OXFORD ENGINEERING SCIENCE SERIES

THE OXFORD ENGINEERING SCIENCE SERIES

10. P. HAGEDORN: *Non-linear oscillations* (Second edition)
13. N.W. MURRAY: *Introduction to the theory of thin-walled structures*
14. R.I. TANNER: *Engineering rheology*
15. M.F. KANNINEN and C.H. POPELAR: *Advanced fracture mechanics*
19. R.N. BRACEWELL: *The Hartley transform*
22. C. SAMSON, M. LeBORGNE, and B. ESPIAU: *Robot control: the task function approach*
23. H.J. RAMM: *Fluid dynamics for the study of transonic flow*
24. R.R.A. SYMS: *Practical volume holography*
25. W.D. McCOMB: *The physics of fluid turbulence*
26. Z.P. BAZANT and L. CEDOLIN: *Stability of structures: elastic, inelastic, fracture, and damage theories*
27. J.D. THORNTON: *Science and practice of liquid-liquid extraction* (Two volumes)
28. J. VAN BLADEL: *Singular electromagnetic fields and sources*
29. M.O. TOKHI and R.R. LEITCH: *Active noise control*
30. I.V. LINDELL: *Methods for electromagnetic field analysis*
31. J.A.C. KENTFIELD: *Nonsteady, one-dimensional, internal, compressible flows*
32. W.F. HOSFORD: *Mechanics of crystals and polycrystals*
33. G.S.H. LOCK: *The tubular thermosyphon: variations on a theme*
34. A. LINAN and F.A. WILLIAMS: *Fundamental aspects of combustion*
35. N. FACHE, D. DE ZUTTER, and F. OLYSLAGER: *Electromagnetic and circuit modelling of multiconductor transmission lines*
36. A.N. BERIS and B.J. EDWARDS: *Thermodynamics of flowing systems: with internal microstructure*
37. K. KANATANI: *Geometric computation for machine vision*
38. J.G. COLLIER and J.R. THOME: *Convective boiling and condensation* (Third edition)
39. I.I. GLASS and J.P. SISLIAN: *Nonstationary flows and shock waves*
40. D.S. JONES: *Methods in electromagnetic wave propagation* (Second edition)
42. G.A. BIRD: *Molecular gas dynamics and the direct simulation of gas flows*
43. G.S.H. LOCK: *Latent heat transfer: an introduction to fundamentals*
44. C.E. BRENNEN: *Cavitation and bubble dynamics*
45. T.C.T TING: *Anisotrophic elasticity: theory, and applications*
46. K. ISHIHARA: *Soil behaviour in earthquake geotechnics*

R

Soil Behaviour in Earthquake Geotechnics

KENJI ISHIHARA

Department of Civil Engineering
Science University of Tokyo

This publication was supported by a generous donation from the Daido Life Foundation

CLARENDON PRESS • OXFORD
1996

Walton Street, Oxford OX2 6DP
New York
d Bangkok Bombay
n Dar es Salaam Delhi
Kong Istanbul Karachi
Kuala Lumpur Madras Madrid Melbourne
Mexico City Nairobi Paris Singapore
Taipei Tokyo Toronto
and associated companies in
Berlin Ibadan

Oxford is a trade mark of Oxford University Press

Published in the United States
by Oxford University Press Inc., New York

A catalogue record for this book is available from the British Library

Library of Congress Cataloging in Publication Data
Ishihara, Kenji.
Soil behaviour in earthquake geotechnics / Kenji Ishihara.
(Oxford engineering science series; 46)
Includes bibliographical references
1. Soil dynamics. 2. Earthquake engineering. I. Title.
II. Series.
TA710.I73 1996 624.1'762–dc20 95-42038

ISBN 0 19 856224 1

Typeset by EXPO Holdings, Malaysia.
Printed in Great Britain by Bookcraft (Bath) Ltd
Midsomer Norton, Avon

PREFACE

Since the late 1960s fairly accurate measurements of soil properties under dynamic loading conditions have been made in the laboratory and in situ. A summary of the then current knowledge on this subject was published by the author in 1976 (in Japanese) entitled *Fundamentals of Soil Dynamics*.

With the further development of new techniques, a vast amount of reliable data has subsequently been produced and is now used for the analysis and design of foundations and earth structures taking into account the effects of earthquake-induced loads. Because of the abundance and diversity of the data it was difficult to review and summarize all of them in a proper perspective. As such, the content included in this book does not necessarily cover all relevant information, and may be biased towards the efforts of the author and other Japanese workers.

The structuring of this book draws basically on the framework of the previous book published in Japanese. Several years have passed since the author started writing the draft. In the meantime, overall encouragement by Professor M. Fukuoka was helpful in advancing the writing work. My colleagues, Professor F. Tatsuoka of the University of Tokyo, Dr T. Kokusho of Central Research Institute of Electric Power Industry, Professor S. Yasuda of Tokyo Denki University, and Dr. S. Iai of the Port and Harbour Research Institute have always been co-operative in providing most useful information and stimulating discussions on the subject matter.

The author wishes to acknowledge the co-operation and assistance by his colleagues at the University of Tokyo, Professor I. Towhata, Mr K. Sugo, and M. Yoshimine. The assistance by Dr R. P. Orense of Kisojiban Consultants is also acknowledged gratefully.

February 1995 K.I.
Tokyo

The Publisher and the Author gratefully acknowledge the generous financial support given for the preparation of the typescript by the Daido Life Foundation.

CONTENTS

1 CHARACTERISTICS OF DYNAMIC PROBLEMS 1

1.1 Range of strain 1
1.2 Differences between static and dynamic loading conditions 2
1.3 Dependency of deformation characteristics upon shear strains 3

2 CHARACTERISTIC CHANGES IN CYCLIC STRESS IN TYPICAL DYNAMIC LOADING ENVIRONMENTS 6

2.1 Cyclic stresses during earthquakes 6
2.2 Traffic loading 10
2.3 Wave-induced loading 13
References 15

3 THE REPRESENTATION OF STRESS–STRAIN RELATIONS IN CYCLIC LOADING 16

3.1 The linear viscoelastic model 17
3.2 The nonlinear cycle-independent model 28
References 39

4 APPARATUS AND PROCEDURES FOR LABORATORY TESTS 40

4.1 Triaxial test apparatus 40
4.2 Simple shear test apparatus 43
4.3 Torsional shear test apparatus 44
4.4 Resonant column test 46
4.5 Wave propagation method in laboratory specimens 54
4.6 Cyclic loading tests with precise measurements of strains 56
References 58

5 IN SITU SURVEY BY WAVE PROPAGATION 60

5.1 Reflection survey 60
5.2 Refraction survey 62
5.3 Uphole and downhole methods 75
5.4 The crosshole method 66
5.5 Suspension sonde 68
5.6 Spectral analysis of surface waves (SASW) 73
References 85

6 LOW-AMPLITUDE SHEAR MODULI 85

6.1 Low-amplitude shear moduli from laboratory tests 85

6.2 Time dependency of low-amplitude shear modulus 107
6.3 Low-amplitude shear moduli from in situ tests 112
6.4 Estimation of shear modulus from the in situ penetration test 119
6.5 Poisson's ratio for saturated soils 120
References 123

7 STRAIN DEPENDENCY OF MODULUS AND DAMPING 127

7.1 Strain-dependent modulus and damping from laboratory tests 127
7.2 Evaluation of strain-dependent soil properties for in situ deposits 142
7.3 Factors affecting modulus reduction and damping characteristics 150
References 152

8 EFFECT OF LOADING SPEED AND STIFFNESS DEGRADATION OF COHESIVE SOILS 152

8.1 Classification of loading schemes 154
8.2 Deformation characteristics of soils under transient loading 162
8.3 Deformation characteristics of soils under cyclic loading 154
8.4 Evaluation of cyclic stiffness degradation 165
8.5 Threshold strains for cyclic degradation 172
8.6 Threshold strains and reference strain 177
References 177

9 STRENGTHS OF COHESIVE SOILS UNDER TRANSIENT AND CYCLIC LOADING CONDITIONS 180

9.1 Load patterns in dynamic loading tests 180
9.2 Definition of dynamic strength of soil 180
9.3 Transient loading conditions 186
9.4 Combined static and cyclic loading 187
9.5 Irregular loading conditions 192
References 206

10 RESISTANCE OF SAND TO CYCLIC LOADING 208

10.1 Simulation of field stress conditions in laboratory tests 208
10.2 The mechanism of liquefaction 208
10.3 Definition of liquefaction or cyclic softening 218
10.4 Cyclic resistance of reconstituted clean sand 221
10.5 Cyclic resistance of in situ deposits of sands 225
10.6 Cyclic resistance of silty sands 227
10.7 Cyclic resistance of gravelly soils 228
10.8 Effects of K_0 conditions on liquefaction resistance of sand 231
10.9 Cyclic resistance of sand under irregular seismic loading 233
10.10 Effects of confining stress and initial shear stress on liquefaction resistance 241
References 242

11 SAND BEHAVIOUR UNDER MONOTONIC LOADING 247

11.1 Flow and non-flow in undrained sand samples 247

11.2 Compression characteristics of sand and the method of sample preparation 250
11.3 Steady state of sand 255
11.4 Quasi-steady state 257
11.5 Quasi-steady state of silty sands 262
11.6 Residual strength of fines-containing sand 268
11.7 Estimate of residual strength 272
11.8 Effects of the fabric on residual strength 275
11.9 Effects of the deformation mode on residual strength 278
References 279

12 EVALUATION OF LIQUEFACTION RESISTANCE BY IN SITU SOUNDINGS 282

12.1 Correlation based on field performances 282
12.2 Correlation based on laboratory tests 287
12.3 The effects of fines on cyclic strength 290
12.4 Correlation for gravelly soils 293
References 299

13 ANALYSIS OF LIQUEFACTION 301

References 306

14 SETTLEMENT IN SAND DEPOSITS FOLLOWING LIQUEFACTION 308

14.1 Basic concepts and procedures 308
14.2 Evaluation of settlement 312
References 315

15 FLOW AND NON-FLOW CONDITIONS AND RESIDUAL STRENGTH 316

15.1 Flow conditions in SPT and CPT 316
15.2 Correlation of residual strength and penetration resistance 325
References 330

16 ONSET CONDITION FOR LIQUEFACTION AND CONSEQUENT FLOW 331

16.1 Interpretation of laboratory tests to assess in situ strength 331
16.2 Onset conditions for liquefaction and consequent flow 333
References 344

APPENDIX: METHODS OF SAMPLE PREPARATION 338

A.1 Moist placement method (wet tamping) 338
A.2 Dry deposition method 339

A.3 Water sedimentation method 340
References 340

INDEX 341

1

CHARACTERISTICS OF DYNAMIC PROBLEMS

There are a variety of problems where the behaviour of soils in dynamic loading is of concern, but because of diversity in the areas and complexity in the conditions under which dynamic loads are applied, it is not easy to establish a hierarchy in the discipline and to put all problems in a proper perspective. However some synthesized considerations may be afforded, if the dynamic problem type is classified in terms of the view point as distinguished from the static problem.

1.1 Range of strain

In the area of classical soil mechanics dealing with static problems, a major concern has been to evaluate the degree of safety of foundations or soil structures against failure. A common approach has been to evaluate available strength of soils and to compare it against the stresses induced by external loading. Thus attention has been centered on the evaluation of soil strength. Settlements of the ground or structures have been another issue of concern associated with the deformation of soils. Consolidation of clay has thus been a major branch area in classical soil mechanics.

Looking back at these two major areas, one can recognize that the attention has been focused on the behaviour of soils which are subjected to a sizable amount of deformation. It is known that the failure in soils takes place normally at a strain level of a few percent and the settlements of engineering concern, due to compression or consolidation, occur in most cases accompanied by a strain in the order of 10^{-3} or greater. Thus it is noted that the phenomena of soils undergoing small strains have not been of concern and thus put out of scope.

In contrast to this, the state of soils in motion is the subject of studies in soil dynamics and therefore inertia force is another kind of agency that cannot be neglected. It is well known that the inertia force plays an increasingly significant role as the time interval at which deformation occurs becomes short. In sinusoidal motion, the inertia force is known to increase in proportion to the square of the frequency at which coils are cyclically deformed. Consequently, even if the level of strain is infinitesimal, the inertia force could become significantly great with increasing rapidity in motion and reach a point where its influence can no longer be disregarded in engineering practice. For this reason, it becomes necessary in soil dynamics to draw attention to the behaviour of soils subjected to a strain level as small as 10^{-6} which is a level completely disregarded in conventional soil mechanics dealing with static problems. The above is one of the most important aspects which distinguish dynamic problems from the static one.

1.2 Differences between static and dynamic loading conditions

It has been established that the void ratio, confining stress, water content, and so forth are factors influencing the mechanical behaviour of soils. Other factors such as stress histories, levels of strains, and temperature have also been known to play a significant role in determining the response of soils. However, these factors are equally important both under dynamic and static loading conditions, and hence are not considered as the elements characterizing the dynamic behaviour as distinguished from the static one. The features distinctive of dynamics would be conceived and derived from the phenomena *impulse*, *vibration*, and *wave* as described below.

1.2.1 *Speed of loading*

The length of time at which a certain level of stress or strain is attained in soils may be defined as the *time of loading*. The rapidity of load application is certainly a feature characterizing the dynamic phenomenon. Several events of engineering significance are classified in accordance with the time of loading and demonstrated accordingly in the horizontal coordinate of Fig. 1.1.

In the case of vibration and waves, events with shorter period or higher frequency are deemed as phenomena with a shorter time of loading, and conversely a longer-period problem is regarded as the one with a longer time of loading. In what follows, the time of loading will be defined approximately as one quarter of the period at

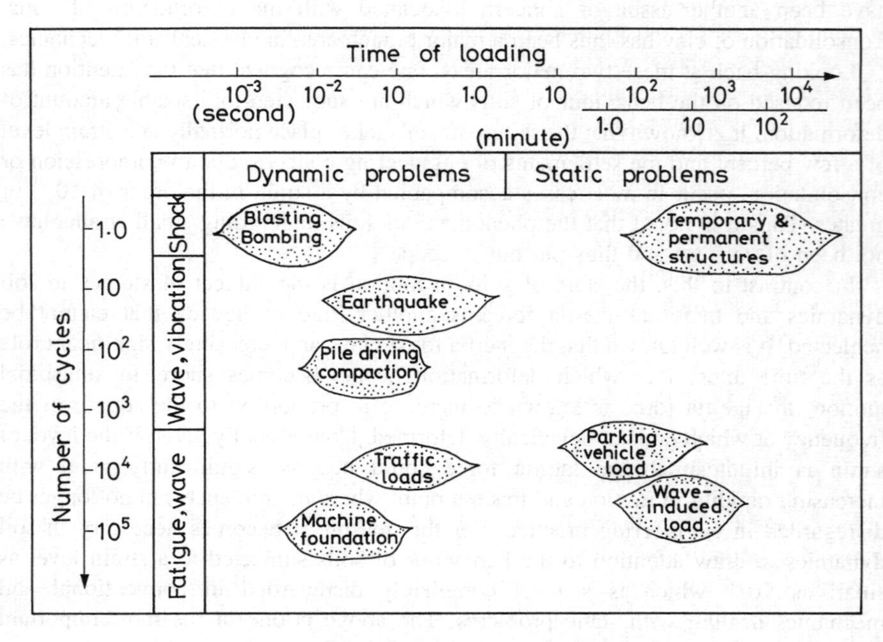

Fig. 1.1 Classification of dynamic problems.

which the load is reciprocated. The problems where the load application lasts for more than tens of second are generally cited as static and those with a shorter time of load application are the target of dynamic problems. The length or shortness of load application may be alternatively expressed in terms of speed of loading or rate of straining and they will be collectively referred to as *speed effect* or *rate effect*.

1.2.2 *Effects of load repetition*

In what is called the dynamic phenomenon, the load is repetitively applied many times with some frequency. Thus repetitiveness in loading is another prime attribute used to classify dynamic problems. The events frequently encountered in engineering practice are classified in this context and accordingly demonstrated in the ordinate of Fig. 1.1.

The problems associated with rapid application of one single impulse are represented by the shock such as that generated by dropping of bombs or blasting of explosives. The duration of loading is as short as $10^{-3}-10^{-2}$ seconds and the load is commonly called an *impulse* or *shock load*. Main shaking during earthquakes involves 10–20 times repetition of loads with differing amplitudes. While the seismic loading is irregular in time history, the period of each impulse, is within the range between 0.1 and 3.0 seconds, giving the corresponding time of loading on the order of 0.02 to 1.0 seconds as accordingly indicated in Fig. 1.1. In the case of pile driving, vibro-compaction, and operation of vibrating rollers, the loads are applied to soils 100 to 1000 times with a frequency 10 to 60 Hz. The foundations on which electric generators or compressors are mounted are also subjected to motions with similar frequencies, but the number of load applications is much larger.

The above events are mainly related to what is termed vibration or wave propagation. Another kind of problem is soil behaviour whilst undergoing the repetitive loads induced by traffic or propagation of water waves. The soils in the subgrade underneath railways or road pavements are subjected to a large number of loads during the life span of their service. Though variable to a large degree, the time of loading may be deemed on the order of 0.1 second to a few seconds. This type of loading is characterized by the number of repetitions being formidably large and, therefore, even though the intensity is trivial, its accumulated effects could be of engineering significance. In the case as above where the number of repetitions is unmeasurably large, the problem may need to be understood as a phenomenon of fatigue. The behaviour of soils as manifested by the load repetitions as above will be referred to as *the effect of repetition*.

1.3 Dependency of deformation characteristics upon shear strains

It has been shown that the deformation characteristics of soils varies to a large extent depending upon the magnitude of shear strains to which soils are subjected. Overall changes in soil behaviour with changes in shear strain are illustrated in Fig. 1.2, in which approximate ranges of the shear strain producing elastic, elasto-plastic, and failure states of stress are indicated. In the infinitesimal strain range below the order of 10^{-5}, the deformations exhibited by most soils are purely elastic and recoverable. The phenomena associated with such small strains would be vibration or wave propagation through soil grounds. Over the intermediate range of strain between 10^{-4} and 10^{-2}, the behaviour of soils is elasto-plastic and produces irrecoverable permanent deformation. The development of cracks or differential settlements in soil structures appears to be associated with the elasto-plastic attribute of soils that is exhibited within such a range of strain. When large strains exceeding a few percent are imposed on soils, the strains tend to become considerably large without a further increase in shear stress and failure takes place in the soils. Slides in slope or compaction and liquefaction of cohesionless soils are associated with failure-inducing large strains.

Another feature to be noted in soil behaviour is the dilatancy, i.e. the tendency of soils to dilate or to contract during drained shear or pore water pressure changes during undrained shear. The dilatancy in the repetition of load does not occur in the infinitesimal and intermediate strain ranges. Its effect begins to appear when the magnitude of shear strain increases above a level of about 10^{-4} to 10^{-3} as indicated in Fig. 1.2. It should be borne in mind that the progressive changes in soil properties during load repetition such as degradation in stiffness of saturated soils or hardening of dry or partially saturated soils can occur as a consequence of the dilatancy effect being manifested during shear.

Still another important aspect under dynamic loading conditions is the influence of the speed with which loads are applied to soils. Laboratory tests have shown that the resistance to deformation of soils under monotonic loading conditions generally increases as the speed of loading is increased, and also that the strength of soils increases with an increase in time to failure. It is to be noted that the effect of the

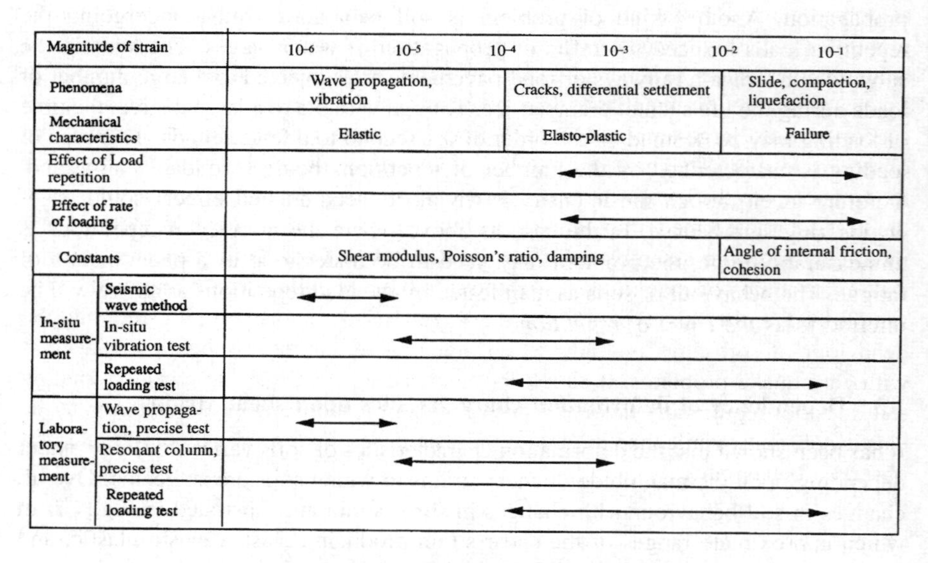

Fig. 1.2 Variation of soil properties with strain.

loading speed does not appear when the shear strain is very small. It has been shown that the threshold shear strain between where the rate effect does or does not occur is in the order of about 10^{-3} as indicated in Fig. 1.2.

In Fig. 1.2 approximate ranges of strain are indicated within which several conventional methods of tests are applicable to evaluate dynamic properties of soils. With respect to in situ measurements, it is difficult to induce strains exceeding 10^{-5} by means of a seismic method because of limitation on the source of energy. This method is therefore useful in obtaining the deformation modulus at infinitesimal strains alone. Somewhat larger strains can be produced in the soil by means of in situ vibration tests, where a large shaker mounted on prototype structures or on foundations is shaken to impart a vibratory force to the surrounding soil ground. In this type of test, the strain induced in the soil is roughly of the order of magnitude of 10^{-5} through 10^{-3}. When it is required to determine the soil properties at strains as large as a few percent, difficulty arises in the vibration test because it requires too much energy to be applied. In such situations, it is preferable to replace the vibration test by what will be called a repeated loading test. If the frequency of vibration is reduced to less than a few cycles per second, the effects of inertia forces are negligible and the test becomes merely repetition of the static test. Since the rate effect is normally small in the frequency range in question, the repeated loading test is a useful tool for investigating soil deposits in the field at intermediate to large strains.

With respect to the tests in the laboratory, wave propagation through columns of soil specimens is the most commonly used technique to determine elastic properties of soils. The resonant column method is also popular for testing soils in the laboratory. In this test, it is possible to impose strains up to about 10^{-4} depending upon the types of soils to be tested. Elastic properties of soils for small strains can also be obtained from laboratory tests on specimens where the deformation is monitored precisely by means of a special device. In order to study soil behaviour at strains as large as a few percent, the influence of frequency must be put aside in favour of the requirement on the part of strain. In general, it is difficult to set specimens into large-amplitude oscillation without sacrificing the accuracy of the observed results. The best way to overcome this difficulty would be to lower the frequency to the level where the vibration test would no longer be feasible. Then the test is transformed into what is called a repeated loading test without any modification in the assemblage of test devices. By means of the repeated loading test, even strains of magnitude large enough to cause failure can be applied to the sample. Recently this type of test has been used extensively in studying the behaviour of soft clays or liquefaction potential of saturated sands in connection with earthquake problems. It should be emphasized here that the repeated loading test does not require any theory on which system response is based to compute the elastic constants, while the vibration test presupposes the response theory to interpret observed behaviour.

2

CHARACTERISTIC CHANGES IN CYCLIC STRESS IN TYPICAL DYNAMIC LOADING ENVIRONMENTS

There are three major areas of engineering interest where environments, natural or artificial, exert excursion of loads on soils which are cyclic in nature. They are earthquakes, traffic, and waves of the sea. The characteristic features of cyclic loads by these external agencies will be described in the following.

2.1 Cyclic stresses during earthquakes

It has been generally accepted that the major part of the ground shaking during an earthquake is due to the upward propagation of body waves from an underlying rock formation. Although surface waves are also involved, their effects are generally considered of secondary importance. The body waves consist of shear waves and compressional (or longitudinal) waves. In the case of level ground, each of the waves produces, respectively, shear stress and compressional stress, as illustrated in Fig. 2.1. During the propagation of compressional waves, normal stress is induced in the vertical as well as horizontal direction, thereby producing the triaxial mode of deformation in an element of soil under level ground.

In the level ground condition, the soil element is not allowed to deform in the horizontal direction and if this condition is imposed, normal stress in the horizontal direction σ_{dh} is shown to be related with vertical normal stress σ_{dv} via the formula

$$\frac{\sigma_{dh}}{\sigma_{dv}} = \frac{\nu}{1-\nu} \tag{2.1}$$

where ν is Poisson's ratio. If the soil is saturated with water, using eqn (6.20) detailed in Chapter 6, Poisson's ratio is given by

$$\nu = \frac{1}{2}(1 - n\,G_0\,C_\ell) \tag{2.2}$$

where n is porosity, G_0 is shear modulus and C_ℓ denotes the compressibility of water. Using the relation of eqn (2.2), eqn (2.1) is rewritten as

$$\frac{\sigma_{dh}}{\sigma_{dv}} = 1 - 2n\,G_0\,C_\ell. \tag{2.3}$$

This relation is displayed in Fig. 2.2. In view of the shear modulus G_0 taking a value less than 50 MPa for soft soil deposits, it is known that the horizontal normal stress σ_{dh} induced by the propagation of the compressional wave is nearly equal to the

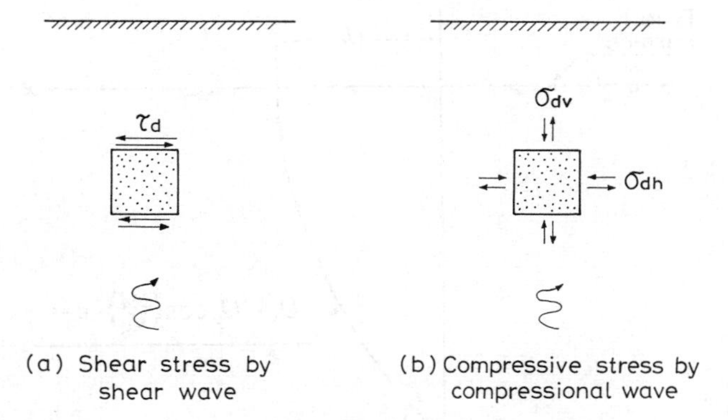

Fig. 2.1 Stresses induced by body wave propagation.

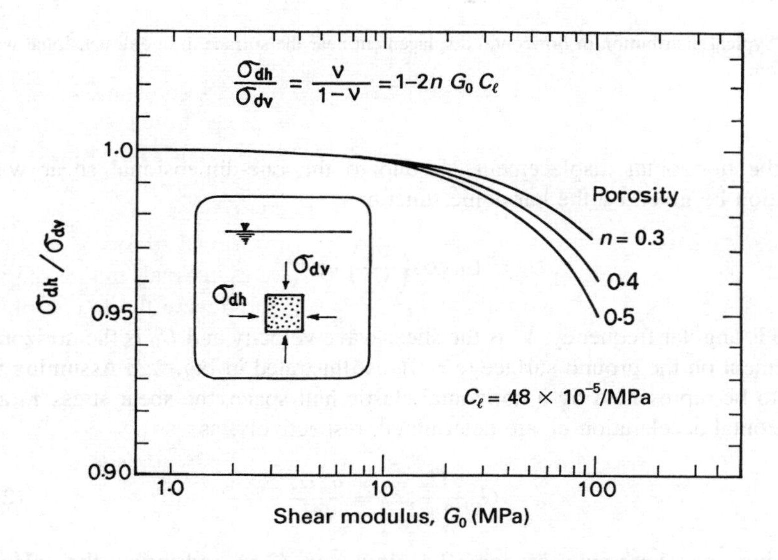

Fig. 2.2 State of stresses induced by propagation of compressional wave.

value of vertical normal stress σ_{dv}. This implies the fact that the propagation of compressional wave through saturated soft soils induces almost purely compressional stress and the component of deviator stress $\sigma_{dv} - \sigma_{dh}$ is practically equal to zero. Since the compressional stress is transmitted through water in the pores, there is no change in the effective stress induced by the compressional wave. For this reason, effects of the compressional wave are disregarded in evaluating the stability of the ground such as liquefaction and consequent settlements in sandy ground. Thus horizontal shear stress due to the propagation of shear waves is the main component of stress that is to be considered in one-dimensional stability analysis of level ground during earthquakes.

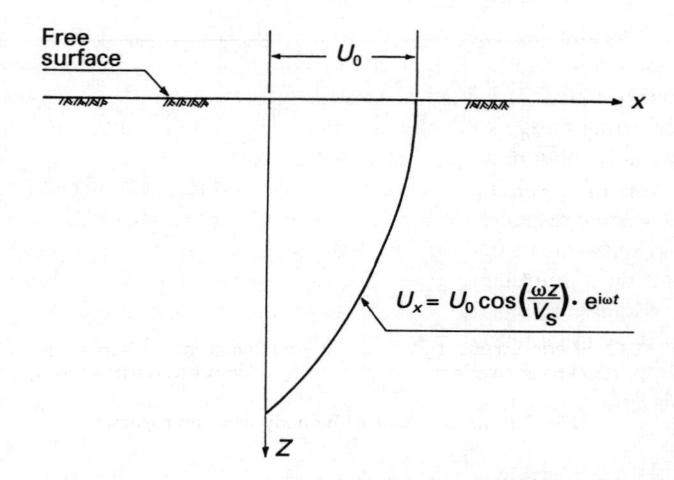

Fig. 2.3 Typical distribution of horizontal displacement near the surface in one-dimensional wave propagation.

Let the horizontal displacement U_x due to the one-dimensional shear wave propagation be given by the harmonic function

$$U_x = U_0 \cos\left(\frac{\omega z}{V_s}\right) \cdot e^{i\omega t} \tag{2.4}$$

where ω is angular frequency, V_s is the shear wave velocity and U_0 is the horizontal displacement on the ground surface ($z = 0$), as illustrated in Fig. 2.3. Assuming the ground to be represented by a horizontal elastic half-space, the shear stress τ_d and the horizontal acceleration a_h are determined, respectively, as

$$\tau_d = G_0 \frac{\partial U_x}{\partial z}, \; a_h = \frac{\partial^2 U_x}{\partial t^2}. \tag{2.5}$$

Introducing the expression of eqn (2.4) into eqn (2.5) and using the relation $G_0 = \rho V_s^2$, one can obtain the ratio $\tau_d/(a_h \rho z)$ as

$$\frac{\tau_d}{a_h \rho z} = \frac{\tan\left(\frac{\omega z}{V_s}\right)}{\left(\frac{\omega z}{V_s}\right)}. \tag{2.6}$$

where ρ is the unit mass of soil.

For a soil element near the surface, the value of $\omega z/V_s$ is very small and $\tan(\omega z/V_s) \doteqdot \omega z/V_s$. Therefore, eqn (2.6) is rendered as

$$\tau_d \doteqdot a_h \, \rho \, z. \tag{2.7}$$

This means that the shear stress acting at depth z is approximately equal to the product of the mass of soil to the depth z and the acceleration on the ground surface. The relation of eqn (2.7) can be alternatively deduced by simply considering the inertia force acting on the soil column which must be in equilibrium with the shear stress acting at its bottom as illustrated in Fig. 2.4.

In the case of a soil body having complex configurations, two-dimensional analyses of seismic response must be conducted in order to determine time histories of change in shear stress. In such a general case, there are two components of shear stress acting on a soil element, i.e. the shear stress τ_d and the stress difference $\sigma_{dv} - \sigma_{dh}$, as shown in Fig. 2.5. As a result of numerous seismic response analyses, it has been shown that one component of shear stress tends to increase or decrease nearly in proportion to the increase or decrease of the other component. This implies the fact that at any instant of time, the ratio between τ_d and $(\sigma_{dv} - \sigma_{dh})/2$ is held

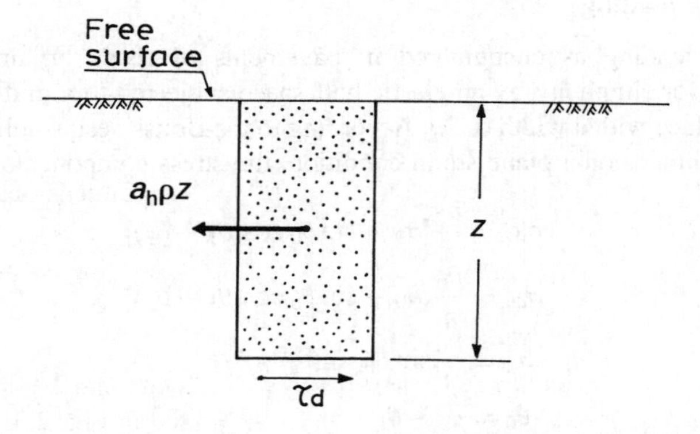

Fig. 2.4 Equilibrium of forces near the surface.

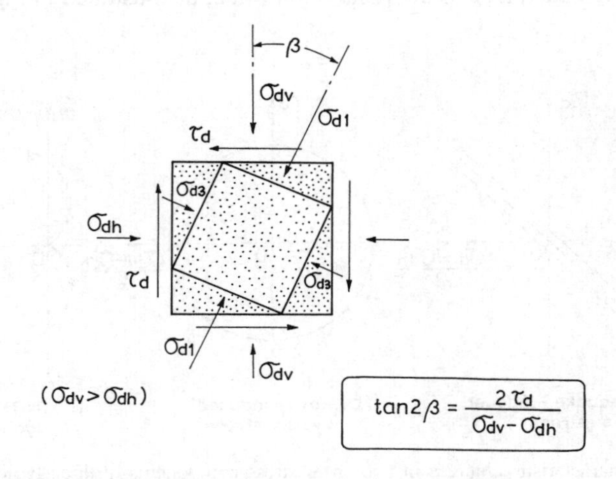

Fig. 2.5 Component of stresses acting on a soil element.

approximately at a constant value. It is to be remembered that this ratio defines the direction of the principal stress axis β as

$$\tan 2\beta = \frac{2\,\tau_d}{\sigma_{dv} - \sigma_{dh}} \tag{2.8}$$

where β is taken as an angle from the vertical to the direction of the axis of major principal stress, as illustrated in Fig. 2.5. Therefore it can be stated that the cyclic change in shear stress during seismic shaking tends to occur so that the direction of the principal stress axis remains almost unchanged throughout the duration of the shaking. This characteristic feature of stress change is schematically illustrated in two-dimensional stress space in Fig. 2.6(a) in which τ_d and $(\sigma_{dv} - \sigma_{dh})/2$ are displayed as variables in two coordinate axes.

2.2 Traffic loading

The traffic loading as encountered in pavements of roads or airfields may be represented for simplicity by an elastic half-space subjected to a uniform load of p_0 over the surface with a width of $2a$. According to the Boussinesq solution for the case of the two-dimensional plane strain condition, the stress components are given by

$$\begin{aligned}
\sigma_{dv} &= \frac{p_0}{\pi}[\sigma_0 + \sin\theta_0 \cos(\theta_1 + \theta_2)] \\
\sigma_{dh} &= \frac{p_0}{\pi}[\sigma_0 - \sin\theta_0 \cos(\theta_1 + \theta_2)] \\
\tau_d &= \frac{p_0}{\pi}\sin\theta_0 \sin(\theta_1 + \theta_2) \\
\theta_0 &= \theta_2 - \theta_1
\end{aligned} \tag{2.9}$$

where θ_1 and θ_2 indicate the angles between the vertical and the lines connecting the edge of the loaded area to the point in question, as illustrated in Fig. 2.7. The major

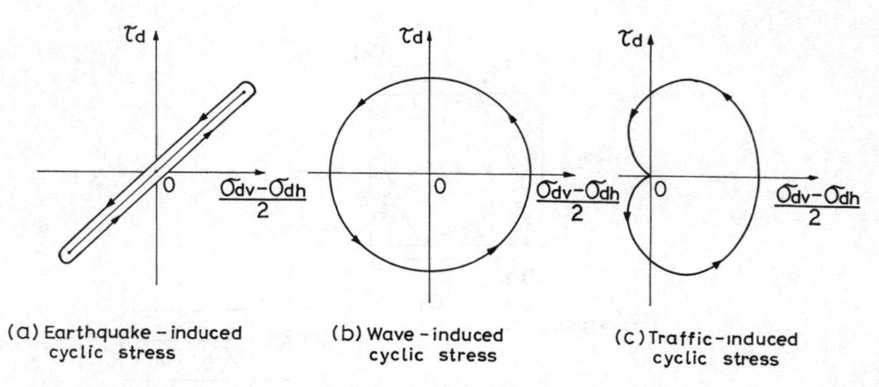

Fig. 2.6 Characteristic changes in two shear stress components in three typical dynamic loading conditions.

and minor principal stresses, σ_{d1} and σ_{d3}, respectively are calculated as follows, by using a well-known formula,

$$\sigma_{d1} = \frac{\sigma_{dv} + \sigma_{dh}}{2} + \sqrt{\left(\frac{\sigma_{dv} - \sigma_{dh}}{2}\right)^2 + \tau_d{}^2} = \frac{p_0}{\pi}(\sigma_0 + \sin\theta_0) \tag{2.10}$$

$$\sigma_{d3} = \frac{\sigma_{dv} + \sigma_{dh}}{2} - \sqrt{\left(\frac{\sigma_{dv} - \sigma_{dh}}{2}\right)^2 + \tau_d{}^2} = \frac{p_0}{\pi}(\sigma_0 - \sin\theta_0).$$

The direction of the major principal stress relative to the vertical is determined by the angle β defined as,

$$\tan 2\beta = \frac{2\tau_d}{\sigma_{dv} - \sigma_{dh}} = \tan(\theta_1 + \theta_2). \tag{2.11}$$

Using the formulae of eqns (2.9), (2.10), and (2.11), distribution of several stress components on the horizontal plane are computed as shown graphically in Fig. 2.8.

The change of stresses induced at a point within the half-space due to the passage of a wheel load can be obtained simply by moving the eyes horizontally from $-\infty$ to $+\infty$ along the x-coordinate in the diagram of Fig. 2.8 and by reading off the stresses at each point on the x-axis. Viewed in this manner, Fig. 2.8 clearly indicates that the component of stress difference $(\sigma_{dv} - \sigma_{dh})/2$ as well as the horizontal shear stress component τ_d change in the course of passage of a wheel load on the ground surface. Accompanied by the change of two components of shear stress, changes in the direction of principal stress are also induced in the ground as shown in Fig. 2.8(c). It is seen that the horizontally directed axis of major principal stress tends to rotate towards the right as the wheel load comes closer to the point in question, but after once becoming vertical, the axis of major principal stress tends to incline in the opposite direction as the load passes by. The characteristic feature of the rotation of principal stress direction can be more clearly visualized by plotting the horizontal shear stress τ_d versus the stress difference $(\sigma_{dv} - \sigma_{dh})/2$ as

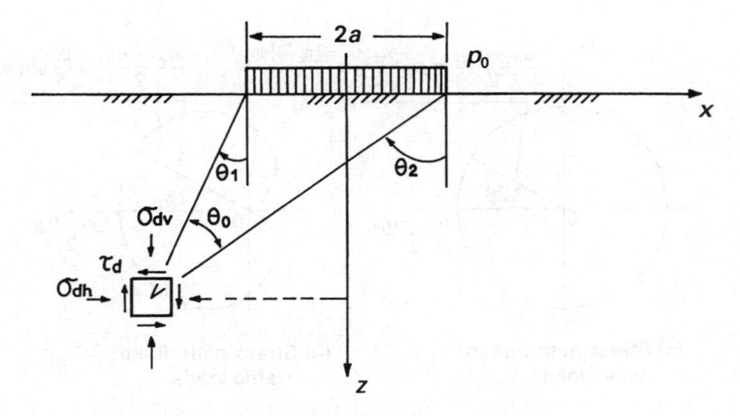

Fig. 2.7 Uniform loads on an elastic half-space.

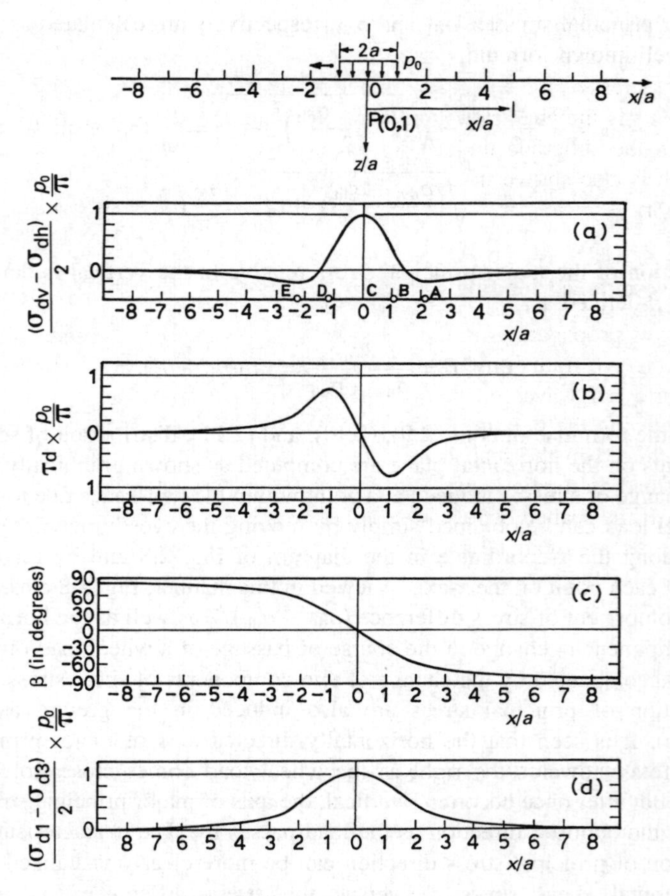

Fig. 2.8 Distribution of several components of stress with in a homogeneous elastic half-space.

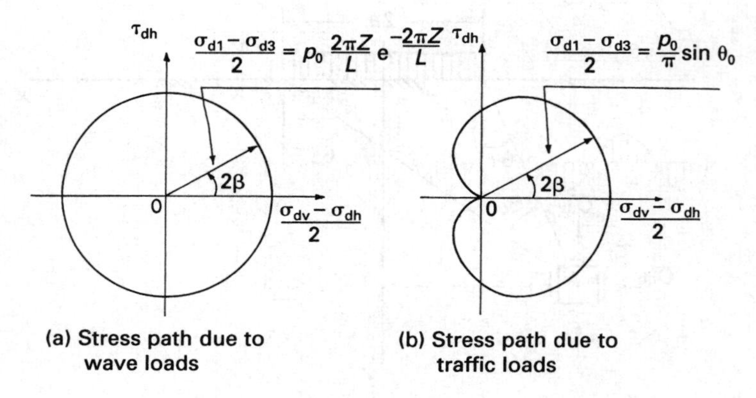

Fig. 2.9 Stress paths due to wave and traffic loads.

demonstrated in Fig. 2.9. It may be seen in Fig. 2.9 that the rotation of principal stress direction takes place first with increasing and then with decreasing magnitude of deviator stress as the wheel load moves from minus to plus infinity on the ground surface. This is the characteristic feature of change in stress that can occur in a soil element in the subgrade underlying pavements or railways (Ishihara, 1983). This stress path is also shown in Fig. 2.6(c).

2.3 Wave-induced loading

Water waves propagating on the ocean may be considered to consist of an infinite number of wave trains having a constant amplitude and wave length. Passage of such an array of waves on the sea creates harmonic pressure changes on the sea floor, increasing the pressure under the crest and reducing it under the trough. The stresses induced in the seabed are, therefore, analyzed by applying a sinusoidally changing load on the horizontal surface from minus to plus infinity, as illustrated in Fig. 2.10. If the seabed deposit is assumed to consist of a homogeneous elastic half space, the stresses can be readily determined by using the classical solution of Boussinesq for the two-dimensional plane strain problem. Assume that a harmonic load

$$p(x) = p_0 \cos\left(\frac{2\pi}{L}x - \frac{2\pi}{T}t\right) \tag{2.12}$$

is distributed on the surface of an elastic half space, where p_0 is the amplitude of the load, L is the wave length and T is the period of waves. The vertical normal stress σ_{dv}, horizontal normal stress σ_{dh} and shear stress τ_d induced in the half-space by this

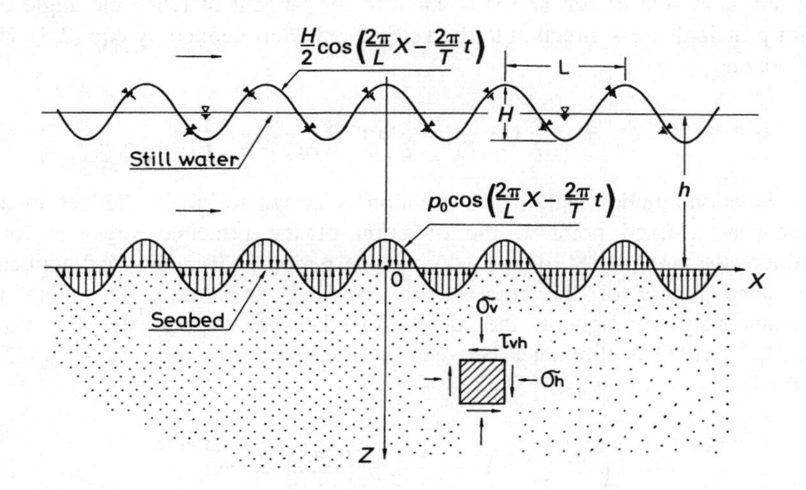

Fig. 2.10 Wave-induced stresses in a seabed deposit.

load are determined (Yamamoto, 1978; Madsen, 1978) as

$$\sigma_{dv} = p_0\left(1+\frac{2\pi z}{L}\right)e^{-\frac{2\pi z}{L}}\cdot\cos\left(\frac{2\pi x}{L}-\frac{2\pi t}{T}\right)$$

$$\sigma_{dh} = p_0\left(1-\frac{2\pi z}{L}\right)e^{-\frac{2\pi z}{L}}\cdot\cos\left(\frac{2\pi x}{L}-\frac{2\pi t}{T}\right) \tag{2.13}$$

$$\tau_d = p_0\frac{2\pi z}{L}e^{-\frac{2\pi z}{L}}\cdot\sin\left(\frac{2\pi x}{L}-\frac{2\pi t}{T}\right)$$

where x and z indicate the spatial coordinates in the horizontal and vertical directions, respectively, as illustrated in Fig. 2.10.

It is well known that the major components of stress associated with the shearing deformation of a solid body are shear stress τ_d and the stress difference defined as $(\sigma_{dv}-\sigma_{dh})/2$. The stress difference is calculated from eqn (2.13) as follows

$$\frac{\sigma_{dv}-\sigma_{dh}}{2} = p_0\frac{2\pi z}{L}e^{-\frac{2\pi z}{L}}\cdot\cos\left(\frac{2\pi x}{L}-\frac{2\pi t}{T}\right). \tag{2.14}$$

Comparison of the expressions for τ_d and $(\sigma_{dv}-\sigma_{dh})/2$ indicated that the amplitude of these two components is identical and they differ only in the time phase of cyclic load application. In order to obtain an insight into the nature of the cyclic loading, let the variable z be eliminated between τ_d and $(\sigma_{dv}-\sigma_{dh})/2$. Then one obtains

$$\frac{2\tau_d}{\sigma_{dv}-\sigma_{dh}} = \tan\left(\frac{2\pi x}{L}-\frac{2\pi t}{T}\right). \tag{2.15}$$

The left-hand side of eqn (2.15) is equal to the tangent of twice the angle of the major principal stress direction to the vertical axis β as defined by eqn (2.8). Hence one obtains

$$\beta = \frac{\pi}{L}x - \frac{\pi}{T}t = \frac{1}{2}\tan^{-1}\left(\frac{2\tau_d}{\sigma_{dv}-\sigma_{dh}}\right). \tag{2.16}$$

This equation implies that, when attention is drawn to stress changes in a soil element at a fixed point x, the direction of the principal stress is rotating continuously through 180 degrees during one period T of cyclic load application. The same nature of the rotation of principal stress axes is taking place simultaneously throughout the depth of the half space. If the variable $(2\pi x/L - 2\pi t/T)$ is eliminated between the shear stress τ_d and $(\sigma_{dv}-\sigma_{dh})/2$, one obtains

$$\left(\frac{\sigma_{dv}-\sigma_{dh}}{2}\right)^2 + \tau_d{}^2 = p_0^2\left(\frac{2\pi z}{L}\right)^2\cdot e^{-\frac{4\pi z}{L}}. \tag{2.17}$$

A textbook of elementary stress analysis shows that the left-hand side of eqn (2.17) is equal to the square of the deviator stress which is defined as the difference between the maximum and minimum principal stresses σ_{d1} and σ_{d3} respectively, divided by two. Hence one obtains

$$\frac{\sigma_{d1}-\sigma_{d3}}{2}=\sqrt{\left(\frac{\sigma_{dv}-\sigma_{dh}}{2}\right)^2+\tau_d{}^2}=p_0\frac{2\pi z}{L}\mathrm{e}^{-\frac{2\pi z}{L}}. \quad (2.18)$$

This equation indicates that, when attention is drawn to stress changes in a soil element at a fixed depth, the deviator stress is kept unchanged at all times and at all points on the horizontal plane during the cyclic load application. The same nature of cyclic alternation of stress is taking place at each depth of the half-space. As can easily be understood from eqn (2.18), the two components of stress τ_d and $(\sigma_{dv}-\sigma_{dh})/2$ are increasing or decreasing alternately so as to keep the deviator stress at a constant value during the entire period of cyclic load application. If the two components of stress are represented in a rectangular coordinate system, eqn (2.18) is an equation of a circle with its radius equal to the deviator stress $(\sigma_{d1}-\sigma_{d3})/2$. Such a plot is presented in Fig. 2.9(a). In this type of plot, the angle which a stress vector makes to the horizontal coordinate axis represents twice the angle of the maximum principal stress axis to the vertical, 2β. It can be stated in summary that the cyclic change of shear stress induced in an elastic half space by a harmonic load moving on its surface is characterized by a continuous rotation of the principal stress direction with the deviator stress being always maintained constant. This is the characteristic feature of the cyclic change in stress induced in the seabed deposit by travelling waves on the sea (Ishihara and Yamazaki, 1984). This type of cyclic loading is also illustrated in Fig. 2.6(b).

Looking back over the three types of cyclic variation of loads as indicated in Fig. 2.6, one can envisage that the earthquake-induced loading is characterized by the jump rotation of the principal stress axes and its continuous rotation is an attribute to the cyclic loads induced by travelling sea waves and traffic.

References

Ishihara, K. (1983). *Soil response in cyclic loading induced by earthquakes, traffic and waves.* Proceedings of the 7th Asian Regional Conference on Soil Mechanics and Foundation Engineering, Haifa, Israel, Vol. 2, pp. 42–66.

Ishihara, K. and Yamazaki, A. (1984). Analysis of wave-oinduced liquefaction in seabed deposits of sand. *Soils and Foundation* **24** (3), pp. 85–100.

Madsen, O.S. (1978). Wave-induced pore pressure and effective stresses in a porous bed. *Geotechnique*, **28**(4), pp. 377–93.

Yamamoto, O.S. (1978). *Seabed instability from waves.* Proceedings of the 10th Annual Offshore Technology Conference, Houston, Texas, Vol. 1, pp. 1819–24.

3

THE REPRESENTATION OF STRESS–STRAIN RELATIONS IN CYCLIC LOADING

In making seismic response analysis of ground or soil structures based on the theory of wave propagation or on the finite element principle, of uttermost importance is to represent the cyclic behaviour of soils in a form of material model correlating shear stress and shear strain. Modelling of soil behaviour under cyclic or random loading conditions must be made so that the model can duplicate the deformation characteristics in the range of strains under consideration. When soil behaviour is expected to stay within the range of the small strain, the use of an elastic model is justified and the shear modulus is a key parameter to properly model the soil behaviour (see Fig. 3.1). When a given problem is associated with the medium range of strain, approximately below the level of 10^{-3}, the soil behaviour becomes elasto-plastic and the shear modulus tends to decrease as the shear strain increases. At the same time, energy dissipation occurs during cycles of load application. The energy dissipation in soils is mostly rate-independent and of hysteretic nature, and the damping ratio can be used to represent the energy absorbing properties of soils. Since the strain level concerned is still small enough not to cause any progressive change in soil properties, the shear modulus and damping ratio do not change with the progression of cycles in load application. This kind of behaviour is called *non-degraded hysteresis type*. Such steady-stage soil characteristics can be represented to a reasonable degree of accuracy by use of the linear viscoelastic theory as shown in Fig. 3.1. The shear modulus and damping ratio determined as functions of shear strain are the key parameters to represent soil properties in this medium strain range. The most useful analytical tool accommodating these strain-dependent but cycle-independent soil properties would be the equivalent linear method based on the viscoelastic concept. Generally, the linear analysis is repeated by stepwise changing the soil parameters until a strain-compatible solution is obtained. The seismic response analysis performed for horizontally layered soil by use of the computer program SHAKE (Schnabel *et al.* 1972) is a typical example of an analytical tool that can be successfully used to clarify the soil response in the medium range of strain.

For the shear strain level larger than about 10^{-2}, soil properties tend to change appreciably not only with shear strain but also with the progression of cycles. This kind of behaviour is termed *degraded hysteresis type*. The manner in which the shear modulus and damping ratio change with cycles is considered to depend upon the manner of change in the effective confining stress during irregular time histories of shear stress application. When the law of changing effective stress is established, it is then necessary to have a constitutive law in which stress–strain relations can be

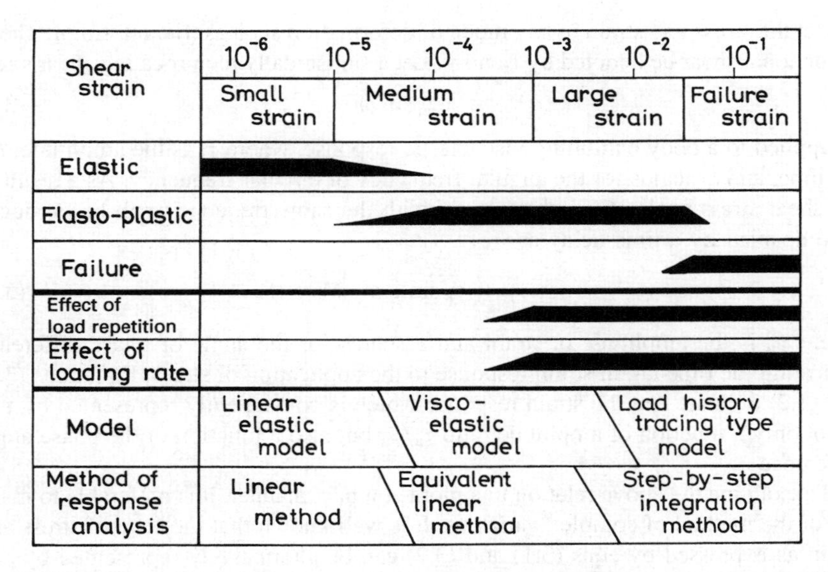

Fig. 3.1 Modelling of soil behaviour in compliance with strain-dependent deformation characteristics.

specified at each step of loading, unloading, and reloading phases. One of the concepts most commonly used at present for this purpose is what is referred to as the Masing law. For analysis of a soil response accommodating such a stress–strain law covering large strain levels near failure, it is necessary to employ a numerical procedure involving the step-by-step integration technique as indicated in Fig. 3.1. In what follows, the methods of modelling soil behaviour will be discussed with emphasis on the non-degraded type.

3.1 The linear viscoelastic model

When the level of cyclic shear strain is still small of the order of 10^{-3} and 10^{-4}, the cyclic behaviour of soils can be represented to a reasonable degree of accuracy by means of a constitutive model based on the classical theory of linear viscoelasticity. In this model, the stress–strain relation is assumed linear, but the energy dissipating characteristics of soils can be logically taken into account. It has been shown that there always exists a certain degree of damping in soils and it plays an important role in determining motions in soil deposits during earthquakes. Thus this model has frequently been used to represent soil behaviour even in the slightly nonlinear range where damping has important effects.

3.1.1 *The general expression of the cyclic stress–strain relationship*

Before discussing the stress–strain law for a specific viscoelastic model, the general form of expression correlating the stress and strain is introduced.

Let the stress and strain in any mode of deformation such as triaxial, simple shear, or torsional shear be denoted by τ and γ. Let a sinusoidally reciprocating shear stress

$$\tau = \tau_a \sin \omega t \tag{3.1}$$

be applied to a body exhibiting viscoelastic response, where τ_a is the amplitude, t is the time, and ω stands for the angular frequency or circular frequency. As a result of the shear stress application, shear strain with the same frequency will be produced accompanied by a time delay as

$$\gamma = \gamma_a \sin(\omega t - \delta) \tag{3.2}$$

where γ_a is the amplitude in strain and δ stands for the angle of phase difference indicating the time lag in strain response to the application of stress. Equations (3.1) and (3.2) indicate that the strain response which is conveniently represented by τ/γ is not only a function of amplitude ratio τ_a/γ_a but also a function of the phase angle difference δ.

To examine the above relation in a more compact manner, it is preferable to make use of the method of complex variables. It is well known that the state of stress and strain as expressed by eqns (3.1) and (3.2) can be alternatively represented by

$$\begin{aligned} \tau_R &= \tau_a \cos \omega t \\ \gamma_R &= \gamma_a \cos(\omega t - \delta) \end{aligned} \tag{3.3}$$

where τ_R and γ_R are the stress and strain, respectively, which keep a conjugate relationship to those of eqns (3.1) and (3.2). In other words, if a viscoelastic body produces a strain response γ when subjected to an input stress τ, then the same body will produce a strain response γ_R when undergoing an input stress τ_R. Thus, it may be stated further that, if the viscoelastic body is subjected to an input stress expressed in a form of complex variables as $\overline{\tau} = \tau_R + i\tau$, then the resulting response in strain would be $\overline{\gamma} = \gamma_R + i\gamma$ where i is the unit imaginary number and $\overline{\tau}$ and $\overline{\gamma}$ are the stress and strain in complex variables.

Introducing eqns (3.1), (3.2), and (3.3) into the definitions of $\overline{\tau}$ and $\overline{\gamma}$, as above, one obtains

$$\begin{aligned} \overline{\tau} &= \tau_a \, e^{i\omega t} \\ \overline{\gamma} &= \gamma_a e^{i(\omega t - \delta)} \end{aligned} \tag{3.4}$$

This is the general form of expressions for the stress and strain in complex variables. In using eqn (3.4), it is stipulated that the real part and imaginary part in the input stress is related only with the corresponding real part and imaginary part in the output strain. The strain versus stress response can be described by the ratio $\overline{\tau}/\overline{\gamma}$ which is explicitly written from eqn (3.4) as

$$\frac{\overline{\tau}}{\overline{\gamma}} = \frac{\tau_a}{\gamma_a} e^{i\delta} = \frac{\tau_a}{\gamma_a}(\cos \delta + i \sin \delta). \tag{3.5}$$

Putting

$$\mu = \frac{\tau_a}{\gamma_a}\cos\delta, \quad \mu' = \frac{\tau_a}{\gamma_a}\sin\delta$$

$$\mu^* = \mu + \mathrm{i}\,\mu' \tag{3.6}$$

eqn (3.5) is rewritten more compactly as

$$\frac{\overline{\tau}}{\overline{\gamma}} = \mu + \mathrm{i}\,\mu' = \mu^* \tag{3.7}$$

where μ and μ' are called *elastic modulus* and *loss modulus*, respectively, and μ^* is named *complex modulus*. The elastic modulus is a parameter indicative of elastic or instantaneous response and loss modulus represents the energy dissipating characteristics of the viscoelastic body. The relationships in eqns (3.6) can be alternatively expressed as,

$$\frac{\tau_a}{\gamma_a} = \sqrt{\mu^2 + \mu'^2} = |\mu^*|$$

$$\tan\delta = \frac{\mu'}{\mu} = \eta \tag{3.8}$$

where η is a parameter called *loss coefficient* which is indicative of energy loss or damping characteristics. From eqn (3.8) the absolute value of the complex modulus μ^* is shown to indicate the shear modulus of the material.

It is to be noted here that the material parameters μ and μ' need not necessarily be real constants but could be a function of angular frequency ω. Therefore the moduli μ and μ' as defined by eqn (3.6) are regarded as the most general form and consequently can take any form expressed as a function of frequency. Once a functional form is specified to these moduli, the viscoelastic behaviour of the material can be described in a more tangible manner. Several methods for specifying these moduli have been proposed either on the basis of direct experiments or on the basis of spring–dashpot models, but this aspect will be described in a later section.

3.1.2 *Hysteretic stress–strain curve*

The stress–strain behaviour of a viscoelastic body as introduced above will be examined in more detail in this section. Equations (3.1) and (3.2) can be viewed as a stress–strain relation expressed by the pair of equations which are mutually correlated through a tracking parameter ωt. Therefore, by eliminating the parameter ωt between these two equations, a single relation as follows is obtained,

$$\left(\frac{\tau}{\tau_a}\right)^2 - 2\cos\delta\left(\frac{\gamma}{\gamma_a}\right)\left(\frac{\tau}{\tau_a}\right) + \left(\frac{\gamma}{\gamma_a}\right)^2 - \sin^2\delta = 0. \tag{3.9}$$

This is regarded as a second order equation with respect to τ/τ_a. Solving with reference to the definition of μ and μ' in eqn (3.6), one obtains,

$$\tau = \mu\gamma \pm \mu'\sqrt{\gamma_a^2 - \gamma^2}. \tag{3.10}$$

This is an alternative expression for the stress–strain relationship derived from the pair of eqns (3.1) and (3.2). An alternative expression is obtained by decomposing the right-hand side of eqn (3.10) into two parts,

$$\begin{aligned} \tau &= \tau_1 + \tau_2 \\ \tau_1 &= \mu\,\gamma \\ \left(\frac{\tau_2}{\mu'\,\gamma_a}\right)^2 &+ \left(\frac{\gamma}{\gamma_a}\right)^2 = 1. \end{aligned} \tag{3.11}$$

The second relation $\tau_1 = \mu\,\gamma$ in eqn (3.11) is depicted in Fig. 3.2 (a) as a straight line with a slope μ. The third relation in eqn (3.11) indicates an ellipse in the plot of τ_2 versus γ having a longer axis at $\tau_2 = \mu'\gamma_a$ and a shorter axis at $\gamma = \gamma_a$. The ellipse is also depicted in Fig. 3.2 (a). The first relation in eqn (3.11) indicates that the addition of two components of shear stress τ_1 and τ_2 plotted in the ordinate should be equal to the actually existing value of shear stress τ. Since the two shear stress components are plotted versus a common coordinate of shear strain γ, the addition can be done easily on the diagram. As a result of this manipulation, an ellipse with an inclined axis is obtained as illustrated in Fig. 3.2(b). The graphical representation of eqn (3.11) as above discloses that, for a given cyclic shear strain, the induced shear stress consists of two parts: one changing linearly with shear strain going back and forth on the straight line in Fig. 3.2(a) and the other changing clockwise on a path along the ellipse in the diagram in Fig. 3.2(a). Thus the actually existing stress composed of these two components traces a path on the inclined ellipse moving clockwise in the course of cyclic loading, as illustrated in Fig. 3.2(b). It is to be noted that the path along the inclined ellipse represents a hysteresis loop which is generally observed in the stress–strain plot in cyclic loading.

As indicated in Fig. 3.2(b), the inclined ellipse cuts the ordinate at a shear stress point of $\mu'\,\gamma_a$. Therefore, the value of μ' may be taken as a measure to indicate a degree of flatness in the shape of the ellipse. The greater the value of μ', the rounder becomes the ellipse indicating that the energy loss or damping is bigger, whereas the smaller the value μ', the thinner the ellipse and, hence, the smaller the damping during the cyclic loading.

To quantify the damping characteristics, it has been customary to draw attention to the amount of energy which is lost during one cycle of load application. The energy loss per cycle is equal to the area enclosed by the hysteresis loop shown in Fig. 3.2(b). Since the area ΔW enclosed by the inclined ellipse should be equal to the area of the upright ellipse shown in Fig. 3.2(a), the enclosed area can be calculated directly using the well known formula of an ellipse,

$$\Delta W = \int \tau\,\mathrm{d}\gamma = \mu'\,\pi\,\gamma_a^2. \tag{3.12}$$

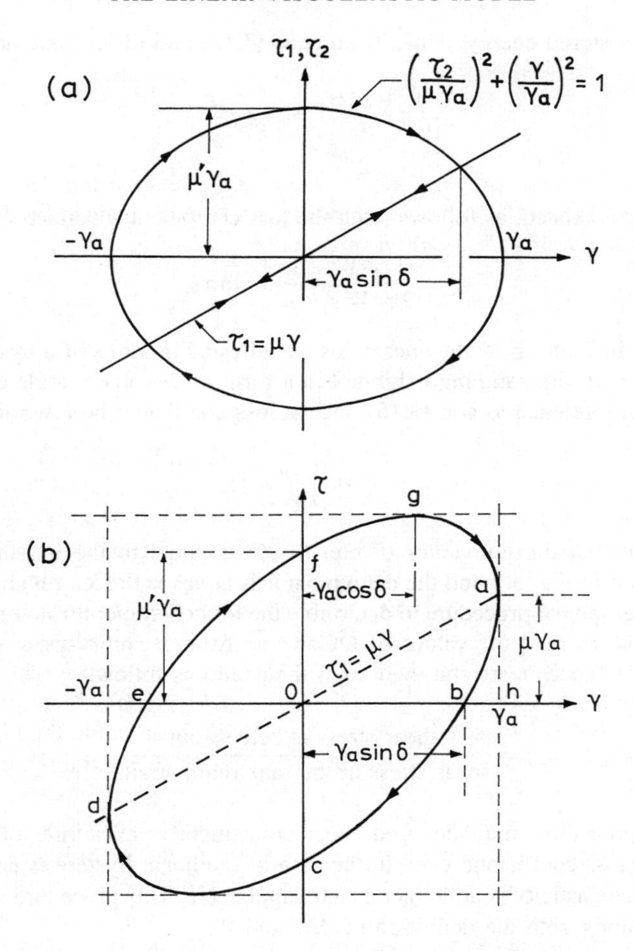

Fig. 3.2 Decomposition of the viscoelastic model into elastic and viscous components.

Next, consider the maximum elastic energy W that can be stored in a unit volume of a viscoelastic body. There are several ways to define the stored energy, but the most logical would be to take up the energy stored by the elastic component of shear stress τ_1. Therefore with reference to the relation $\tau_1 = \mu' \gamma_a$ in eqn (3.11), the energy is expressed as

$$W = \frac{1}{2}\,\tau_1\,\gamma_a = \frac{1}{2}\,\mu\,\gamma_a^2. \tag{3.13}$$

As a measure of the damping characteristics, the energy loss ΔW may be used, but since it is a function of strain amplitude γ_a, the energy loss itself is not an appropriate quantity indicative of an intrinsic material property. In view of this, it is customary to take up a quantity defined as the ratio of the energy loss per cycle to the maximum stored energy. Thus, from eqns (3.12) and (3.13), one obtains

$$\frac{\Delta W}{W} = \frac{\mu' \pi \gamma_a^2}{\frac{1}{2}\mu \gamma_a^2} = 2\pi \frac{\mu'}{\mu}. \tag{3.14}$$

This quantity is related, as follows, with the loss coefficient η defined by eqn (3.8),

$$\eta = \frac{1}{2\pi}\frac{\Delta W}{W} = \frac{\mu'}{\mu} = \tan\delta. \tag{3.15}$$

This relation indicates how the energy loss manifested in terms of a hysteresis loop is correlated with the damping exhibited in a form of the phase angle difference.

Now, with reference to eqn (3.15), let the loss coefficient be rewritten as,

$$\eta = \frac{\mu' \gamma_a}{\mu \gamma_a}. \tag{3.16}$$

It is apparent that the numerator of eqn (3.16) is equal to the length $\overline{\text{Of}}$ of the hysteresis loop in Fig. 3.2 and the denominator is equal to the length $\overline{\text{ah}}$. Therefore, the simplest graphical procedure to determine the loss coefficient from the hysteresis loop would be to read off values of $\overline{\text{Of}}$ and $\overline{\text{ah}}$ from a cyclic stress–strain curve obtained from experiments and then takes their ratio as follows:

$$\eta = \frac{\text{shear stress at zero strain}}{\text{shear stress at the maximum strain}}. \tag{3.17}$$

The above procedure may be used most conveniently, as a rule of thumb, to calculate the loss coefficient, even in the case of nonlinear hysteresis curves where the linear viscoelasticity is no longer exactly applicable. This procedure is illustrated in Fig. 3.3, along with the definition of ΔW and W.

3.1.3 *Model representation by the spring–dashpot system*

The viscoelastic behaviour of a body discussed above from a general point of view can be examined in a more specific and physically understandable manner by introducing spring–dashpot models. In this type of model, the elastic property is represented by a spring and the damping characteristics is expressed by a dashpot, and they are connected in parallel or in series, as illustrated in Fig. 3.4. It is well known that the energy loss can occur from different internal mechanisms in a deforming body, but what is important to be noted here is that the dashpot can represent the energy loss characteristics due only to the viscosity, that is, the damping generated in proportion to the velocity or time rate of deformation. This kind of energy loss will be referred to as *rate-dependent damping*. In the case of cyclic loading, the *rate dependency* is manifested in such a way that the deformation depends upon frequency. Therefore the frequency-dependent characteristics are

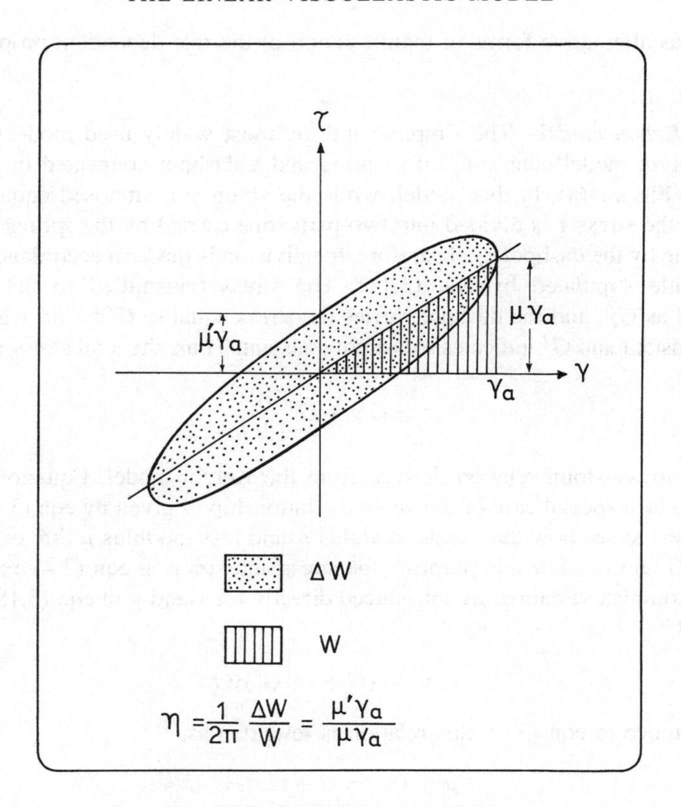

Fig. 3.3 Definition of loss coefficient.

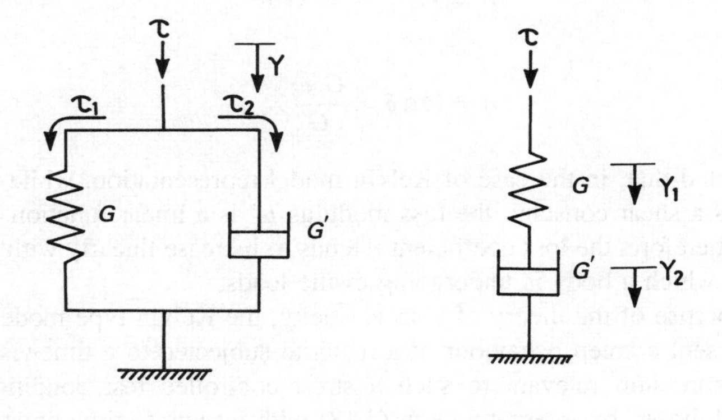

Fig. 3.4 Typical viscoelastic model.

regarded as alternative forms of manifestation of the rate-dependent properties of a material.

3.1.3.1 *Kelvin model* The simplest and the most widely used model is what is called Kelvin model consisting of a spring and a dashpot connected in parallel as shown in Fig. 3.4(a). In this model, while the strain γ is imposed equally to two elements, the stress τ is divided into two parts: one carried by the spring τ_1 and the other borne by the dashpot τ_2. Therefore, the division is made in accordance with the general rule stipulated by eqn (3.11). The stress transmitted to the spring is calculated as $G\gamma$, and the dashpot carries the stress equal to $G'\mathrm{d}\gamma/\mathrm{d}t$, where G is a spring constant and G' indicates a dashpot constant. Thus, the total stress $\tau = \tau_1 + \tau_2$ is given by

$$\tau = G\gamma + G' \frac{\mathrm{d}\gamma}{\mathrm{d}t}. \tag{3.18}$$

This is a stress–strain relation derived from the Kelvin model. Equation (3.18) is deemed to be a special case of the general relationship as given by eqn (3.5). Thus it is of interest to see how the elastic modulus μ and loss modulus μ' are expressed in terms of G and G'. For this purpose, the stress and strain in eqn (3.4) expressed in terms of complex variables are introduced directly for τ and γ in eqn (3.18) with the result that

$$\tau_{\mathrm{a}}\,\mathrm{e}^{\mathrm{i}\delta} = (G + \mathrm{i}\,\omega G')\gamma_{\mathrm{a}}. \tag{3.19}$$

With reference to eqn (3.6), this relation is rewritten as,

$$\mu + \mathrm{i}\,\mu' = G + \mathrm{i}\,\omega G'. \tag{3.20}$$

Comparing the real part and imaginary part separately, one immediately obtains

$$\mu = G, \quad \mu' = G'\omega$$

$$\eta = \tan\delta = \frac{G'\omega}{G}. \tag{3.21}$$

It is to be noted that, in the case of Kelvin model representation, while the elastic modulus μ is a shear constant, the loss modulus μ' is a linear function of angular frequency. Therefore, the loss coefficient η tends to increase linearly with increasing frequency at which a body is undergoing cyclic loads.

In the expertise of the theory of viscoelasticity, the Kelvin type model has been used to represent a creep behaviour of a material subjected to a timewise constant load. The expression relevant to such a stress-controlled test condition can be obtained, as follows, by integrating eqn (3.18) with respect to time under an initial condition, $\tau = \tau_0$ at $t = 0$,

$$\gamma = \frac{\tau_0}{G}(1 - \mathrm{e}^{-t/\bar{t}}) \tag{3.22}$$

where $\bar{t}$ is defined as $\bar{t} = G'/G$ and called the *retardation time*. If $\bar{t}$ is put equal to t in eqn (3.22), one obtains $\gamma = 0.632\ \tau_0/G$. Therefore the retardation time in turn means the length of time required to achieve 63.2% of the total strain under the condition of sustained shear stress.

3.1.3.2 *Maxwell model* This model consists of the spring and dashpot connected in series as shown in Fig. 3.4(b). In this model, the stress τ is carried commonly but the strain γ is of two parts; γ_1 coming from the deformation of the spring and γ_2 resulting from the dashpot deformation. Each component of the strain is correlated to the stress through the relations, $\tau = G\gamma_1$ and $\tau = G'\mathrm{d}\gamma_2/\mathrm{d}t$. Therefore, the stress–strain relationship for the Maxwell model is obtained through the relation $\gamma = \gamma_1 + \gamma_2$, as

$$\frac{\tau}{G'} + \frac{1}{G}\frac{\mathrm{d}\tau}{\mathrm{d}t} = \frac{\mathrm{d}\gamma}{\mathrm{d}t}. \tag{3.23}$$

In the manner similar to the above, the expressions for the elastic modulus μ and loss modulus μ' are obtained by directly replacing τ and γ in eqn (3.23) by the stress $\bar{\tau}$ and strain $\bar{\gamma}$ in eqn (3.4). Introducing eqn (3.4) into eqn (3.23), one obtains

$$\left(\frac{1}{G} + \frac{1}{\mathrm{i}\,\omega G'}\right)\tau_{\mathrm{a}} = \mathrm{e}^{-\mathrm{i}\delta}\,\gamma_{\mathrm{a}}. \tag{3.24}$$

With reference to eqn (3.6), the expressions for μ and μ' are obtained as

$$\begin{aligned} \mu &= \frac{\dfrac{1}{G}}{\left(\dfrac{1}{G}\right)^2 + \left(\dfrac{1}{G'\,\omega}\right)^2} \\ \mu' &= \frac{\dfrac{1}{G'\,\omega}}{\left(\dfrac{1}{G}\right)^2 + \left(\dfrac{1}{G'\,\omega}\right)^2} \\ \eta &= \tan\delta = \frac{G}{\omega G'}. \end{aligned} \tag{3.25}$$

It can be seen in eqn (3.25) that the loss coefficient η is inversely proportional to the angular frequency. The Maxwell model has been used to represent a relaxation behaviour of a material undergoing an imposed strain of constant magnitude. By integrating eqn (3.23) under an initial condition, $\gamma = \gamma_0$ at $t = 0$, the expression for the stress can be obtained as

$$\tau = \gamma_0\, G\mathrm{e}^{-t/\bar{t}} \tag{3.26}$$

where $\bar{t} = G'/G$ is called the *relaxation time*. Putting $\bar{t} = t$, one obtains $\tau = 0.368\ G\gamma_0$. Thus the retardation time implies the time needed to relax the initially induced shear stress by 63.2% under the condition of sustained shear strain.

As shown later, the loss coefficient is equal to twice the value of the damping

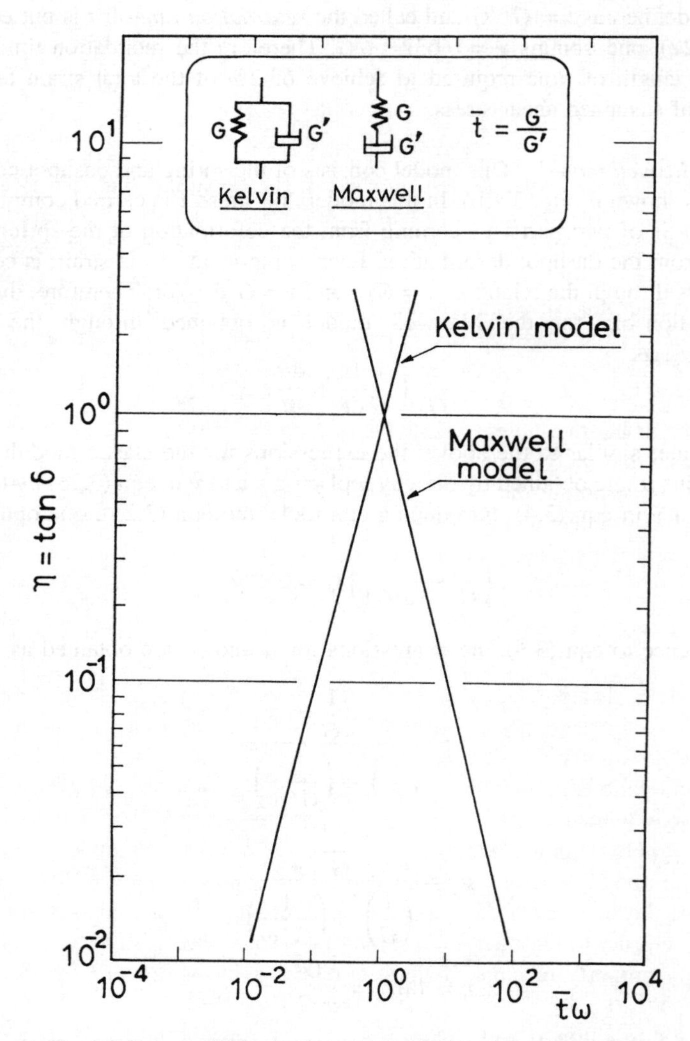

Fig. 3.5 Loss coefficients of two models as functions of frequency.

ratio which is of prime importance in seismic response analysis. Therefore it is of interest to examine the nature of the loss coefficient defined in the Kelvin and Maxwell type models. To visualize the frequency dependency, the loss coefficient η defined in eqns (3.21) and (3.25) is plotted versus the non-dimensional parameter $\omega\bar{t}$ in Fig. 3.5, where it can be obviously seen that the loss coefficient in Kelvin model increases and that in Maxwell model decreases with increasing frequency in cyclic loading.

As shown later, a vast number of laboratory experiments on soils have generally shown that the damping properties are practically independent of frequency in the range normally encountered in the seismic loading environments. Thus the application of any spring–dashpot model to practical problems should be restricted to some special cases where the frequency of loading varies within a narrow range, or where the damping itself is inconsequentially small.

3.1.3.3 *Non-viscous type Kelvin model* It is apparent that the frequency-dependent nature of the loss coefficient in any of the spring–dashpot models emerges from the use of the viscous dashpot which correlates the stress with the time rate of strain. Therefore, in order to eliminate this shortcoming, it is necessary to adopt a particular kind of dashpot which is rate independent. It is to be recalled here, however, that in the classical theory of thermodynamics, the change in entropy indicative of the energy dissipation has always been treated in terms of time rate of change in variables on the basis of the established laws of physics. Therefore even the existence of the rate-independent dashpot itself is very much in question and its physical basis may not be justified. Notwithstanding such a dilemma, it would be expedient to incorporate the rate-independent dashpot and to evolve a convenient model which can reflect actual behaviour of soils to a good degree of accuracy. The simplest model satisfying this requirement would be what may be termed the *non-viscous type Kelvin model* which is written as

$$\tau = (G + \mathrm{i}\,G'_0)\gamma \tag{3.27}$$

where G'_0 is a dashpot constant. This model consists of a spring and a rate-independent dashpot connected in parallel as shown in Fig. 3.6 where a different sign is used to signify the dashpot. Equation (3.27) can be interpreted as being a stress–strain relationship in which the stress is composed of two parts, one occurring simultaneously with strain, $\tau_1 = G\gamma$, and the other occurring 90 degree out of phase, $\tau_2 = G'_0\gamma$. It is to be noted that the stress–strain relation by this model contains the imaginary term $\mathrm{i}\,G'_0\gamma$ which is not allowed to enter into any kind of equation describing a physical law in the real world. However, the presence of this term is mandatory in order to reasonably represent the phenomenon of phase lag and hence the damping property of soils, if the rate-independent loss coefficient were to be utilized.

The elastic modulus μ and loss modulus μ' corresponding to this model can be obtained following the same procedure as before. Introducing eqn (3.4) into eqn (3.27), one obtains

$$\tau_\mathrm{a}\,\mathrm{e}^{\mathrm{i}\delta} = (G + \mathrm{i}\,G'_0)\gamma_\mathrm{a}. \tag{3.28}$$

With reference to eqn (3.6), the expression of μ and μ' for the non-viscous type Kelvin model is derived as,

$$\begin{gathered} \mu = G \qquad \mu' = G'_0 \\ \eta = \tan\delta = \frac{G'_0}{G}. \end{gathered} \tag{3.29}$$

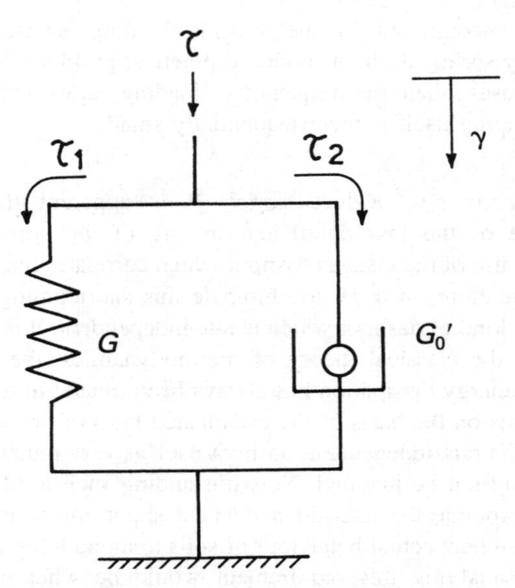

Fig. 3.6 Non-viscous type Kelvin model.

Needless to say, all the moduli take constant values independent of frequency of cyclic loading.

3.2 The nonlinear cycle-independent model

When the amplitude of shear strain is still small, the response of soils does not change with the progression of cycles and therefore the modulus and damping properties remain the same throughout the duration of cyclic stress application. However, the level of shear strain is assumed to be large enough to produce a nonlinear hysteresis loop in the cyclic stress–strain relationship. This type of soil behaviour seems to be manifested when the induced shear strain is within the range approximately between 10^{-5} and 10^{-3} as demonstrated in Fig. 3.1. Unlike the well-arranged viscoelastic model, there is no nonlinear model of any kind established on a sound physical basis. Most of the models hitherto proposed are of formal nature in favour of practical usage enabling the best fit to observed data.

3.2.1 *Framework for a nonlinear stress–strain model*

When the cyclic stress with a fairly large amplitude is applied to soils, the stress–strain curve constitutes a closed hysteresis loop as demonstrated in Fig. 3.7(b). Suppose the load is increased first to a level indicated by point a and then cycled through points bcdefa, where it is assumed that the stress reversal from unloading to reloading occurs at point d which is located symmetrically to the first reversal point a. In this stress–strain curve, it is recognized that there are two types of curves: one

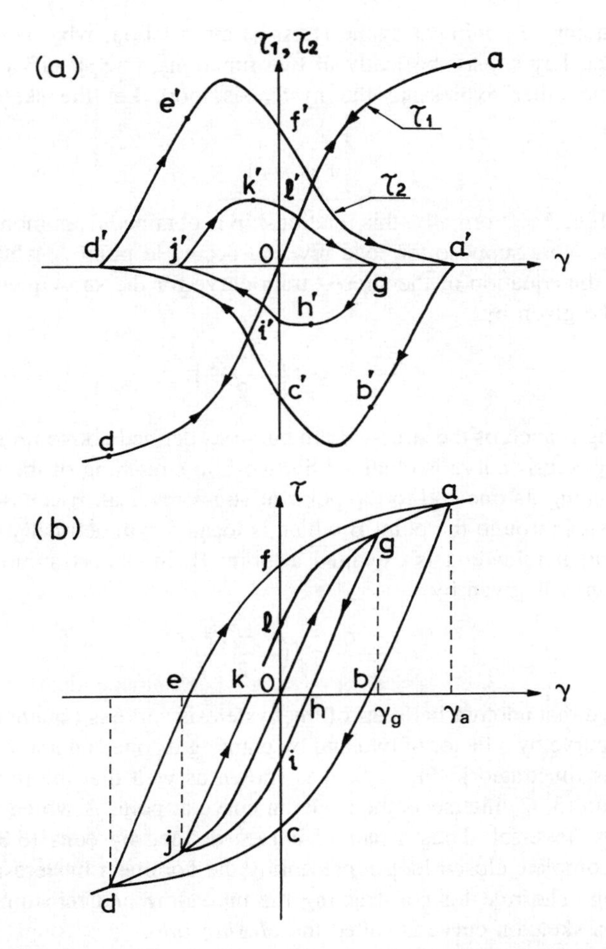

Fig. 3.7 Decomposition of nonlinear hysteresis curve into elastic and energy dissipating components.

associated with monotonic loading doa, and the other constituting a cyclic loop acdef. The former is called a *backborne curve* or *skeleton curve* and the latter is called a *hysteresis loop*. If the stress value in the hysteresis curve is subtracted on the diagram from that of the skeleton curve, two separate curves are obtained as displayed in Fig. 3.7(a). These two curves are viewed as bearing the same physical meanings as those shown in Fig. 3.2(a) which are the indication of a nonlinear behaviour of soils. As a logical extension to the nonlinear case, it may as well be mentioned that the skeleton curve and the hysteresis loop shown in Fig. 3.7(a) indicate, respectively, the elastic property and energy dissipating characteristics which are nonlinear. Because of the nonlinearity, the skeleton curve is not a straight line, nor does the hysteresis loop have rounded corners.

In constructing a nonlinear cyclic stress–strain relation, what is common to all models is that they consist basically of two functions, one specifying the skeleton curve and the other expressing the hysteresis loop. Let the skeleton curve be expressed by

$$\tau = f(\gamma) \tag{3.30}$$

as shown in Fig. 3.8. Normally, this relationship is obtained from monotonic loading tests on soils. Now suppose the load reversal occurs at point A where $\gamma = \gamma_a$ and $\tau = \tau_a$, then the equation of the stress–strain curve for the subsequent unloading is assumed to be given by

$$\frac{\tau - \tau_a}{2} = f\left(\frac{\gamma - \gamma_a}{2}\right). \tag{3.31}$$

The unloading branch of the stress–strain curve as defined above implies that a half part of the hysteresis curve is obtained by two-fold stretching of the skeleton curve and by translating its one end to the point of stress reversal. It is easily shown that the curve passes through the point B which is located symmetrically to the point of stress reversal. If reloading is executed at point B, the stress–strain curve for the reloading branch is given by

$$\frac{\tau + \tau_a}{2} = f\left(\frac{\gamma - \gamma_a}{2}\right). \tag{3.32}$$

It is also noted that another half part of the hysteresis curve is obtained by enlarging the skeleton curve by a factor of two and by planting its one end at the point of stress reversal B, as illustrated in Fig. 3.8. It is shown as well that the reloading branch defined by eqn (3.32) intersects the skeleton curve at point A which was the initial point of stress reversal. Thus a pair of curves defined by eqns (3.31) and (3.32) constitute a complete closed loop representing the nonlinear hysteresis curve in the cyclic loading. The rule for constructing the unloading and reloading branches as above using a skeleton curve is called the *Masing rule*.

Having established the basic framework of the nonlinear stress–strain relation, the next step would be to derive a series of formulae to calculate the shear moduli and loss coefficient or damping ratio. By analogy with the reasoning in the linear viscoelastic model, the nonlinear deformation characteristic is normally represented by the secant modulus which is defined as a slope of a line connecting the origin and the point of strain amplitude on the skeleton curve, as illustrated in Fig. 3.9. Thus the secant modulus G is determined merely through eqn (3.30) as

$$G = \frac{\tau_a}{\gamma_a} = \frac{f(\gamma_a)}{\gamma_a} \tag{3.33}$$

where τ_a and γ_a denote the amplitude of shear stress and shear strain, respectively, being currently used in the cyclic loading scheme.

In the same fashion as in the viscoelastic model, the energy dissipation per cycle is represented by the area enclosed by the hysteresis loop ΔW. Thus by analogy to

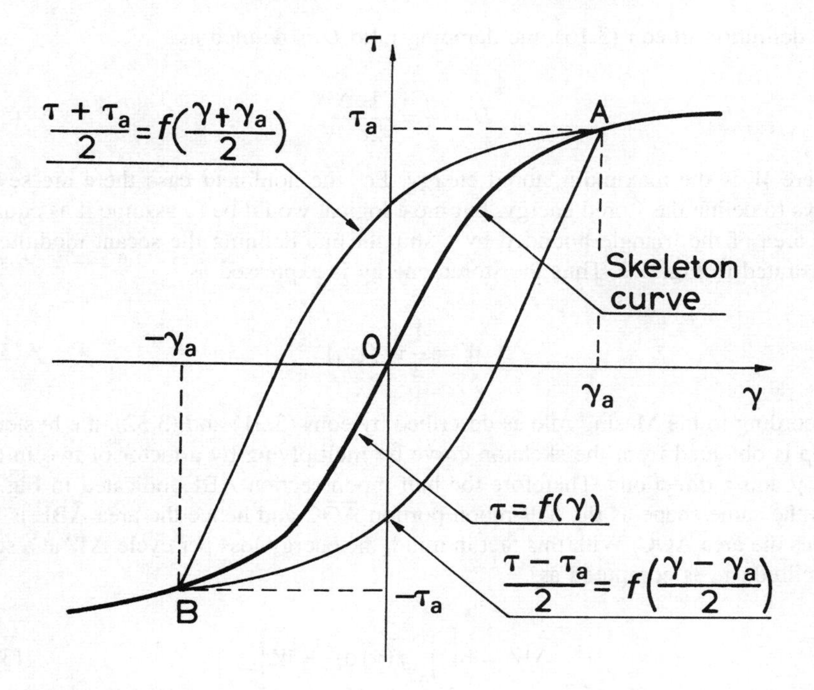

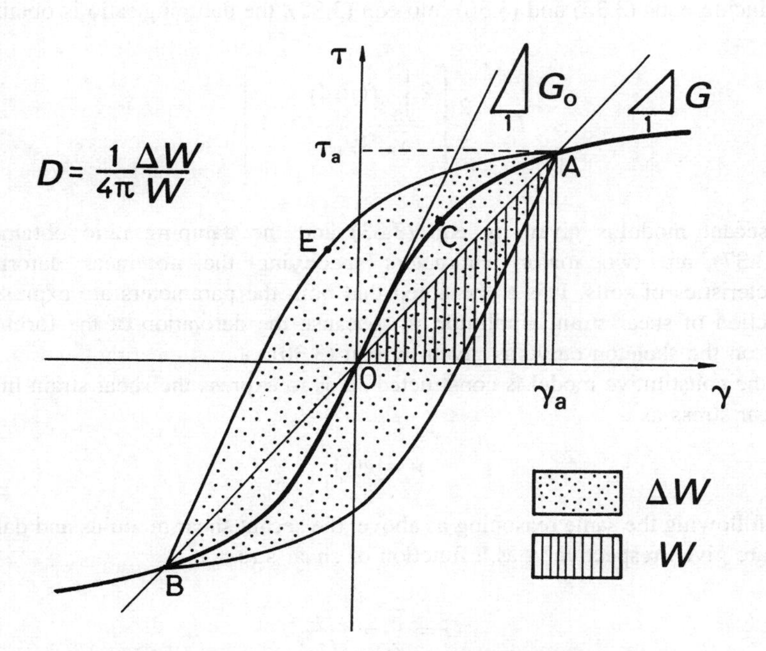

Fig. 3.9 Definition of elastic stored energy and dissipation of energy.

the definition in eqn (3.15), the damping ratio D is defined as

$$D = \frac{\eta}{2} = \frac{1}{4\pi}\frac{\Delta W}{W} \tag{3.34}$$

where W is the maximum stored energy. For the nonlinear case there are several ways to define the stored energy, but most logical would be to assume it as equal to the area of the triangle bounded by a straight line defining the secant modulus, as illustrated in Fig. 3.9. Thus the stored energy is expressed as

$$W = \frac{1}{2}\gamma_{\mathrm{a}} f(\gamma_{\mathrm{a}}). \tag{3.35}$$

According to the Masing rule as described by eqns (3.31) and (3.32), the hysteresis loop is obtained from the skeleton curve by multiplying by a factor of two in both the γ and τ directions. Therefore the half-moon section ABE indicated in Fig. 3.9 has the same shape as the half-moon portion $\overline{\mathrm{AOC}}$ and hence the area $\overline{\mathrm{ABE}}$ is four times the area $\overline{\mathrm{AOC}}$. With this fact in mind, the energy loss per cycle ΔW at a strain amplitude γ_{a} is computed as

$$\Delta W = 8\left[\int_0^{\gamma_{\mathrm{a}}} f(\gamma)\,\mathrm{d}\gamma - W\right]. \tag{3.36}$$

Introducing eqns (3.35) and (3.36) into eqn (3.32), the damping ratio is obtained as

$$D = \frac{2}{\pi}\left[\frac{2\int_0^{\gamma_{\mathrm{a}}} f(\gamma)\,\mathrm{d}\gamma}{\gamma_{\mathrm{a}} f(\gamma_{\mathrm{a}})} - 1\right]. \tag{3.37}$$

The secant modulus given by eqn (3.33) and the damping ratio obtained by eqn (3.37) are two major parameters specifying the nonlinear deformation characteristics of soils. It is to be noted that both the parameters are expressed as a function of shear strain amplitude γ_{a}, because the derivation of the formulae is based on the skeleton curve as given by eqn (3.30).

If the constitutive model is constructed so as to express the shear strain in terms of shear stress as

$$\gamma = g(\tau) \tag{3.38}$$

then, following the same reasoning as above, the secant shear modulus and damping ratio are given respectively, as a function of shear stress τ_{a},

$$G = \frac{\tau_{\mathrm{a}}}{\gamma_{\mathrm{a}}} = \frac{\tau_{\mathrm{a}}}{g(\tau_{\mathrm{a}})} \tag{3.39}$$

$$D = \frac{2}{\pi}\left[1 - \frac{2\int_0^{\tau_a} g(\tau)\,d\tau}{\tau_a\, g(\tau_a)}\right]. \tag{3.40}$$

3.2.2 *Models for nonlinear stress and strain relations*

Two kinds of material models have been used to describe the nonlinear stress–strain relations of soils. The first is a two-parameter model as represented by the hyperbolic or exponential functions. The second is a four-parameter model known as the *Ramberg–Osgood model* (R–O model). In the following, the basic aspect of these models will be discussed somewhat in detail.

3.2.2.1 *Hyperbolic model* It is logical to assume that any stress–strain curve of soils is bounded by two straight lines which are tangential to it at small strains and at large strains, as illustrated in Fig. 3.10. The tangent at small strains denoted by G_0 represents the elastic modulus at small strains and the horizontal asymptote at large strains indicates the upper limit of the stress τ_f, namely the strength of soils. The stress–strain curve bounded by these two straight lines may be expressed in differential form as

$$\frac{d\tau}{d\gamma} = G_0\left(1 - \frac{\tau}{\tau_f}\right)^n \tag{3.41}$$

where n is an arbitrary number. This expression indicates that the tangent to the stress–strain curve takes a value of G_0 at $\tau = 0$ and tends to decrease with increasing stress until it becomes equal to zero at $\tau = \tau_f$. Except for the case of $n = 1$, eqn (3.41) can be integrated, as follows, so as to satisfy the condition $\gamma = 0$ when $\tau = 0$,

$$\gamma = \frac{\gamma_r}{n-1}\left[\frac{1}{(1 - \tau/\tau_f)^{n-1}} - 1\right] \tag{3.42}$$

where a new parameter, γ_r, called reference strain is defined as

$$\gamma_r = \frac{\tau_f}{G_0}. \tag{3.43}$$

The reference strain indicates a strain which would be attained at failure stress, if a soil were to behave elastically, as illustrated in Fig. 3.10.

One of the interesting features of the stress–strain curve as given by eqn (3.41) is that it produces a constant damping ratio of $2/\pi$ as a limit when the strain becomes large. This can be proved by modifying eqn (3.37) as

$$\left(1 + \frac{\pi}{2}D\right)\gamma_a f(\gamma_a) = 2\int_0^{\gamma_a} f(\gamma)\,d\gamma. \tag{3.44}$$

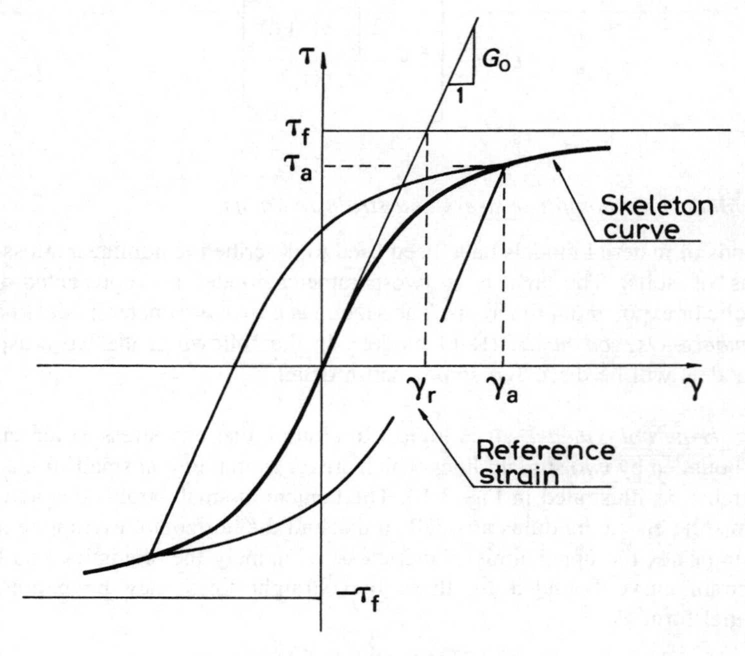

Fig. 3.10 Definition of reference strain.

If the damping ratio is assumed to take a constant value, D_0, at large strains, eqn (3.44) is differentiated as,

$$\left(1 - \frac{\pi}{2} D_0\right) f(\gamma_a) = \left(1 + \frac{\pi}{2} D_0\right) \gamma_a f'(\gamma_a). \tag{3.45}$$

Introducing eqns (3.41) and (3.42) into eqn (3.45) with $\tau = \tau_a$, $\gamma = \gamma_a$, one obtains

$$1 - \frac{\pi}{2} D_0 = \frac{\gamma_r}{n-1} \frac{G_0}{\tau_0} \left(1 + \frac{\pi}{2} D_0\right) \left[\left(1 - \frac{\tau_a}{\tau_f}\right) - \left(1 - \frac{\tau_a}{\tau_f}\right)^n\right] \tag{3.46}$$

where $f(\tau_a) = \tau_a$ and $f'(\gamma_a) = d\tau/d\gamma$ by definition. When the strain is very large, τ_a becomes equal to τ_f and thus the right-hand side of eqn (3.46) vanishes. Therefore

$$D_0 = \frac{2}{\pi} = 0.637. \tag{3.47}$$

The same conclusion can also be obtained for the case of $n = 1$ by following a similar procedure.

The stress–strain curve for the hyperbolic model can be obtained directly from eqn (3.42) by putting $n = 2$:

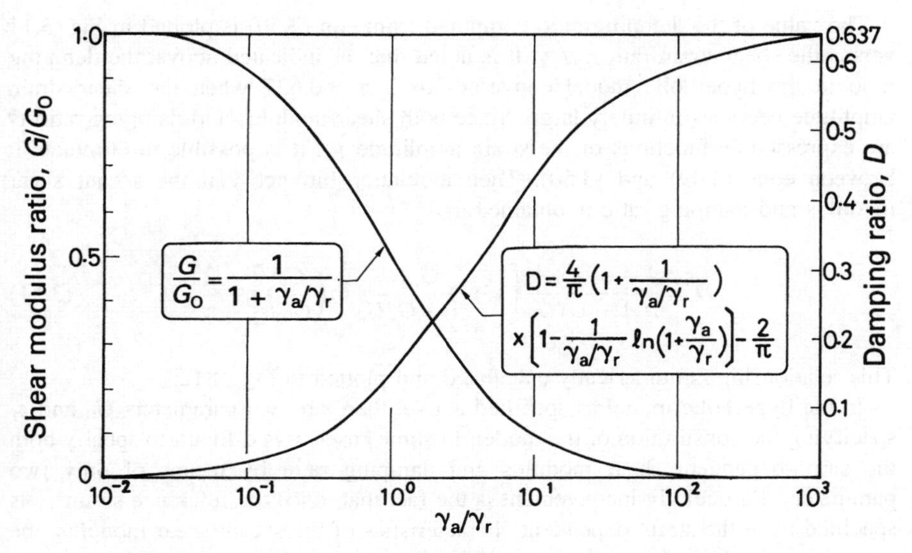

Fig. 3.11 Shear modulus ratio and damping ratio for hyperbolic model.

$$\tau = \frac{G_0 \gamma}{1 + \dfrac{\gamma}{\gamma_r}}. \tag{3.48}$$

This equation has been extensively used for representing the stress–strain relations of a variety of soils since it was used by Kondner and Zelasko (1963) and Duncan and Chang (1970). The hyperbolic model has also been used to specify the hardening rule in the theory of soil plasticity (Vermeer, 1978).

The expression for the secant modulus in the cyclic loading is obtained from eqn (3.48) as

$$\frac{G}{G_0} = \frac{1}{1 + \gamma_a/\gamma_r}, \qquad G = \frac{\tau_a}{\gamma_a}. \tag{3.49}$$

The value of secant shear modulus ratio G/G_0 computed by eqn (3.49) is plotted versus the shear strain ratio γ_a/γ in Fig. 3.11. It is noted that the secant shear modulus is reduced to half the initial shear modulus when shear strain becomes equal to the reference strain.

The equation for the damping ratio of the hyperbolic model can be derived by applying the Masing rule to the skeleton curve given by eqn (3.48). By introducing eqn (3.48) into eqn (3.37), the damping ratio is obtained as

$$D = \frac{4}{\pi}\left[1 + \frac{1}{\gamma_a/\gamma_r}\right]\left[1 - \frac{\ln(1 + \gamma_a/\gamma_r)}{\gamma_a/\gamma_r}\right] - \frac{2}{\pi}. \tag{3.50}$$

The value of the damping ratio computed from eqn (3.50) is plotted in Fig. 3.11 versus the shear strain ratio γ_a/γ_r. It is noted that, as indicated above, the damping ratio in the hyperbolic model converges to $2/\pi = 0.637$ when the shear strain amplitude becomes infinitely large. Since both shear modulus and damping ratio D are expressed as functions of the strain amplitude γ_a, it is possible to eliminate it between eqns (3.49) and (3.50). Then a relationship between the secant shear modulus and damping ratio is obtained as

$$D = \frac{4}{\pi}\frac{1}{1 - G/G_0}\left[1 + \frac{G/G_0}{1 - G/G_0}\ln\left(\frac{G}{G_0}\right)\right] - \frac{2}{\pi}. \tag{3.51}$$

This relationship is numerically calculated and plotted in Fig. 3.12.

In the hyperbolic model as specified above, there are two parameters G_0 and τ_f specifying the constitution of the model. In some cases, it is difficult to specify both the strain-dependent shear modulus and damping ratio by means of only two parameters. Particularly inconvenient is the fact that, once the reference strain γ_r is specified from the strain-dependent characteristics of the secant shear modulus, the value of strain-dependent damping ratio is automatically determined, and there is no choice for any parameter to be adjusted to achieve a good fit to experimentally obtained damping data.

Indicated as dotted area in Fig. 3.12 is the approximate range in which lie a majority of test data thus far obtained. It may be seen that, while the model

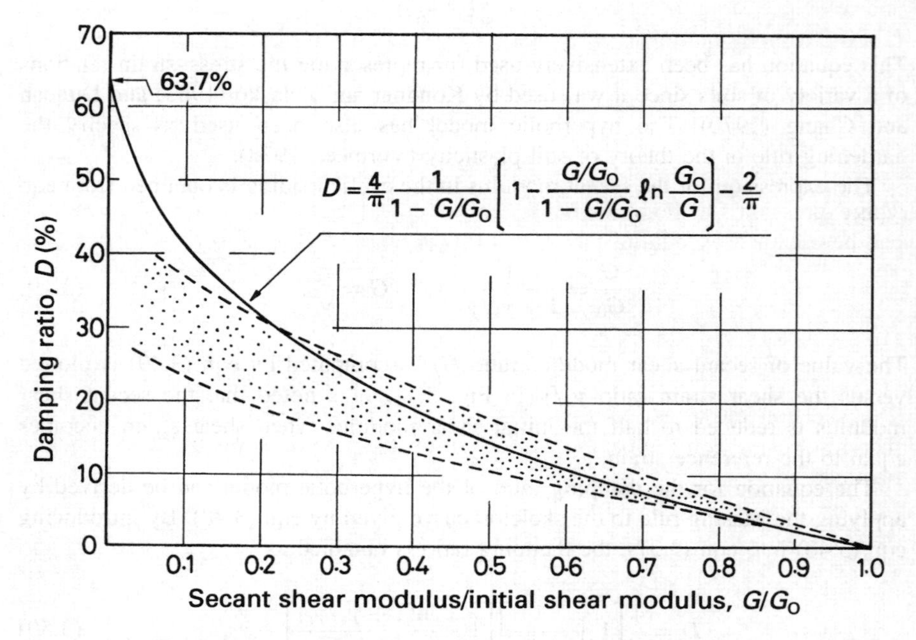

Fig. 3.12 Relation between damping ratio and shear modulus ratio.

representation is satisfactory in the range of small strains, it tends to deviate from actual behaviour of soils with increasing shear strains, thereby overestimating the damping ratio.

Another model sometimes used in the theory of plasticity is the exponential function. It is derived by integrating eqn (3.41) with $n = 1$ to give

$$\tau = \tau_f(1 - e^{-\gamma/\gamma_r}). \tag{3.52}$$

The expression for the secant modulus for the cyclic loading is obtained as

$$\frac{G}{G_0} = \frac{\gamma_r}{\gamma_a}(1 - e^{-\gamma_a/\gamma_r}). \tag{3.53}$$

The same argument as above holds true as well for the exponential model.

3.2.2.2 *Ramberg–Osgood model* The original form of the stress–strain relation for the skeleton curve is given by

$$\frac{\gamma}{\gamma_y} = \frac{\tau}{\tau_y}\left[1 + \alpha\left|\frac{\tau}{\tau_y}\right|^{r-1}\right] \tag{3.54}$$

where τ_y and γ_y are shear stress and shear strain, respectively, to be appropriately chosen, and α and r are constants. Thus, the R–O model contains four parameters that can be adjusted to achieve a best fit to experimental data. The most widely used way to specify the quantities τ_y and γ_y are to equate them to the shear strength τ_f and reference strain γ_r respectively, as suggested by Idriss *et al.* (1978) and Hara (1980). Thus by putting $\tau_y = \tau_f$ and $\gamma_y = \gamma_r$, eqn (3.54) is rewritten as

$$\tau = \frac{G_0\,\gamma}{1 + \alpha\left|\dfrac{\tau}{\tau_f}\right|^{r-1}}. \tag{3.55}$$

The expression for the strain-dependent modulus for the cyclic loading condition can be obtained by putting $\tau = \tau_a$, $\gamma = \gamma_a$ in eqn (3.55), as

$$\frac{G}{G_0} = \frac{1}{1 + \alpha\left|\dfrac{G}{G_0}\cdot\dfrac{\gamma_a}{\gamma_r}\right|^{r-1}}. \tag{3.56}$$

The skeleton curve as expressed by eqn (3.55) has a form of eqn (3.38) and, therefore, using eqn (3.40), the expression of damping ratio for the R–O model is obtained as

$$D = \frac{2}{\pi}\frac{r-1}{r+1}\cdot\alpha\cdot\frac{\left|\dfrac{G}{G_0}\cdot\dfrac{\gamma_a}{\gamma_r}\right|^{r-1}}{1 + \alpha\left|\dfrac{G}{G_0}\cdot\dfrac{\gamma_a}{\gamma_r}\right|^{r-1}}. \tag{3.57}$$

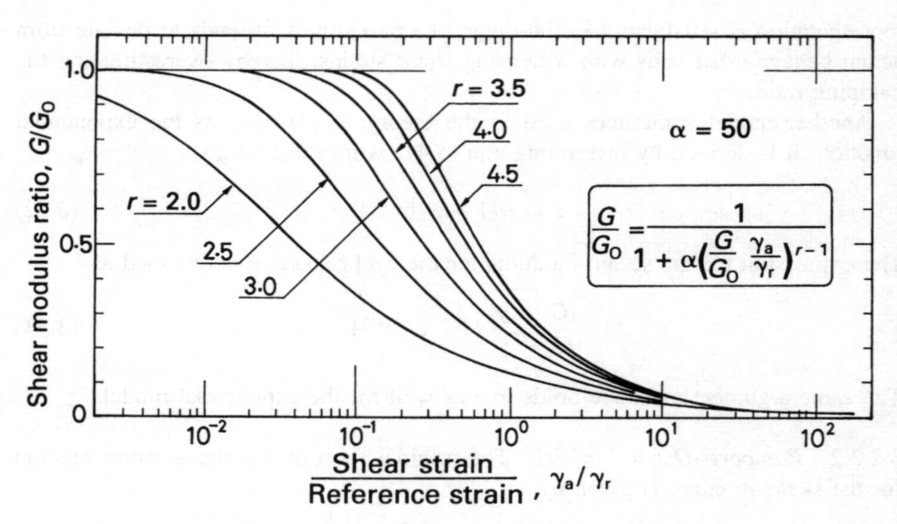

Fig. 3.13 Numerical example of Ramberg–Osgood model.

For typical sets of values for α and r, the relations of eqns (3.56) and (3.57) are presented numerically in Fig. 3.13, where it can be seen that G/G_0 value tends to decrease and the damping ratio to increase with increasing shear strain ratio γ_a/γ_r.

As in the case of the hyperbolic model, the secant shear modulus as well as the damping ratio is expressed as a function of shear strain ratio γ_a/γ_r, and therefore by eliminating γ_a/γ_r between eqn (3.56) and eqn (3.57), it becomes possible to derive a relationship between the shear modulus and damping ratio as

$$D = \frac{2}{\pi}\frac{r-1}{r+1}\left(1 - \frac{G}{G_0}\right). \tag{3.58}$$

One of the drawbacks of the R–O model is the fact that the shear strain γ_a increases in proportion to τ_a when the shear strain becomes large, as easily proven by eqn (3.55). In view of the possible range of the parameter r supposedly taking a value between 2 and 4, the quantity τ_a tends to increase indefinitely with increasing shear strain, which is inconsistent with the real behaviour of soils. One of the ways to overcome this contradiction would be to set a rule for determining the parameter α so that in no case the shear stress τ_a exceeds a value τ_f corresponding to failure. This method, suggested by Hara (1980), consists of utilizing a shear strain at failure as a new parameter. By putting $\tau = \tau_f$ when $\gamma = \gamma_f$ in eqn (3.55), the following relation is obtained,

$$\alpha = \frac{\gamma_f}{\gamma_r} - 1. \tag{3.59}$$

An advantage of determining the parameter α in the above fashion is that the soil property can be represented with a good accuracy particularly in the large strain range, and if the failure strain γ_f is chosen large enough so as not to be exceeded in any case, the potential drawback of the R–O model can be circumvented.

With respect to the determination of the parameter r, it would be reasonable to take into account the damping characteristics of soils. Inasmuch as the damping ratio D in the R–O model is correlated with the shear modulus ratio G/G_0 through eqn (3.51), the value of r can be determined, if the values of D and G/G_0 at a certain strain level is known. When the damping ratio at the failure state, D_0 is known, the value of r is determined from eqn (3.51) as

$$r = \frac{1 + \dfrac{\pi D_0}{2} \dfrac{1}{1 - G_f/G_0}}{1 - \dfrac{\pi D_0}{2} \dfrac{1}{1 - G_f/G_0}} \tag{3.60}$$

where G_f is the secant shear modulus at failure and defined as

$$G_f = \frac{\tau_f}{\gamma_f}. \tag{3.61}$$

In addition to the constants α and τ, two other parameters, namely the initial shear modulus G_0 and shear strength τ_f ought to be determined in order to completely establish the model, whether it is of the hyperbolic type or of R–O type. This aspect will be discussed in details in later chapters.

References

Duncan, J.M. and Chang, C.Y. (1970). Nonlinear analysis of stress and strain in soils. *Proceedings ASCE, SM5*, **96**, pp. 1629–53.

Hara, A. (1980). Dynamic deformation characteristics of soils and seismic response analyses of the ground. Dissertation submitted to the University of Tokyo.

Idriss, I.M., Dorby, R. and Singh, R.M. (1978). Nonlinear behaviour of soft clays during cyclic loading. *Proceedings ASCE, GT12*, **104**, pp. 1427–47.

Kondnor, R.L. and Zelasko, J.S. (1963). *A hyperbolic stress–strain formulation of sands*. Proceedings 2nd Pan American Conference on Soil Mechanics and Foundation Engineering, pp. 289–324.

Schnabel, P.B., Lysmer, J. and Seed, H.B. (1972). *SHAKE, a computer program for earthquake response analysis of horizontally layered sites*. Report No. 72–12, University of California, Berkeley.

Vermeer, P.A. (1978). A double hardening model for sand. *Geotechnique*, **28**, pp. 413–33.

4

APPARATUS AND PROCEDURES FOR LABORATORY TESTS

4.1 Triaxial test apparatus

The triaxial test apparatus has been widely in use for testing cohesionless soils in the laboratory under both monotonic and cyclic loading conditions. While there are several models with some variations in details, the basic design is the same as the conventional triaxial apparatus used for testing cohesive soils. In the conventional type developed in the UK, the rods or columns to support the top cap and piston are posted outside the triaxial cell. In this type, after the cap is mounted on top of the specimen, the cell, the top cap and the columns are assembled together and then the top cap is fastened to the column. At this stage, the specimens cap is perching by itself within the cell without any side support. Then the loading piston is lowered to touch down on top of the specimens cap. In this operation, the specimen needs to stand upright in order to achieve a precise connection. To elude this difficulty, another type of triaxial cell as shown in Fig. 4.1 has been in use particularly in Japan. In this type of equipment, the cell cap is first put in place supported by the columns. Then the loading piston is lowered and can be connected to the cap of the specimen by hand, because there is no cell yet at this stage. The load cell and displacement gauge are assembled manually. Then the triaxial cell is put in place and fixed to the top cap of the cell. This type of manual handling is helpful particularly when the piston and specimens cap is to be connected for transmitting extensional force. This upward tensional force is induced in the vertical piston when the specimen undergoes the triaxial extension. Returning to the test assembly in Fig. 4.1, the load cell for monitoring the axial load is encased in the triaxial chamber to make it free from the piston friction. To provide cushion for impulsive loading, air is normally retained in the upper part of the cell, particularly when cyclic loading is executed.

One of the most important features required of the cyclic triaxial test is that it should be capable of applying extensional loads to the specimens cap so that a state of triaxial extension can be produced cyclically in the specimen without changing the chamber pressure. This is necessary to achieve the so-called two-way loading in which cyclic stresses reverse its direction between the triaxial compression and extension. For this purpose, the vertical piston should be firmly connected to the specimens cap.

An example of the assemblage for the cyclic triaxial test is shown in Fig. 4.2. In this system, air pressure is generated by a compressor and transmitted to two different static loading paths. The first is the chamber pressure system. The air

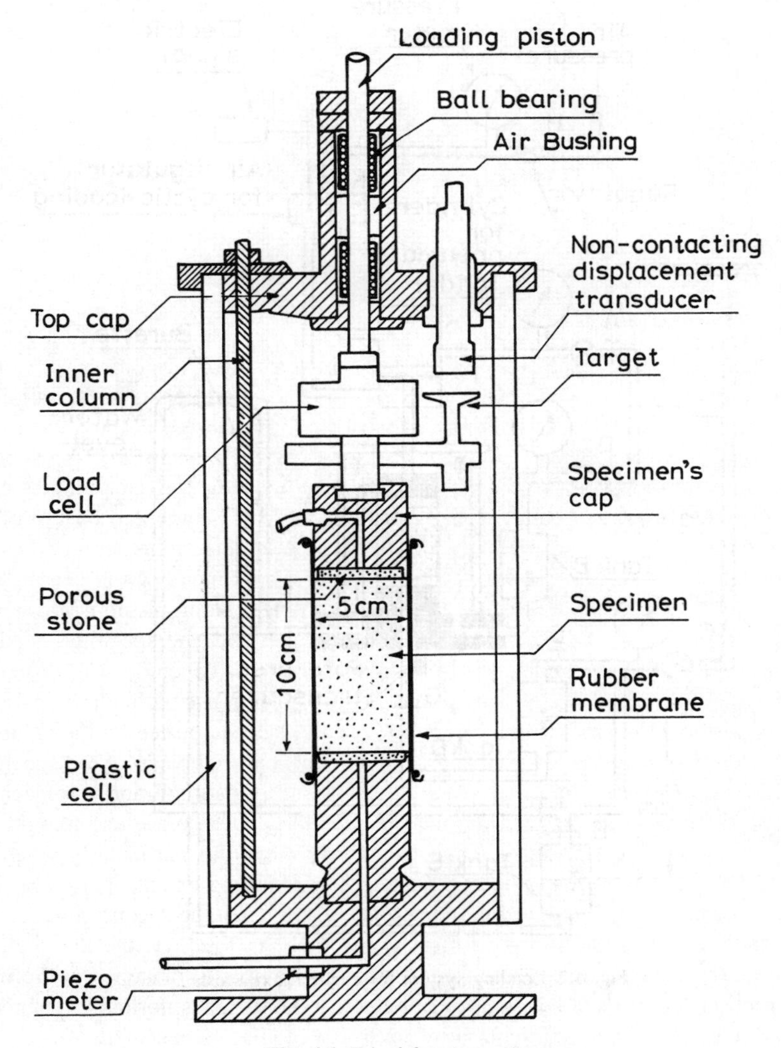

Fig. 4.1 Triaxial test apparatus.

pressure controlled by B is transmitted to a tank B after being reduced to a desired pressure by a pressure regulator. This air pressure is transmitted to water leading to the triaxial chamber. Since the vertical stress induced by the chamber pressure is smaller than the horizontal stress due to the existence of the vertical rod within the cell, it is generally necessary to have another system to apply an additional vertical stress which can be achieved by the regulator A indicated in Fig. 4.2.

The second loading system is the one for supplying water into the specimen and

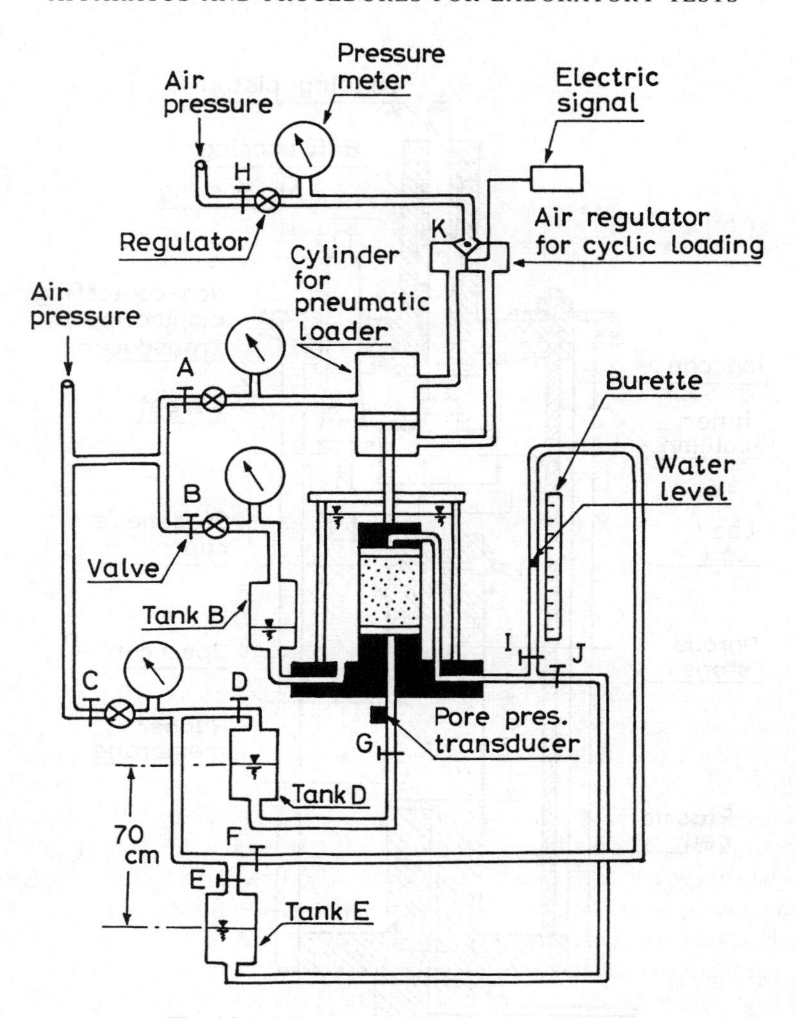

Fig. 4.2 Loading system for cyclic triaxial tests.

for applying the back pressure. The valve–regulator connections indicated by D and E are the intakes of air pressure for these purposes. The tank D in Fig. 4.2 is to supply pressurized water into the specimen through the base pedestal and tank E is used for applying back pressure through the top drainage line. The air–water interface in the burette pipe serves for monitoring the volume of water coming out of the specimen during consolidation. If the tanks D and E are placed at the same elevation, the back pressure applied at the top and bottom of the specimen can be made equal. If the tank D is placed on a table with a higher elevation, say, 70 cm, than the table of the tank E, then a differential head is generated which can be used to circulate water through the specimen. To expel air bubbles from the specimen,

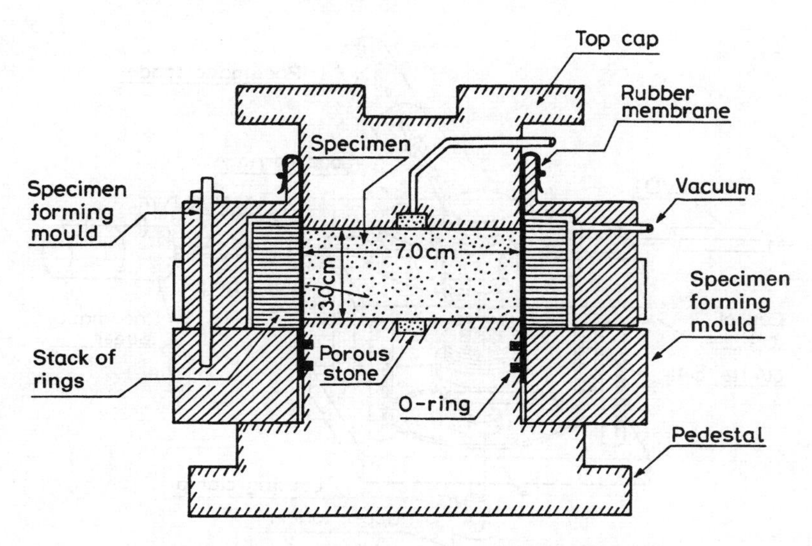

Fig. 4.3 Specimen forming mould for simple shear tests.

circulation of water under a small differential head becomes necessary. The water circulation can be achieved by connecting the pipe from the tank E directly to the top drainage line.

The cyclic axial stress is produced either by means of a pneumatic loader or an electro-hydraulic system. In the pneumatic system, an air regulator is used to control the movement of the loading piston with a desired amplitude and frequency. A function generator is used to feed signals to the regulator. Signals of irregular time sequence retrieved from a magnetic tape can also be fed to the air regulator. The loading ram is driven by the compressed air which is controlled by the air regulator K in Fig. 4.2. The cyclic loader connected to the piston of the triaxial apparatus is thus capable of producing any irregular wave form to be transmitted to the soil specimen within the cell. Any time sequence of irregular wave forms stored on a magnetic tape or computer diskette can be retrieved as an analog command and transmitted to the loader which is driven by air pressure.

4.2 Simple shear test apparatus

If connected to the cyclic loader, any type of simple shear test apparatus can be used for testing soil specimens under cyclic or dynamic loading conditions. One of the models as it sits with the forming mould is shown in Fig. 4.3. The specimen enclosed by the rubber membrane is flanked by a stack of ring-shaped Teflon rings so that the simple shear mode of deformation can be produced by inhibiting the displacement in the lateral direction. When the vertical stress is applied at the stage of consolidation the lateral stress is induced in accordance with the K_0 condition. When the cyclic stress is applied undrained, the pore pressure builds upon within the

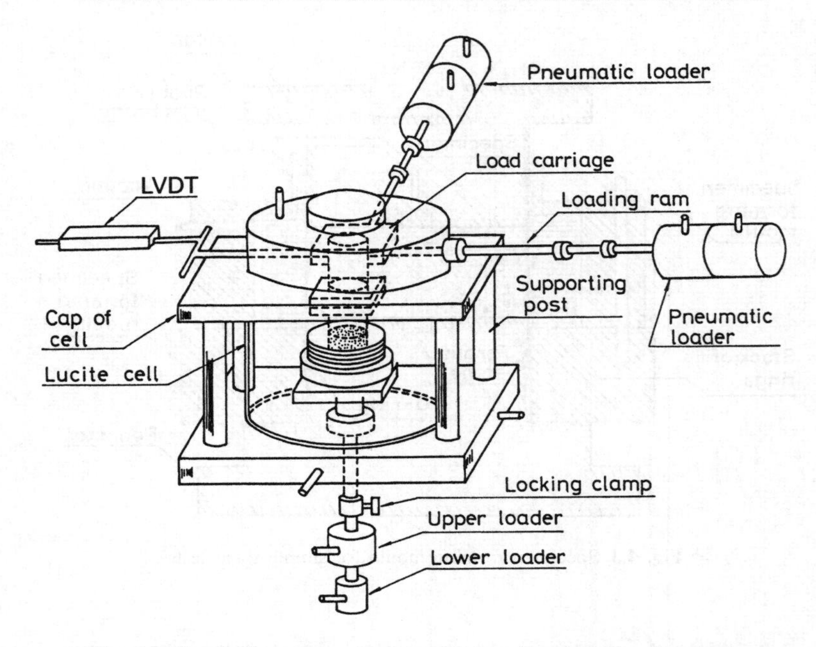

Fig. 4.4 Simple shear test apparatus with two-directional loading device.

saturated specimen and reaches a value in excess of the initial lateral stress induced under the K_0 condition. In this situation, the rubber membrane tends to expand and even to extrude from a thin space at the upper periphery of the specimen. To avoid this undesirable situation, the simple shear device as shown in Fig. 4.3 needs to be put inside the chamber. If a cell pressure equal to the vertical stress is applied to produce initially an isotropic state of consolidation the simple shear test can be performed without difficulty. The conduct of such tests is described by Ishihara and Yamazaki (1980). It is generally difficult to perform simple shear tests in ideal conditions because of several other disadvantages such as non-uniformity of strain distribution. If the cyclic loaders are connected to the specimens top cap in two horizontal mutually perpendicular directions, multidirectional simple shear tests can be conducted. This type of arrangement is shown in Fig. 4.4.

4.3 Torsional shear test apparatus

A solid cylindrical or hollow cylindrical specimen can be tested in the torsional test apparatus. The tests on solid cylindrical specimens have a shortcoming in that the strain distribution is not uniform in the radial direction in the horizontal plane of the sample. To minimize this effect, the use of the hollow cylindrical samples has been preferred in recent years.

There are several design models in use, and one of the pieces developed in Japan is displayed in Fig. 4.5. In the this type of apparatus, four components of stress,

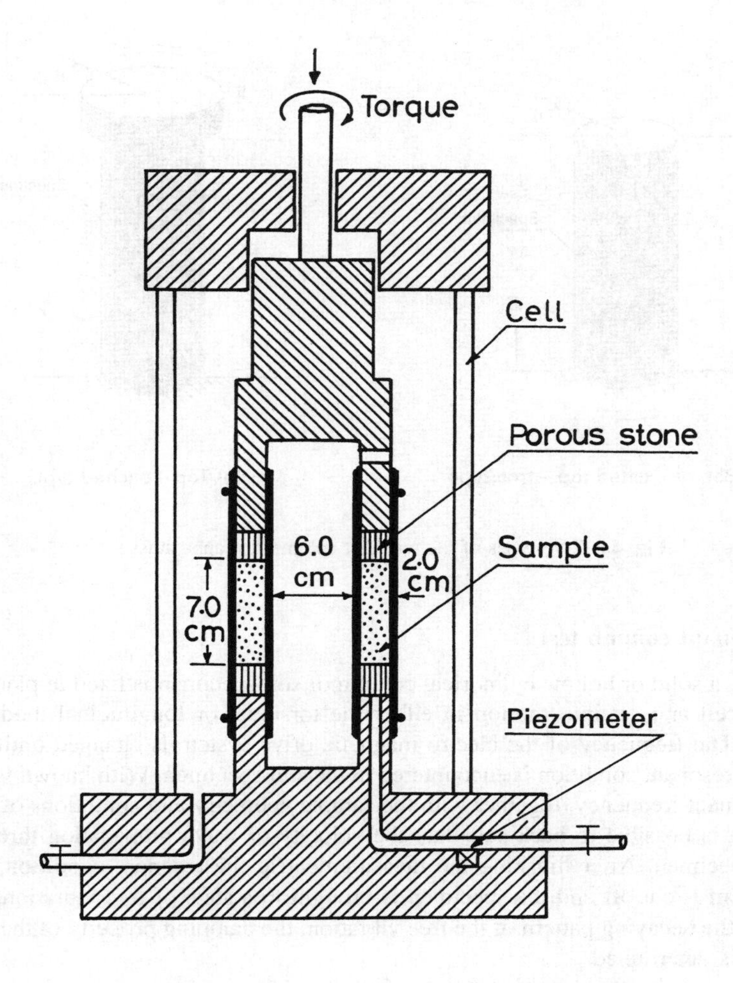

Fig. 4.5 Hollow-cylindrical torsional test apparatus.

namely vertical, torsional, and two lateral stresses (inner and outer cell pressures) can be applied to the specimen under controlled conditions. By connecting the inner and outer cells, the tests are commonly performed under the condition of equal lateral stress. By means of what is called the torsional test apparatus, the triaxial loading tests can also be performed. It is further possible to have a specimen subjected to any combined application of torsional and triaxial shear stresses. Thus complex stress change involving rotation of principal stress axes can be produced in the test specimen (Ishihara and Towhata, 1983; Towhata and Ishihara, 1985). The torsional apparatus is versatile and useful for investigating basic aspects of deformation characteristics of soils, but for practical purposes it may not be suitable.

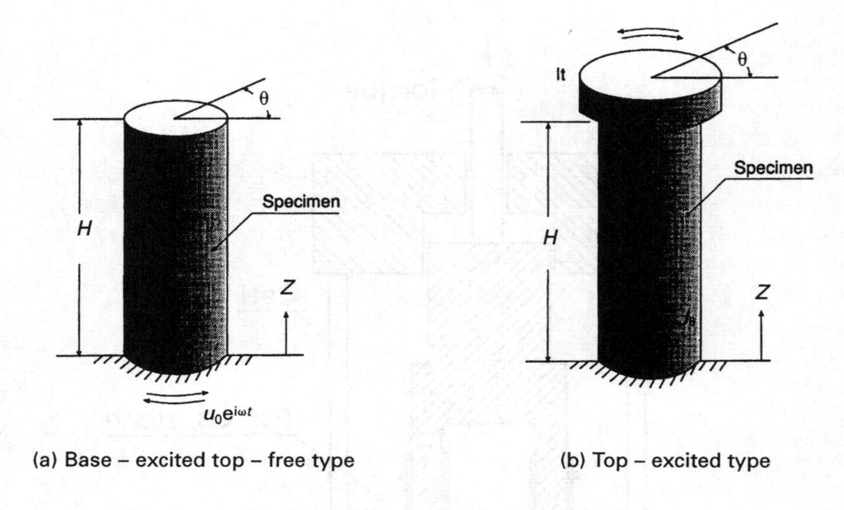

Fig. 4.6 Two types of the resonant column test apparatus.

4.4 Resonant column test

In this test, a solid or hollow cylindrical column of soil specimen is fixed in place in a triaxial cell and set into motion in either the torsional or longitudinal mode of vibration. The frequency of the electro-magnetic drive system is changed until the first mode resonant condition is encountered in the soil specimen. With known value of the resonant frequency, together with the sample geometry and conditions of end restraint, it is possible to back-calculate the velocity of wave propagation through the soil specimen. After finishing the measurement at the resonant condition, the drive system is cut off and the specimen is brought to a state of free vibration. By observing the decaying pattern of the free vibration, the damping property of the soil specimen is determined.

The above procedure is repeated several times with stepwise increased power of the driving force. As the driving force is increased, the specimen is tuned to resonate with a lower frequency because of the reduction in stiffness caused by increased level of induced shear strain. In the phase of the free vibration test that follows, an increased damping ratio would be obtained because of higher level of nonlinearity of the specimen due to the increased shear strain. As a result of several sequences of the tests, a set of data will be obtained for the velocity and damping ratio as functions of shear strains.

There are several versions in the resonant column test apparatus employing different conditions to constrain the specimen's deformation at the top and bottom ends. The most commonly used conditions at the ends are displayed in Fig. 4.6 for the case of the tests employing the torsional mode of vibration.

In the model shown in Fig. 4.6(a), the specimen is excited at the bottom and the response is picked up at the top end in terms of either velocity or acceleration. The apparatus of this type is called *fixed–free* and its details are described by Hall and Richart (1963). The other type of *fixed–free* apparatus developed by Hardin (1965) is shown in Fig. 4.6(b) in which the driving force is applied at the top of the specimen with the response pickup also placed at the top. When the driving force is applied at the top end, a somewhat massive device needs to be mounted at the top. Therefore the moment inertia of this mass in the torsional mode ought to be considered in the process of back-analysis for determining the velocity of the specimen. The theoretical background for making the analysis for this specimen–apparatus system is described below with reference to the notations indicated in Fig. 4.6(b). For simplicity of presentation, the soil tested is assumed to be elastic. The general solution for the wave equation is given as,

$$\theta = E\,\mathrm{e}^{\mathrm{i}\omega(t-z/v_\mathrm{s})} + F\,\mathrm{e}^{\mathrm{i}\omega(t+z/v_\mathrm{s})} \tag{4.1}$$

where θ is the angle of rotation and ω denotes angular frequency. The first term on the right-hand side indicates the angle of rotation due to the upward propagation of the shear wave, i.e. the positive direction of the coordinate z, and the second term indicates the component produced by the wave propagating downwards in the negative direction of z. The values of E and F indicate respectively the amplitude of upward and downward propagating shear waves. The value of E and F are determined so that the general solution of eqn (4.1) satisfies the boundary conditions

$$\begin{aligned} &\theta = 0 && \text{at } Z = 0 \\ &\theta = I_\mathrm{t}\frac{\partial^2\theta}{\partial t^2} + GJ_\theta\frac{\partial\theta}{\partial Z} = T_0\,\mathrm{e}^{\mathrm{i}\omega t} && \text{at } Z = H \end{aligned} \tag{4.2}$$

where I_t is the rotational moment of inertia of the mass mounted at the top, J_θ is the area polar moment of inertia of the cross section, T_0 is the amplitude of driving torque, and G denotes the shear modulus of the specimen. The second condition in eqn (4.2) indicates that the resistance in torsional mode due to the mass inertia plus the rigidity of the specimen must be balanced at the top by the externally applied driving force. By introducing eqn (4.1) into the boundary conditions as expressed by eqn (4.2), it is possible to determine the values of E and F. Substituting these constants back into eqn (4.1), and putting $z = H$, one can obtain the expression for the rotational angle at the top θ_t as,

$$\frac{\theta_t}{\theta_\mathrm{s}} = \frac{\sin\left(\dfrac{\omega H}{V_\mathrm{s}}\right) / \dfrac{\omega H}{V_\mathrm{s}}}{\cos\left(\dfrac{\omega H}{V_\mathrm{s}}\right) - \dfrac{I_\mathrm{t}}{I_\theta}\cdot\dfrac{\omega H}{V_\mathrm{s}}\cdot\sin\left(\dfrac{\omega H}{V_\mathrm{s}}\right)} \tag{4.3}$$

where $I_\theta = \rho H J_\theta$ and θ_s denotes the rotational angle or angular displacement which is produced at the top when the force equal to the amplitude T_0 is applied statically to the specimen. The angular frequency at resonance can be readily obtained by imposing a condition $\theta_\mathrm{t} = \infty$ in eqn (4.3) which is realized when the denominator becomes equal to zero,

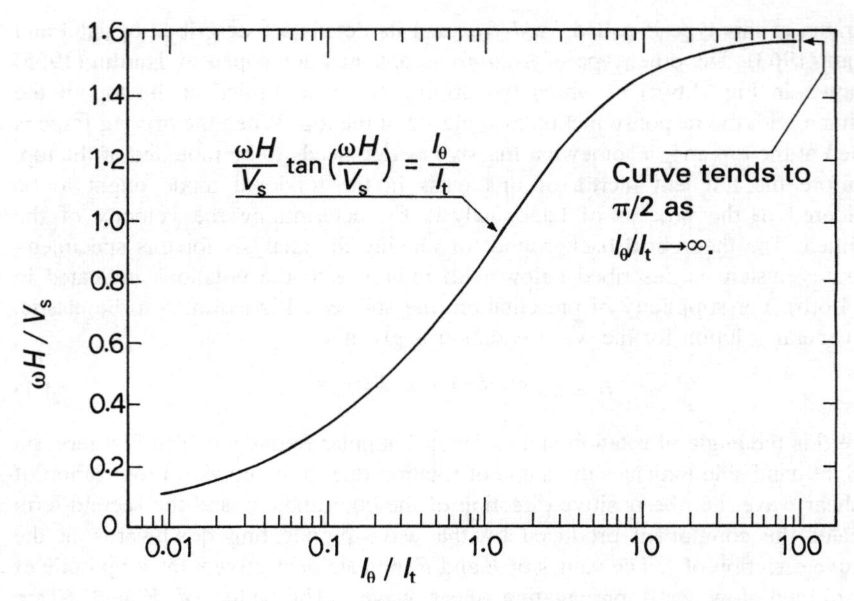

Fig. 4.7 Numerical chart for I_θ/I_t (Richart, Hall, and Woods, 1970).

$$\frac{\omega H}{V_s}\tan\left(\frac{\omega H}{V_s}\right)=\frac{I_\theta}{I_t}. \tag{4.4}$$

The relation given by eqn (4.4) is expressed in a graphical form in Fig. 4.7 for the first-mode frequency in terms of $\omega H/V_s$ plotted versus the value of I_θ/I_t. If the mass at the top is very small, as compared to the rigidity of the specimen, then by putting $I_\theta/I_t \rightarrow \infty$, the first-mode angular frequency ω_1 is obtained as

$$\omega_1=\frac{\pi}{2}\frac{V_s}{H}. \tag{4.5}$$

This extreme condition is found to be equal to the fixed–free test condition illustrated in Fig. 4.6(a). In the general case where I_θ/I_t takes a finite value, the first-mode angular frequency is determined as a function of I_θ/I_t. Suppose a certain value φ_1 is read off from Fig. 4.7 for a given value of I_θ/I_t, then the resonant angular frequency is determined as

$$\omega_1=\varphi_1\frac{V_s}{H}. \tag{4.6}$$

In terms of the resonant frequency $f_1=\omega_1/(2\pi)$, eqn (4.6) is rewritten as

$$V_s=\sqrt{\frac{G}{\rho}}=\frac{2\pi}{\varphi_1}f_1\,H \tag{4.7}$$

where ρ denotes the unit mass of the soil specimen. If a value of f_1 is obtained from the results of the resonant column test, then the shear modulus G is determined from eqn (4.7) as

$$G = \left(\frac{2\pi}{\varphi_1}\right)^2 H^2 f_1^2 \rho. \tag{4.8}$$

In the above discussion, the soil material has been assumed to behave as an elastic body. Actual soils are known, however, to exhibit more or less nonlinear behaviour while dissipating energy during the application of cyclic loads. Non-elastic behaviour may be represented, to a reasonable level of accuracy, by the linear viscoelastic model, if the loading is limited within a certain level not to induce large shear strain. By incorporating a viscoelastic model to represent the stress–strain relation, it is possible to obtain the general solution such as that given by eqn (4.1) for the waves propagating through the viscoelastic medium. By introducing an additional term in the second boundary condition of eqn (4.2) to allow for the effect of viscosity, it is also possible to derive the expression for the rotational angle θ_t at the top of the specimen when it is subjected to a harmonic excitation $T_0\,e^{i\omega t}$ at the top. Because of the complexity in the mathematical derivation, the solution for the viscoelastic case is not presented here. However it should be noted that the influence of viscosity is known to act mainly to suppress the amplitude of the response and its effect on the value of the resonant frequency is minor and negligible. As such, the frequency at resonance determined by eqn (4.7) is deemed to hold valid for the viscoelastic medium as well. When the column of the specimen is set into vibration at the resonant frequency, the amplitude of resulting motion is governed mainly by the viscosity or damping property of the material. However, the distribution of the rotational angle (or displacement) through the height of the specimen is not influenced significantly by the damping characteristics. The pattern of distribution of the rotational angle can be derived by examining the wave length at which the shear wave is travelling in the specimen when it is oscillating at a resonant frequency. Using the relation of eqn (4.7), the wave length L is given by

$$L = \frac{V_s}{f_1} = \frac{2\pi}{\varphi_1} H. \tag{4.9}$$

Suppose there is an apparatus, for example, designed to have a value of $I_\theta / I_t = 0.75$. From the chart in Fig. 4.7, a value of $\varphi_1 = \pi / 4$ is obtained. Thus, eqn (4.9) gives the relation

$$L = 8H.$$

This implies that at the resonant state, the column of the specimen is cyclically rotated with the mode of deformation such that the specimen's length is equal to $1/8$ of the wave length. The distribution of deformation in such a case is displayed in Fig. 4.8(b) in comparison to the specimen's height.

In the case of a fixed–free apparatus shown in Fig. 4.6(a) where there is no mass at the top, the value of I_θ / I_t becomes infinitely large and φ_1 takes a value of $\pi/2$ as indicated in Fig. 4.7.

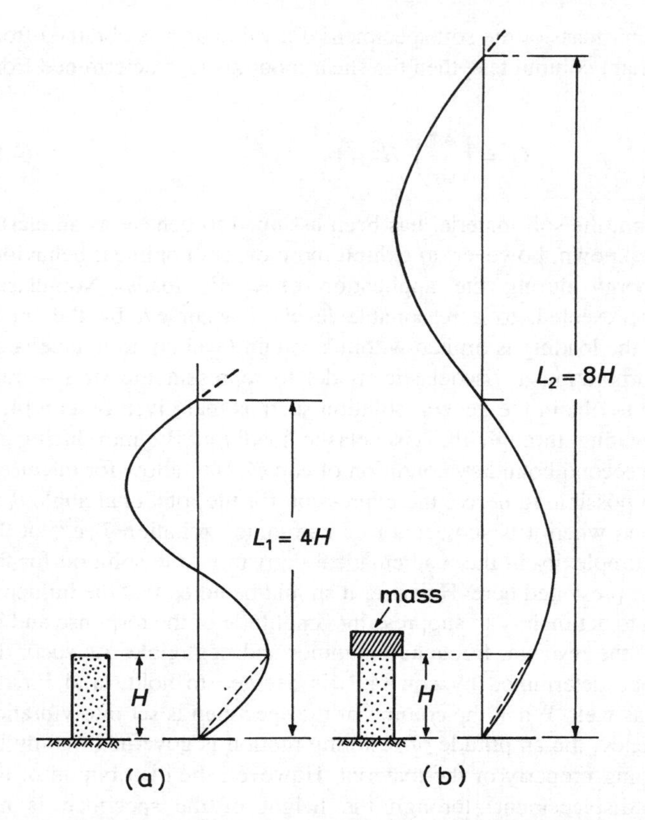

Fig. 4.8 Distribution of deformation through the height of a specimen in the first mode of vibration.

Therefore, eqn (4.9) yields

$$L = 4H.$$

The distribution of displacement for this case along the length of the specimen is shown in Fig. 4.8(a). Comparison of the distribution between the two cases indicates that the presence of the top mass in the resonant column device tends to act towards reducing the resonant frequency and hence increasing the wave length. Also shown by dashed lines in Fig. 4.8 is the linear pattern of distribution of the deformation within the specimen. It should be recalled that, if the deformation is linearly distributed, then the shear strain is uniform through the length of the specimen which is a great advantage for the testing conditions. Thus it can be mentioned that with increasing weight of the mass at the top, the shear strain induced in the resonant column test tends to become increasingly uniform through the length of the specimen.

The damping ratio of the soil specimen is determined for each stage of stepwise increased input driving force by observing an amplitude-decay curve obtained in the

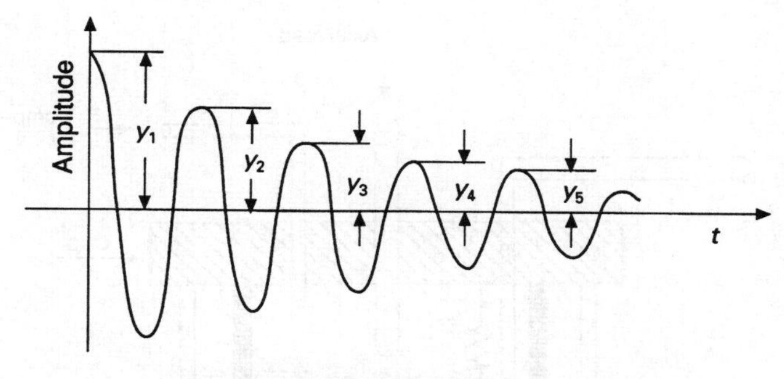

(a) Decay of free vibration

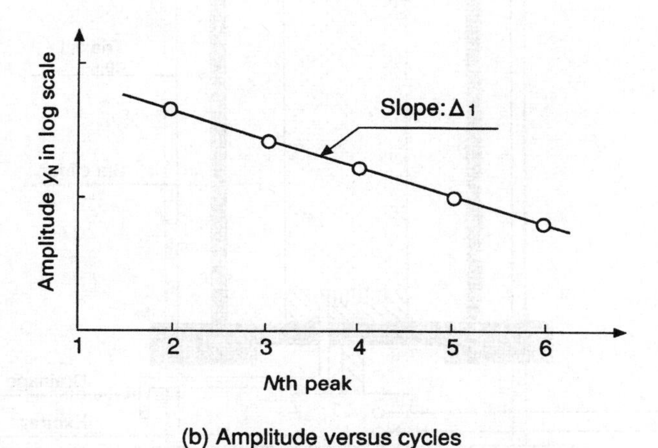

(b) Amplitude versus cycles

Fig. 4.9 Method of determining the damping ratio from the free vibration phase.

phase of free vibration following the forced vibration test. Suppose the amplitude in the free vibration decays with time as illustrated in Fig. 4.9(a). The logarithmic decrement Δ_1 is defined as

$$\Delta_1 = \log\frac{y_1}{y_2} = \log\frac{y_2}{y_3} = \cdots \qquad = \log\frac{y_{N-1}}{y_N} \tag{4.10}$$

where N denotes the number of cycles. Therefore

$$\begin{aligned} \log y_1 &= \Delta_1 + \log y_2 \\ \log y_2 &= \Delta_1 + \log y_3 \\ &\dots \\ \log y_{N-1} &= \Delta_1 + \log y_N. \end{aligned} \tag{4.11}$$

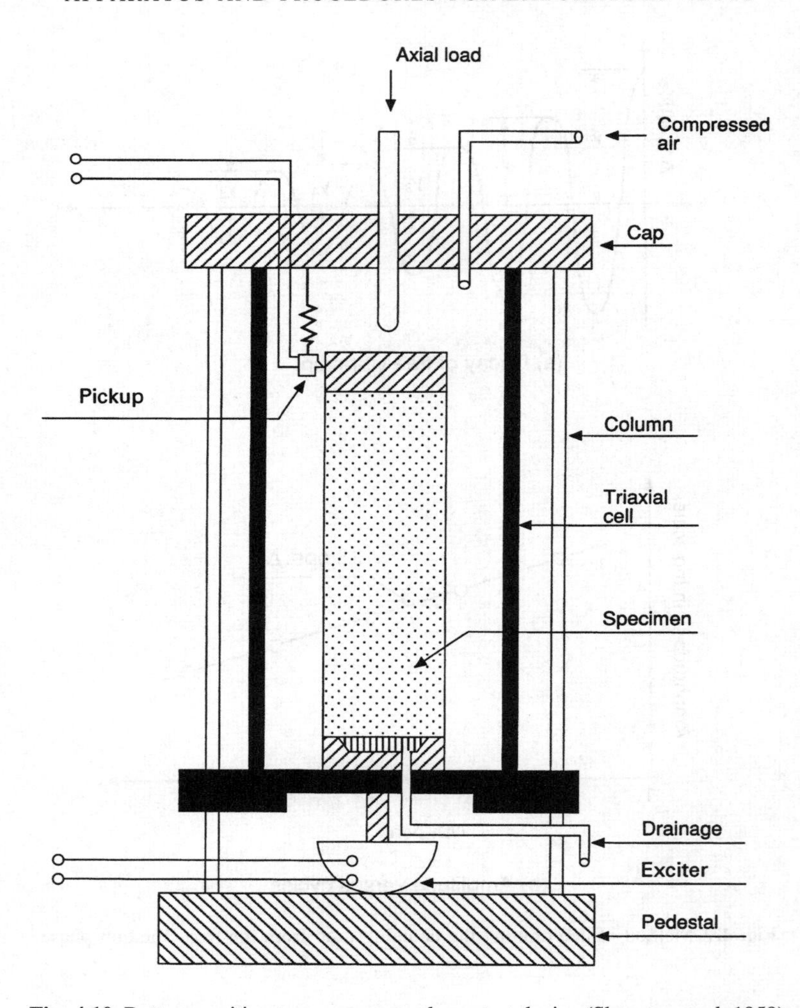

Fig. 4.10 Bottom exciting type resonant column test device (Shannon *et al.* 1959).

Adding all these relations, one obtains

$$\Delta_1 = \frac{1}{N-1} \log \frac{y_1}{y_N}. \tag{4.12}$$

If the logarithm of the amplitude is plotted versus the number of cycles as illustrated in Fig. 4.9(b), the slope of the straight line can be taken as equal to the value of the logarithmic decrement Δ_1.

From the theory of vibration, the logarithmic decrement is known to be correlated with the damping ratio D as

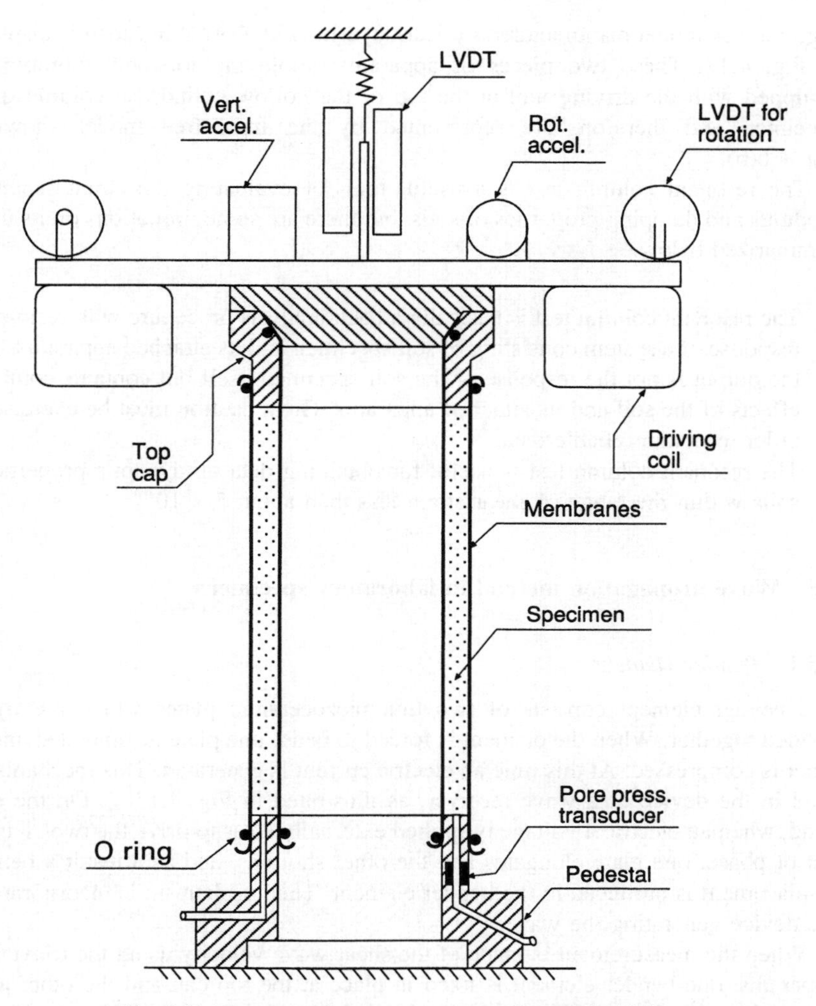

Fig. 4.11 Resonant column test apparatus (Drnevich, 1972).

$$D = \frac{1}{2\pi}\Delta_1. \tag{4.13}$$

In the resonant column test, the damping ratio D is determined at each sequence of tests with varying levels of shear strain amplitude. A model of the resonant column apparatus developed by Shannon *et al.* (1959) is shown in Fig. 4.10. In the model, the excitation is applied in the longitudinal mode at the base of a solid column specimen and the vertical displacement is monitored by a crystal pickup at the top. The specimens behaviour in this apparatus is represented by the fixed–free model shown in Fig. 4.6(a). Another model designed by Drnevich (1972) is shown in

Fig. 4.11. A model manufactured by Kuribayashi *et al.* (1974) in Japan is displayed in Fig. 4.12. These two pieces of apparatus employing torsional vibration are equipped with the driving unit at the top of the hollow cylindrical column of the specimen and therefore are represented by the fixed–free model shown in Fig. 4.6(b).

The resonant column test is a useful tool for evaluating the strain-dependent modulus and damping properties of soils, but there are some limitations on its use as summarized below.

1. The resonant column test is basically a back-analysis procedure with recourse to response of a system consisting of soil specimen and its attached apparatus. Thus the output is not the response of the soil specimen itself but contains combined effects of the soil and its attached apparatus. Great caution must be exercised in order to obtain reliable data.
2. The resonant column test is useful for obtaining data on dynamic properties of soils within the range of shear strain less than about 5×10^{-4}.

4.5 Wave propagation method in laboratory specimens

4.5.1 *Bender element*

The bender element consists of two thin piezoceramic plates which are rigidly bonded together. When the element is forced to bend, one plate is elongated and the other is compressed. At this time an electric current is generated. This mechanism is used in the device of a wave receiver, as illustrated in Fig. 4.13(a). On the other hand, when an electrical voltage is applied externally so as to drive the two elements out of phase, one plate elongates and the other shortens, and as a result a bending displacement is produced in the bender element. This mechanism is incorporated in the device generating the wave.

When the measurement is made of the shear wave velocity using the triaxial test apparatus, one bender element is fixed in place at the top cap and the other at the pedestal, as illustrated in Fig. 4.13. As shown in the details of Fig. 4.14, the element typically has a dimension of 1 mm in thickness, 12 mm in width and about 15 mm in length. In the setup of the triaxial cell, the bender element at both ends protrudes into the specimen as a cantilever, as shown in Fig. 4.14(b). When the bender element at the top is set into motion, the soil surrounding it is forced to move back and forth horizontally and its motion propagates downwards as a shear wave through the soil specimen. The arrival of this wave is picked up by the other bender element fixed at the pedestal. The arrival signal is recorded on the oscilloscope along with the signal from the source element, making it possible to know the time of travel of the shear wave. The length of the wave travel is taken to be equal to the specimen's length minus the protrusion of the two bender elements at the both ends. The bender element can also be incorporated in the oedometer test apparatus and simple shear test device. The details of the mechanism and installation of this device are

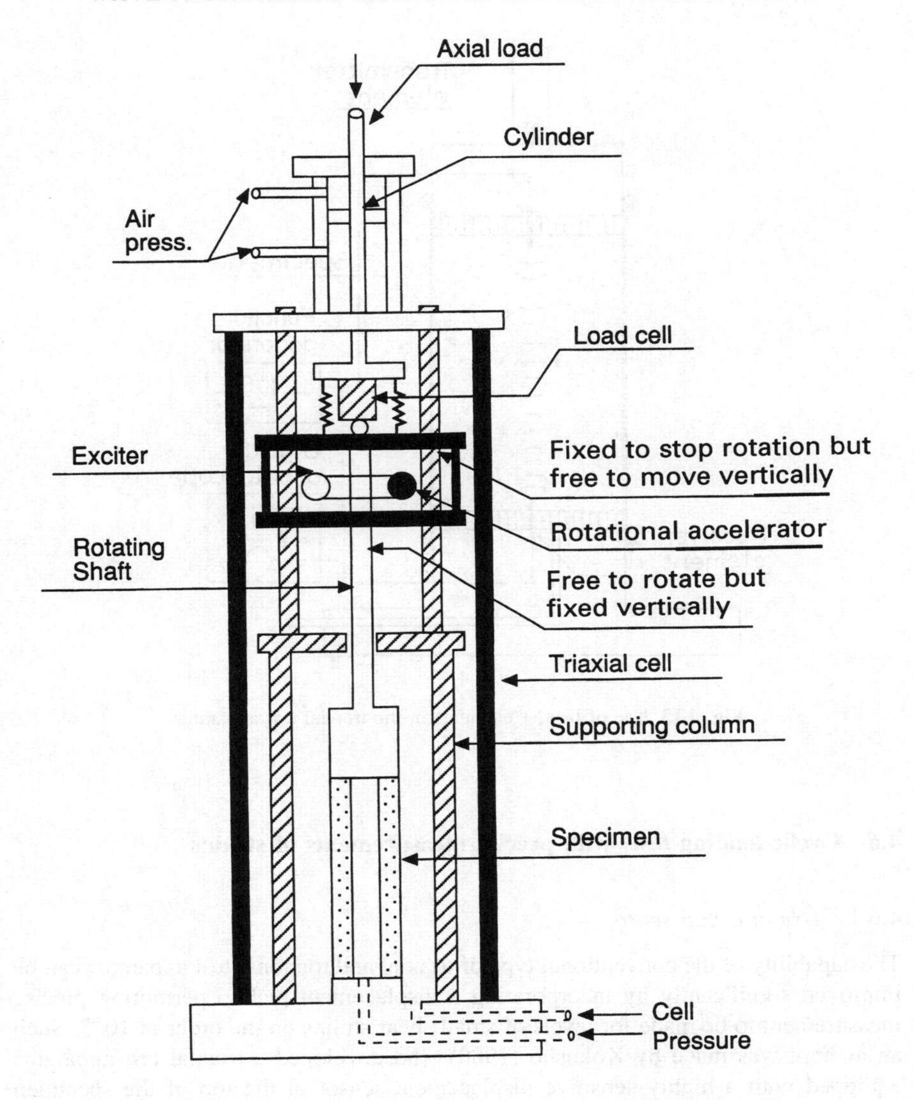

Fig. 4.12 Resonant column tests apparatus (Kuribayashi *et al.* 1974).

described by Dyvik and Madshus (1985). One of the disadvantages is that the bender element must be waterproofed to prevent short circuits, which is often difficult to achieve when testing dense or hard saturated materials. Insertion of the bender element into such hard materials can easily damage the sealing for waterproofing. The outcome of the tests is described in the works by Viggiani (1991), and De Alba and Baldwin (1991).

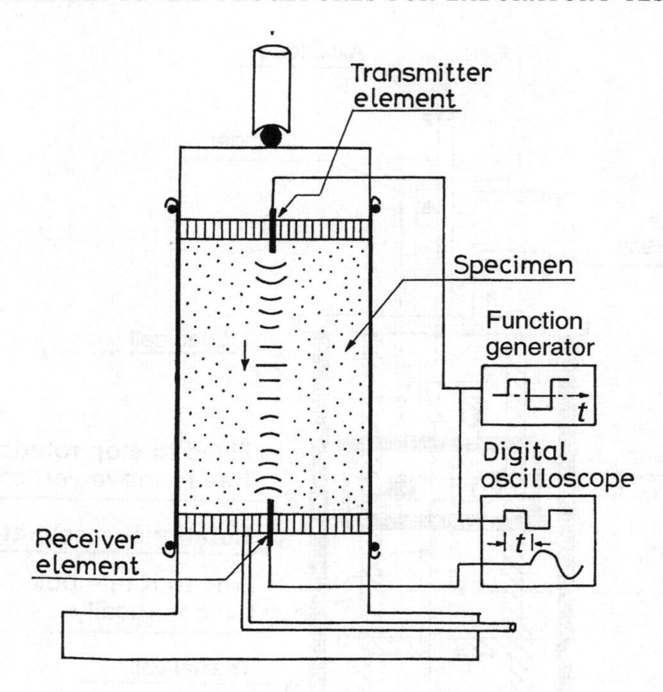

Fig. 4.13 Use of bender elements in the triaxial test apparatus.

4.6 Cyclic loading tests with precise measurements of strains

4.6.1 *Use of a gap sensor*

The capability of the conventional type of triaxial and torsional test apparatus can be improved significantly by incorporating a displacement pickup permitting precise measurement to be made for levels of small shear strains on the order of 10^{-6}. Such an attempt was made by Kokusho (1980) who developed a triaxial test apparatus equipped with a highly sensitive displacement sensor at the top of the specimen within the triaxial chamber. This sensor called a *gap sensor* or *proximity transducer* consists of two small non-contacting discs each encasing an electromagnetic coil. Even a small change in the distance between these two discs can be sensed by virtue of a change in the inductance. By installing two pieces of this sensor at two mutually opposite locations above the specimen and by taking an average of measured data, it is possible to monitor the displacement to a desired level of resolution. By means of this sensor, measurements are made possible to cover a wide range of shear strain from 10^{-6} to 10^{-3}. To ensure preciseness in measuring the axial load, a load cell is installed inside the triaxial chamber. The setup of this type of equipment developed by Kokusho (1980) is shown in Fig. 4.1.

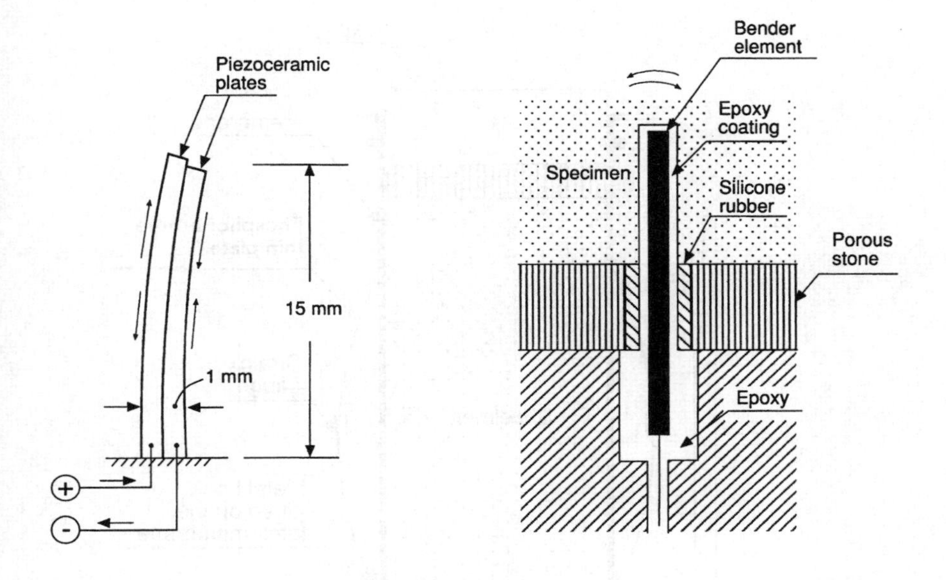

Fig. 4.14 Details of the bender element (Dyvik and Madshus, 1985).

4.6.2 *Use of a local deformation transducer*

It is likely that the use of the gap sensor tends to overestimate the shear strain because of the difficulty in providing perfect contact between a test specimen and the end plattens (top cap and pedestal at the bottoms). The measured error due to the imperfect contact is generally called the *bedding error*. The imperfect bedding tends to overestimate the strains measured by the gap sensors. It was pointed out by Tatsuoka and Shibuya (1991) that the effects of bedding error on strain measurements may be insignificant when performing tests on soft clays, but otherwise its effect would be vital particularly for testing stiff material such as soft rocks.

As an alternative technique, a device called a *local deformation transducer* (LDT) was developed by Goto *et al.* (1991) for measuring small strains of sedimentary rocks tested in the triaxial chamber. The LDT consists of a thin plate of phosphor bronze onto both sides of which a strain gauge is glued. This plate is initially bent and put in place into the metal hinges which are attached to the flank of the specimen. A pair of LDTs sit around the specimen shown in Fig. 4.15. This device is claimed to permit precise measurements of shear strain to be made even to an infinitesimally small strain on the order of 10^{-7}. The use of the LDT has been limited to the tests under monotonic loading conditions where the stress is applied slowly. However there is evidence by Tatsuoka and Shibuya (1991) indicating the coincidence of small-strain moduli between those obtained by the resonant column

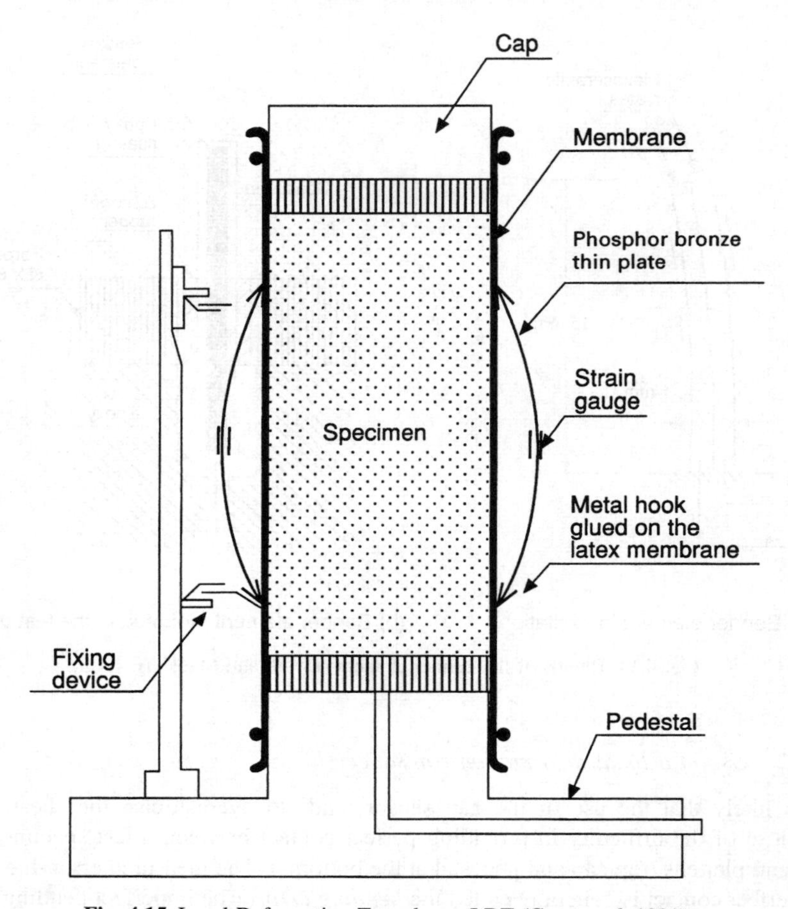

Fig. 4.15 Local Deformation Transducer, LDT (Goto *et al.* 1991).

tests and by the slow monotonic loading tests using the LDT. Thus it would be appropriate to assert that the dynamic modulus of soils at small strains can be obtained by means of a static monotonic loading, if accurate measurements are made of shear strain by using the LDT or any other device with equal capability.

References

De Alba, P. and Baldwin, K.C. (1991). *Use of bender elements in soil dynamics experiments.* Recent Advances in Instrumentation, Data Acquistion and Testing in Soil Dynamics, Geotechnical Special Publication No. 29, ASCE, pp. 86–101.

Drnevich, V.P. (1972). Undrained cyclic shear of saturated sand. *Journal of Soil Mechanics and Foundations Division*, ASCE, **98**, SM8, pp. 807–25.

Dyvik, R. and Madshus, C. (1985). *Lab measurements of G_{max} using bender elements.* Proceedings ASCE Annual Convention, Advances in the Art of Testing Soils under Cyclic

Conditions, Detroit, pp. 186–96.

Goto, S., Tatsuoka, F., Shibuya, S., Kim, Y.S., and Sato, T. (1991). A simple gauge for local small strain measurements in the laboratory. *Soils and Foundations* **31**(1), pp. 169–80.

Hall, J.R. and Richart, F.E. (1963). Dissipation of elastic wave energy in granular soils. *Journal of Soil Mechanics and Foundations Division*, ASCE, **89**, SM6, pp. 27–56.

Hardin, B.O. (1965). The nature of damping in sands. *Journal of Soil Mechanics and Foundations Division*, ASCE, **91**, SM1, pp. 63–97.

Ishihara, K. and Yamazaki, F. (1980). Cyclic simple shear tests on saturated sand in multi-directional loading. *Soils and Foundations*, **20** (1), pp. 45–59.

Ishihara, K. and Towhata, I. (1983). Sand response to cyclic rotation of principal stress directions as induced by wave loads. *Soils and Foundations*, **23**(4), pp. 11–26.

Kokusho, T. (1980). Cyclic triaxial test of dynamic soil properties for wide strain range. *Soils and Foundations*, **20**(2), pp. 45–60.

Kuribayashi, E., Iwasaki, T., Tatsuoka, F., and Horiuchi, S. (1974). *Dynamic deformation characteristics of soils—measurements by the resonant column test device*. Report of the Public Works Research Institute, Japan, No. 912 (in Japanese).

Richart, F.E., Hall, J.R., and Woods, R.D. (1970). *Vibration of soils and foundations*. Prentice Hall.

Shannon, W.L., Yamane, G., and Dietrich, R.J. (1959). *Dynamic triaxial tests on sand*. Proceedings of the 1st Panamerican Conference on Soil Mechanics and Foundation Engineering, Mexico City, Vol. 1, pp. 473–86.

Tatsuoka, F. and Shibuya, S. (1991). *Deformation characteristics of soils and rocks from field and laboratory tests*. Proceedings of the 9th Asian Regional Conference on Soil Mechanics and Foundation Engineering, Bangkok, Vol. 2, pp. 101–77.

Towhata, I. and Ishihara, K. (1985). Undrained strength of sand undergoing cyclic rotation of principal stress axes. *Soils and Foundations*, **25**(2), pp. 135–47.

Viggiani, G. (1991). *Dynamic measurements of small strain stiffness of fine grained soils in the triaxial apparatus*. Experimental Characterization and Modelling of Soils and Soft Rocks, Proceedings of the Workshop in Napoli, pp. 75–97.

5

IN SITU SURVEY BY WAVE PROPAGATION

5.1 Reflection survey

When a body wave comes across an interface of two media with different stiffnesses, the direction of its propagation is deflected in accordance with Snell's law as illustrated in Fig. 5.1. If the angle of incidence ψ_1 is smaller than a critical angle ψ_c, the wave is refracted and propagates in the second medium with an angle ψ_2 which is larger than ψ_1 (where we have assumed that the wave travels faster in the second medium). If the incident wave encounters the interface with an angle larger than ψ_c, the wave is reflected into the first medium with the same angle ψ_R to the vertical, as illustrated in Fig. 5.2.

The reflection survey takes advantage of the reflection of the P-wave (longitudinal or compressional wave) which is the fastest and so the first to arrive and hence the easiest to identify at a point of monitoring on the ground surface. In the reflection survey illustrated in Fig. 5.3, an arrival time t_d of the P-wave through a direct path from the source A to a point such as B′, C′, or D′ is monitored. Such a

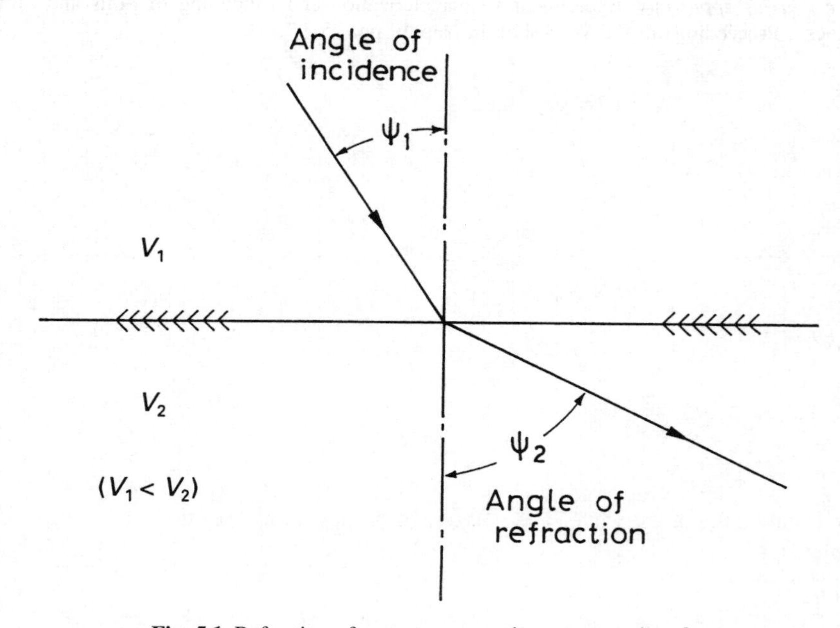

Fig. 5.1 Refraction of a wave propagation across an interface.

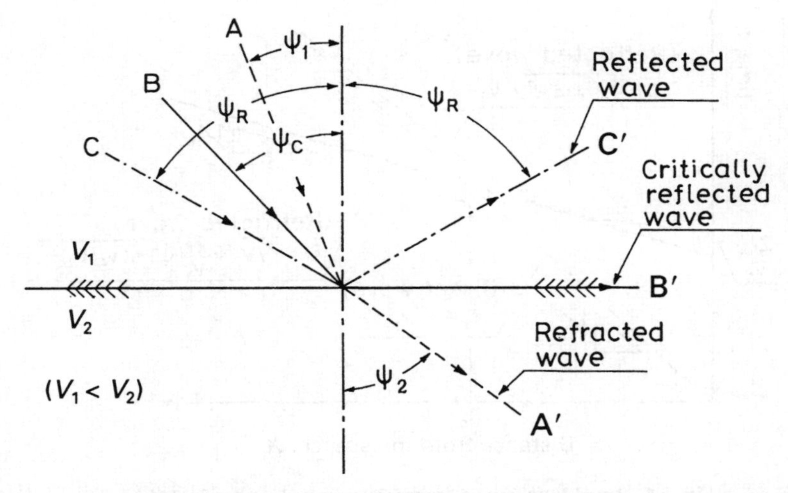

Fig. 5.2 Critical angle of incidence ψ_c differentiating between reflection and refraction.

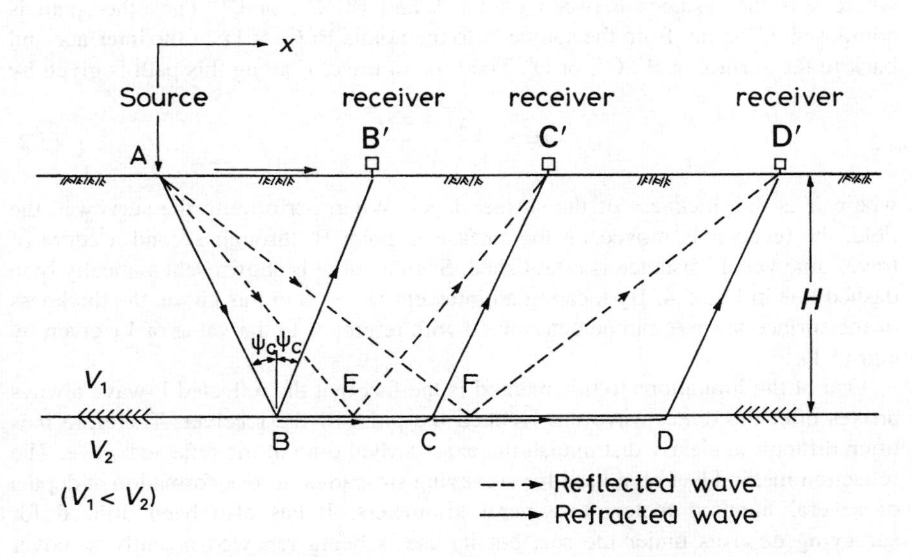

Fig. 5.3 Propagation paths for reflected and refracted waves (Richart *et al.* 1970).

wave is called the direct wave. The velocity of propagation V_1 of the surface layer is calculated by

$$V_1 = \frac{x}{t_d} \tag{5.1}$$

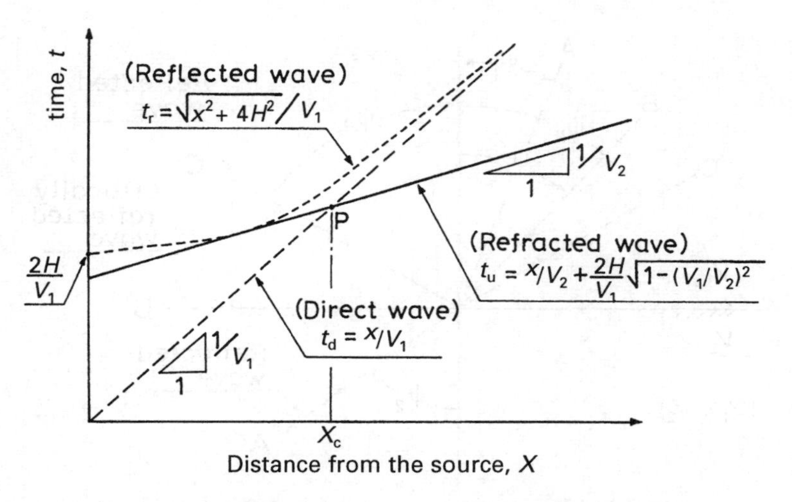

Fig. 5.4 Travel time curves for direct, reflected, and refracted waves.

where x is the distance between points A and B′, C′, or D′. The other path is composed of the ray from the source A to the points B, C, or D, at the interface and back to the surface at B′, C′, or D′. The time of travel t_r along this path is given by

$$t_r = \frac{\sqrt{x^2 + 4H^2}}{V_1} \tag{5.2}$$

where H is the thickness of the surface layer. When performing the survey in the field, the receiver is moved on the surface at point B′ through D′ and a curve of travel time versus distance is established. Such a curve is shown schematically by a dashed line in Fig. 5.4. By locating an intercept at $x = 0$ in this curve, the thickness of the surface layer H can be determined with reference to the value of V_1 given by eqn (5.1).

One of the limitations to this method is the fact that the reflected P-wave always arrives after the direct wave has reached the point of the receiver. Therefore it is often difficult to clearly distinguish the exact arrival time of the reflected wave. The reflection method has been used for surveying structures of rock formation at depths of several hundred meters to several kilometers. It has also been utilized for surveying deposits under the sea, but its use is being renewed recently to cover shallow-depth deposits of soils.

5.2 Refraction survey

This method is the simplest and most conventionally used procedure to map out the overall picture of soil and rock profiles over a wide area. In this method, an impact or explosive energy is applied on the ground surface and the front of its propagation is captured at a point some distant from the source. The distance and time of arrival

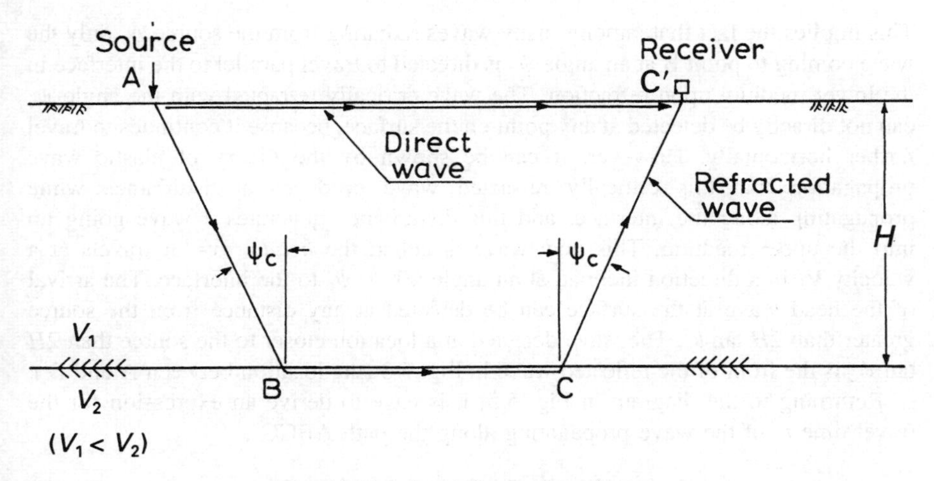

Fig. 5.5 Travel path for a refracted wave.

make it possible to calculate the velocity of wave propagation (Ballard and Mclean, 1975). Since refracted waves are utilized, it is a prerequisite for this method to have a soft surface layer underlain by a stiffer soil or rock deposit at which the waves are refracted.

The basic concept on this methodology is illustrated in Fig. 5.5 where V_1 and V_2 are the velocities of wave propagation through the surface and underlying layers, respectively. The variable H denotes the thickness of the surface layer. When an excitation is given at point A, waves are generated and propagate in all directions, but the most important are two waves: one travelling on the surface and arriving at point C′, and the other going down to point B and coming up to point C′ after travelling along the interface and passing through point C. The former is called the *direct wave* and the latter the *refracted wave*. The travel time of the direct wave t_d is given by eqn (5.1).

To compute the travel time of the refracted wave, it is necessary to specify the critical angle of incidence ψ_c with which the wave is refracted at point B and directed to propagate towards point C along the path B to C. Let the direction of incident and refracted wave be denoted by ψ_1 and ψ_2 as illustrated in Fig. 5.1. According to the Snell's law, the angles ψ_1 and ψ_2 are related with the ratio between the wave propagation velocity in two mutually contacting media as

$$\frac{\sin\psi_1}{\sin\psi_2} = \frac{V_1}{V_2}. \tag{5.3}$$

For the refracted wave propagating in the path along the interface, the angle of refraction is given as $\psi_2 = 90°$. Then the critical angle of incidence is given by

$$\sin\psi_c = \frac{V_1}{V_2}. \tag{5.4}$$

This implies the fact that, among many waves radiating from the source A, only the wave coming to point B at an angle ψ_c is directed to travel parallel to the interface in the lower medium upon refraction. The wave critically refracted with the angle ψ_c can not directly be detected at any point on the surface, because it continues to travel farther horizontally. However, it can be shown by the theory of elastic wave propagation that this critically refracted wave produces a disturbance while propagating along the interface, and this disturbance generates a wave going up into the upper medium. This new wave is called the *head wave*. It travels at a velocity V_1 in a direction inclined at an angle $90° - \psi_c$ to the interface. The arrival of the head wave at the surface can be detected at any distance from the source greater than $2H \tan\psi_c$. The wave detected at a location closer to the source than $2H \tan \psi_c$ is the front of the reflected wave in Fig. 5.3 (the location between A and B′).

Returning to the diagram in Fig. 5.5, it is easy to derive an expression for the travel time t_u of the wave propagating along the path ABCC′,

$$t_u = \frac{H}{V_1 \cos \psi_c} + \frac{x - 2H \tan \psi_c}{V_2} + \frac{H}{V_1 \cos \psi_c} \tag{5.5}$$

where V_2 is the velocity of shear wave propagation through the lower medium.

It is to be noticed that the velocity of the critically refracted wave should be taken as V_2, because it travels in the lower medium parallel to the interface. Since the velocity of propagation is assumed to be faster in the lower medium than in the surface layer, the time of travel from the point B to C would become significantly shorter. This is the reason, as discussed below, why the head wave could arrive at the receiver point earlier than the direct wave. With reference to eqn (5.4), it is possible to express the critical angle ψ_c as a function of V_1/V_2, and hence to eliminate ψ_c entering in eqn (5.5). The travel time t_u is thus rewritten as

$$t_u = \frac{x}{V_2} + 2H\sqrt{\frac{1}{{V_1}^2} - \frac{1}{{V_2}^2}}. \tag{5.6}$$

If expressed in the travel time diagram, eqn (5.6) represents a straight line with a slope of $1/V_2$ and with an intercept at $2H\sqrt{1/{V_1}^2 - 1/{V_2}^2}$, as shown in Fig. 5.4. Also shown in this diagram is the expression for the direct wave as given by eqn (5.1). It can be seen that, when the distance from the source is short enough, the direct wave arrives first, followed by the reflected wave. However, the arrival of the refracted head wave is preceded by that of the direct wave, if the distance from the source is large. The distance x_c from the source to the point at which the direct wave and refracted head wave arrive at the same time is called crossover distance. By setting $t_d = t_u$ in eqns (5.1) and (5.6), the expression for x_c can be obtained as

$$x_c = 2H\sqrt{\frac{V_2 + V_1}{V_2 - V_1}}. \tag{5.7}$$

In the practice of the refraction survey, monitoring of wave arrival can be performed by successively moving a receiver along a linear array on the ground surface, while holding the source at a fixed point. By plotting the arrival time of the direct and refracted waves against the distance from the source, two travel time curves such as those shown schematically in Fig. 5.4 can be obtained. By reading off the slopes of these two lines, values of V_1 and V_2 are determined. At the same time the value of the crossover distance x_c can be read off. Using the value of x_c in eqn (5.7), the thickness of the surface layer H is determined.

The above concept holds valid for both shear wave and longitudinal wave propagation. As is well known, the longitudinal wave always propagates at a velocity faster than that of the shear wave. Whatever means may be used for exciting the source, the two body waves are always generated and therefore, on the recorded data from the receiver there are always traces of arrival of these waves plus noises of unknown origins. Under these circumstances, it is often difficult to distinguish clearly the exact time of the reflected wave arrival for the shear wave propagation, because it always arrives after the longitudinal wave. It is rather easier to identify the travel time of the longitudinal wave because it always arrives first at the receiver point. For this reason, the method of refraction survey as described above is generally applied in practice for P-wave logging only.

5.3 Uphole and downhole methods

Uphole and downhole surveys are performed by monitoring longitudinal or shear waves propagating vertically in soil deposits in the vicinity of a borehole. The uphole method consists of generating waves at a point in the borehole and monitoring its arrival on the surface. Explosives are fired generally as a source in the borehole, and both shear wave (S-wave) and longitudinal wave (P-wave) are generated simultaneously. The arrivals of these two waves are monitored by several receivers placed in an array on the ground surface. In soil deposits with low to medium stiffness, the propagation of the P-wave is sufficiently faster than that of the shear wave and, therefore, the later arrival of the shear wave can be discerned on the monitored record. In the case of stiff soils and rocks, the difference in the propagation velocity of these two waves is not so pronounced and so it becomes difficult to discern the trace of the shear wave arrival.

In the case of the downhole logging, a geophone or hydrophone is clamped to the wall of a borehole, as illustrated in Fig. 5.6, to monitor the arrival of the wave front propagating downward from the source on the ground surface. As the source, a wooden plate clamped on the surface is hit manually by a hammer. If the plate is hit horizontally, it generates a shear wave polarized in the horizontal direction. The longitudinal wave (P-wave) is generated by hitting the plate vertically or by dropping a weight onto it. In the downhole method, the geophone is lowered to the desired depths successively while generating the wave each time on the surface. The downhole survey can be conducted effectively in crowded city areas where available space is limited. The use of this method has been prevalent particularly in Japan, because it can be combined with the drilling for the standard penetration test (SPT). The data are normally plotted in a form of time versus distance from the source. Figure 5.7 is a typical example of S-wave and P-wave data obtained from downhole

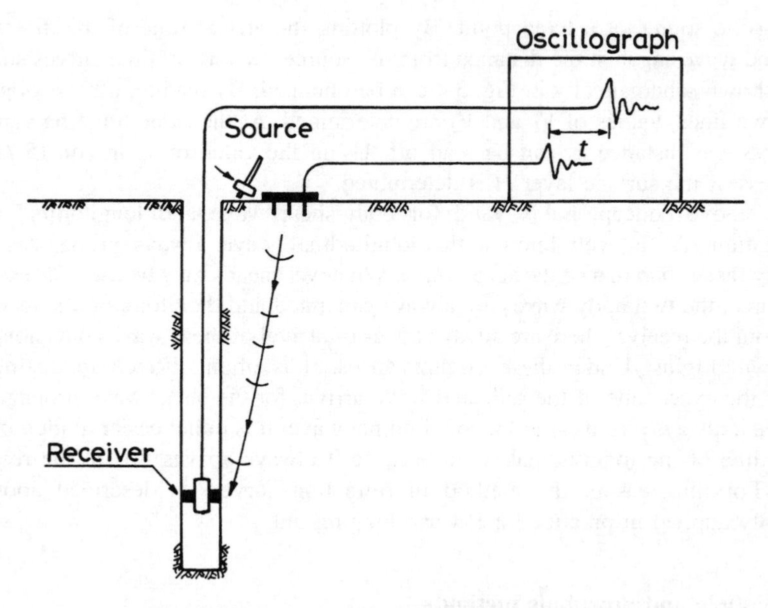

Fig. 5.6 Velocity logging by the downhole method.

logging. By connecting the data points on several segments of straight lines, the velocity of propagation and thickness of each layer can be delineated. Soil stratigraphy and SPT profiles established in the same hole during the logging are also shown in Fig. 5.7. As seen in the figure, the minimum thickness of a layer that can be identified by the downhole method would be on the order of 2–3 m and most of the data are gross averages of the velocities over several thin layers.

5.4 The crosshole method

In this method, a shear wave or a compressional wave is generated in a source borehole and its propagation in the horizontal direction is detected by means of receivers placed in two or three other adjacent boreholes in a linear array. The layout of the test is shown in Fig. 5.8. The impulse energy in the source borehole is applied by various methods. When the logging is performed with SPT, the drop of the hammer may be utilized to generate a compressional wave at the bottom of the borehole. When a shear wave is to be generated, a specially designed in-hole anchor and hammer assembly is used. This assembly is lowered into the source borehole to a desired depth by a tension cable and wedged to the wall of the borehole by expanding the clamps. By dropping a hammer on top of the clamped anchor, a downward shearing motion is generated. A special device can also be attached to this anchor so that hitting can be made upward from the bottom. Thus the vertically oriented impulsive force is applied to the borehole wall both downwards and upwards. In the adjacent boreholes, geophones sensing the vertical velocity are

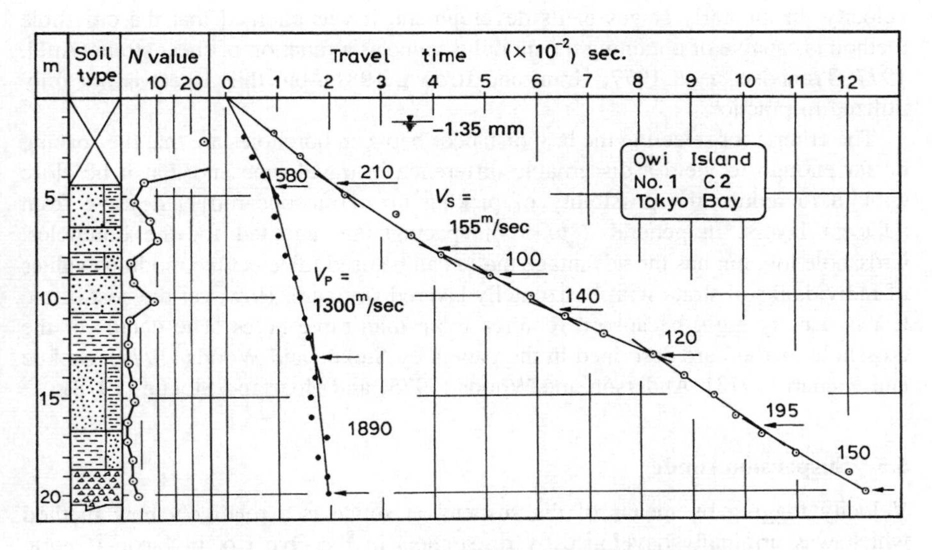

Fig. 5.7 An example of velocity logging by the downhole method.

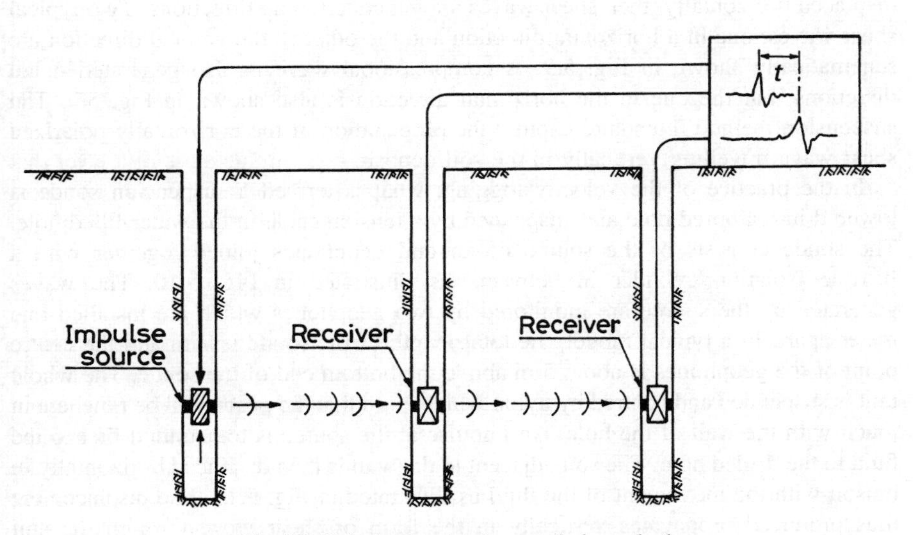

Fig. 5.8 Velocity logging by crosshole method.

placed tightly against the walls at the same elevation as the clamped anchor in the source hole. Once in place, the hammer is dropped on the anchor and the signals from the geophones are monitored and stored in the oscilloscope. The difference in travel time between the two adjacent geophones is used to compute the shear wave velocity. In the early stages of its development, it was claimed that the crosshole method is capable of obtaining the modulus value as a function of shear strain (Auld, 1977; Troncoso *et al.* 1977; Grant and Brown, 1981), but this aspect is not fully utilized in practice.

The criteria for selecting the best distances between boreholes are that the spacing be far enough to yield a discernable difference in travel time and that it be close enough to reduce the possibility of picking up extraneous refracted wave from adjacent layers. In general 2 to 5 m spacings are adopted for the boreholes. Crosshole logging has the advantage that it can be used to detect the modulus values of individual soil strata with horizontally layered structure. However the cost of the test is usually high, because it requires more than three holes. The details of the crosshole method are described in the papers by Stokoe and Woods (1972), Stokoe and Richart (1973), Anderson and Woods (1975) and Hoar and Stokoe (1978).

5.5 Suspension sonde

Velocity logging by means of the suspension sonde is a relatively new method which was originally developed by researchers of the Oyo Co. in Japan (Ogura, 1979; Ogura, 1988; Kitsunezaki, 1982). The basic concept of this technique is illustrated in Fig. 5.9. It is known that if an element of medium in a half-space is displaced horizontally, then shear waves are generated in all directions. Two typical shear waves, one in a horizontal direction and the other in the vertical direction are schematically shown in Fig. 5.9. A compressional wave is also generated in all directions, but the one in the horizontal direction is also shown in Fig. 5.9. The suspension method intends to capture the propagation of the horizontally polarized shear wave travelling vertically in the soil deposit.

In the practice of the velocity logging, what is termed a suspension sonde is lowered into a bored hole and suspended by a tension cable in the water-filled hole. The sonde consists of the source driver and geophones joined together with a flexible isolation cylinder in between, as illustrated in Fig. 5.10. The waves generated by the source are monitored by two geophones which are installed one meter apart. In a typical model, the total length of the sonde is 7 m and the centre point of the geophones is about 5 m above the bottom end of the sonde. The whole unit is suspended and centred by nylon whiskers so that the probe will be nowhere in touch with the wall of the hole. An impulse at the source is transmitted first to the fluid in the drilled hole. The soil adjacent to the wall is then displaced horizontally in unison with the movement of the fluid as illustrated in Fig. 5.10. The displacement thus produced propagates vertically in the form of shear wave through the soil stratum. At this time the fluid is forced to move horizontally in unison to the adjacent soil and, hence, the fluid wave travels at the same speed as the wave in the soil. Then when the shear wave reaches a receiver point, the horizontal movement of the fluid also arrives at the same point simultaneously. Thus by monitoring the motion of the fluid by the geophones, it is possible to measure the travel time of the shear wave and hence its velocity of propagation.

When an impulse is applied at the source, it is practically difficult to produce the

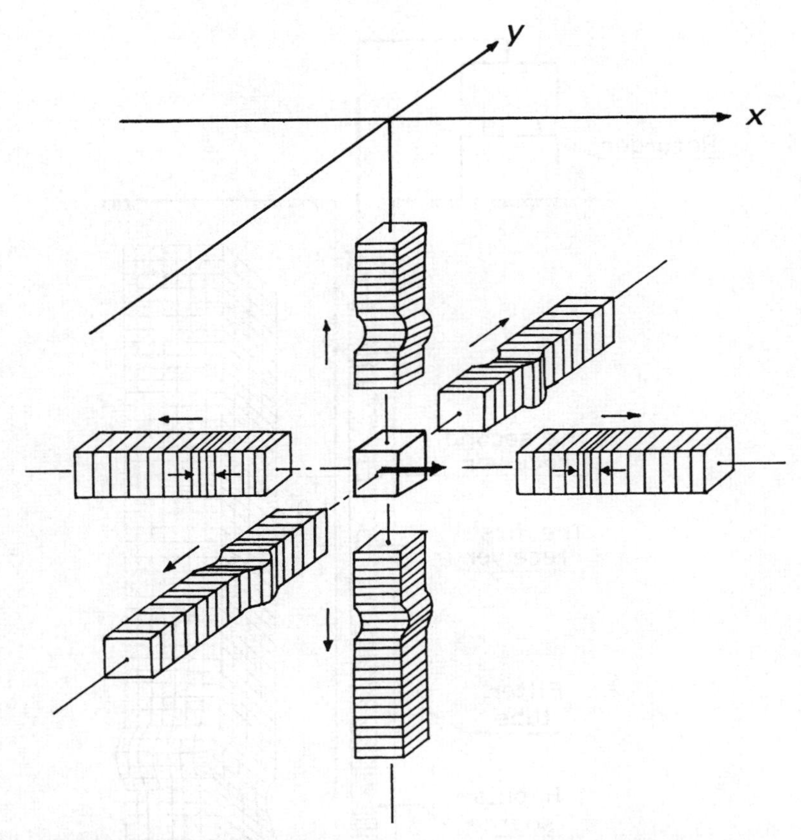

Fig. 5.9 Mode of soil deformation within a half-space due to a horizontal impulse.

shear wave alone and a compressional wave is always generated and tends to travel vertically through the fluid in the borehole. This is called the tube wave and could be a source of noise for the receivers. The generation of the tube wave can be minimized by utilizing a special device for the impulse source, but no less effective is the presence of an obstacle in the path of tube wave propagation. If there is such an obstacle, it becomes difficult for the wave to travel through it because most of the wave energy is reflected upon hitting the obstacle. To implement this principle, an isolation cylinder called a filter tube is installed in the middle of the suspension sonde as shown in Fig. 5.10. This filter tube is composed of a rubber tube containing compressed air. Experiments have shown that, even if it does not cover the entire cross section of the borehole, the presence of such a filter tube could be very effective in prohibiting the passage of the tube wave and hence in reducing noise. Thus by the time the shear wave arrives at the receivers, most of the noise has already been wiped out and a clear shear wave signal can be transmitted and monitored by the two geophones at the receiver points. In addition, the geophone

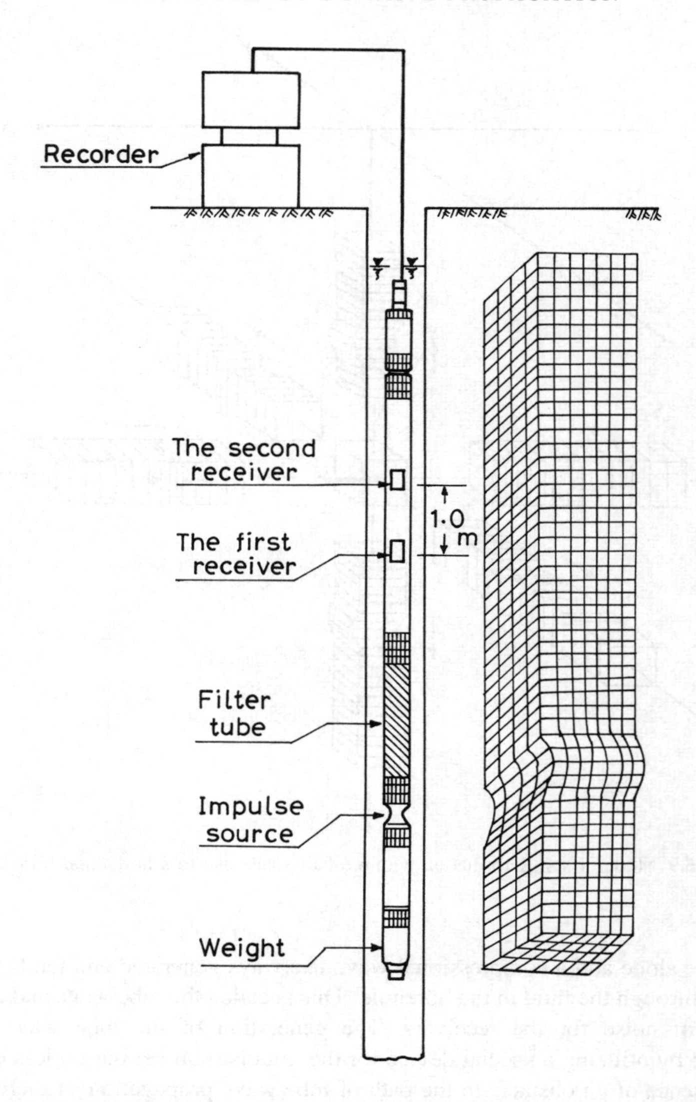

Fig. 5.10 Velocity logging by suspension method (Nigbor and Imai, 1994).

unit is designed so as to have the same density as the surrounding fluid in the borehole. Also the geophone itself can move freely without developing inertia interaction with the fluid.

In practical operation, an impulse is generated by an electrical device in one horizontal direction and the motion in the same horizontal direction is monitored by the geophones. Then the impulsive source is activated in the opposite direction and the horizontal output signal is monitored. The mode of deformation of the soil

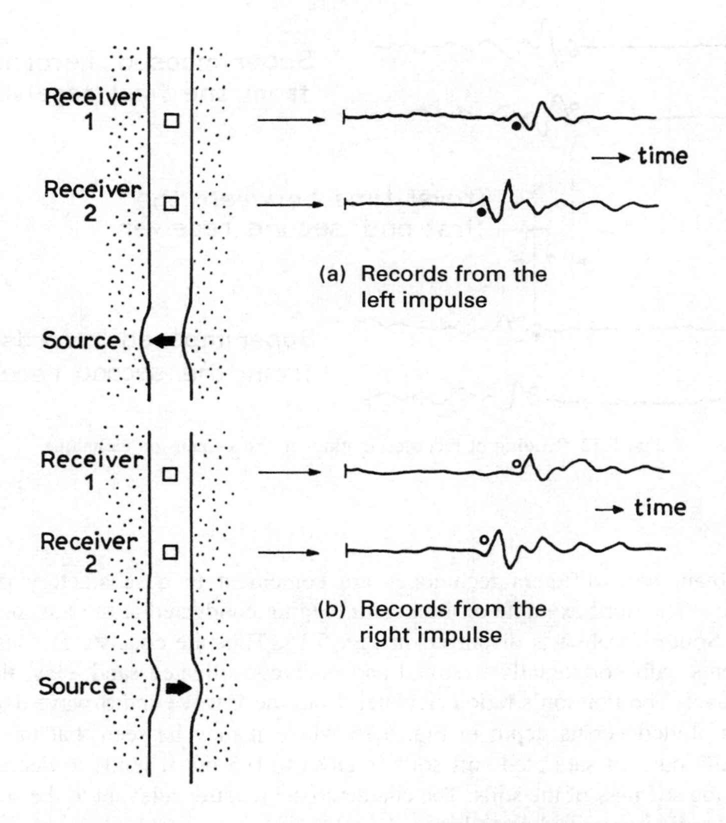

Fig. 5.11 Monitoring of fronts of wave propagation by the suspension method (Kitsunezaki, 1982).

deposit produced by the shear wave and an example of the data set recorded by the two receivers one meter apart are schematically shown in Fig. 5.11. The data pair are arranged together in Fig. 5.12, where it can be seen that the pair of records in one and the other oppositely polarized wave propagation is clearly visible, helping to precisely identify the time of wave arrival at two receiver points.

By using a special device for exciting the impulsive source, it is possible to also produce a compressional wave (P-wave) and to detect its arrival at the points of two receivers by means of vertical sensors encased in the same geophone. Figure 5.13 shows a typical example of velocity logging by means of the suspension sonde for a site at the east end of the San Francisco Bay Bridge (Nigbor and Imai, 1994). It may be seen that the measured shear wave velocities ranges from $V_s = 230$ ms^{-1} in the upper bay mud deposit to a value of 1400 ms^{-1} in the bedrock. Another example (Ishihara *et al.* 1989) is demonstrated in Fig. 5.14 in which the shear wave velocity obtained by the suspension sonde is compared with that established by the downhole method employing a hammer blow on the surface. It may be seen that the two sets of

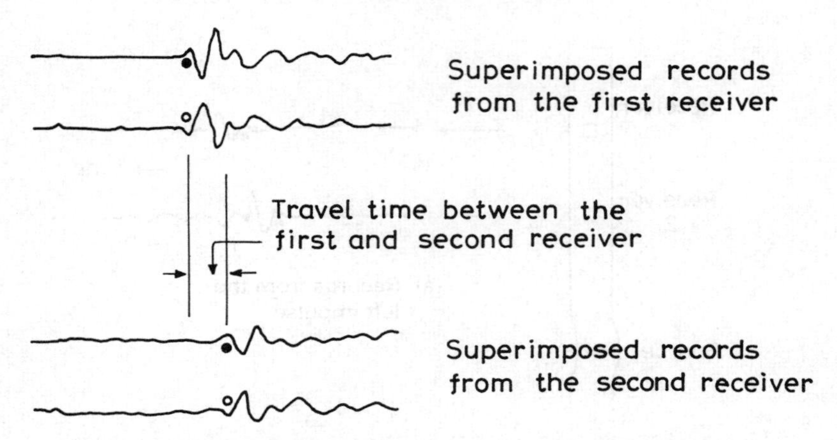

Fig. 5.12 Reading of recorded motions in the suspension technique.

data obtain with different techniques are coincident to a satisfactory degree of accuracy. The third example of the in situ logging conducted at the Savannah River site in South Carolina is displayed in Fig. 5.15. This site consists of coastal plain sediments with horizontally stratified and underconsolidated sand, clay, limestone, and gravel. The Poisson's ratio calculated from the P-wave and S-wave data in Fig. 5.15 is plotted versus depth in Fig. 5.16 where it may be seen that the value of Poisson's ratio of saturated soft soils is close to 0.5 but it tends to decrease with increasing stiffness of the soils. The characteristic features relevant to the suspension method are summarized as follows.

1. This method offers definite advantages in that it permits the measurement at a shorter spacing of 1 m to be made through the depth of the deposit as compared to other techniques and therefore a high level of resolution can be achieved in mapping up soil profiles.
2. The velocity logging by this method can be executed at a greater depth than ever been possible because of the proximity of the source to the receivers. The maximum depth ever tried is 300 m.

3. The energy from the source is always transmitted through the fluid in boreholes. Therefore, if there is no water in the boreholes, it becomes impossible to implement the logging.

4. In uncased portion of the borehole, the velocity logging can be effectively worked out. In the presence of stiff steel casing, it becomes difficult to detect the arrival of shear wave through soil deposits. However, if a flexible casing such as vinyl chloride is used, the use of the suspension sonde is still effective in monitoring the shear wave propagation.

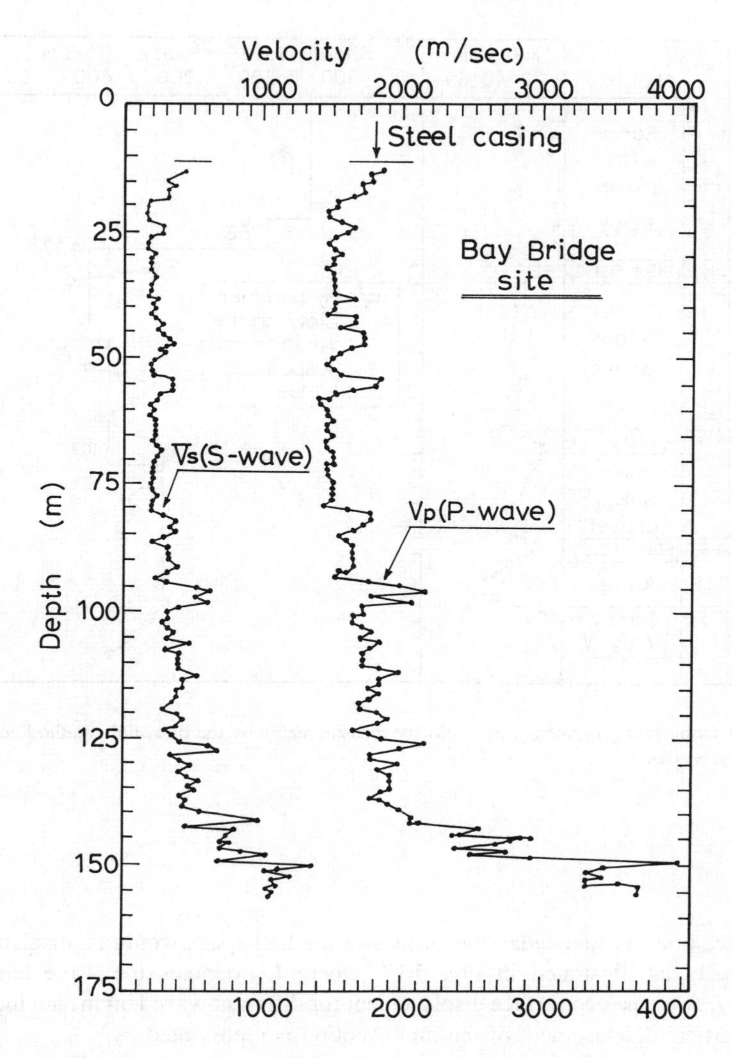

Fig. 5.13 An example of at-depth velocity logging by the suspension method (Nigbor and Imai, 1994).

5.6 Spectral analysis of surface waves (SASW)

The method consists of monitoring the propagation of surface wave of the Rayleigh type on the ground surface. It is known that the waves generated by a vertically oscillating load at the surface of an elastic half-space are predominantly Rayleigh waves (R-waves). Thus if a transducer sensing vertical motions is placed on the ground surface at some distance from a source point, the vertical component of the motions due to R-wave propagation can be monitored. When the input oscillation at

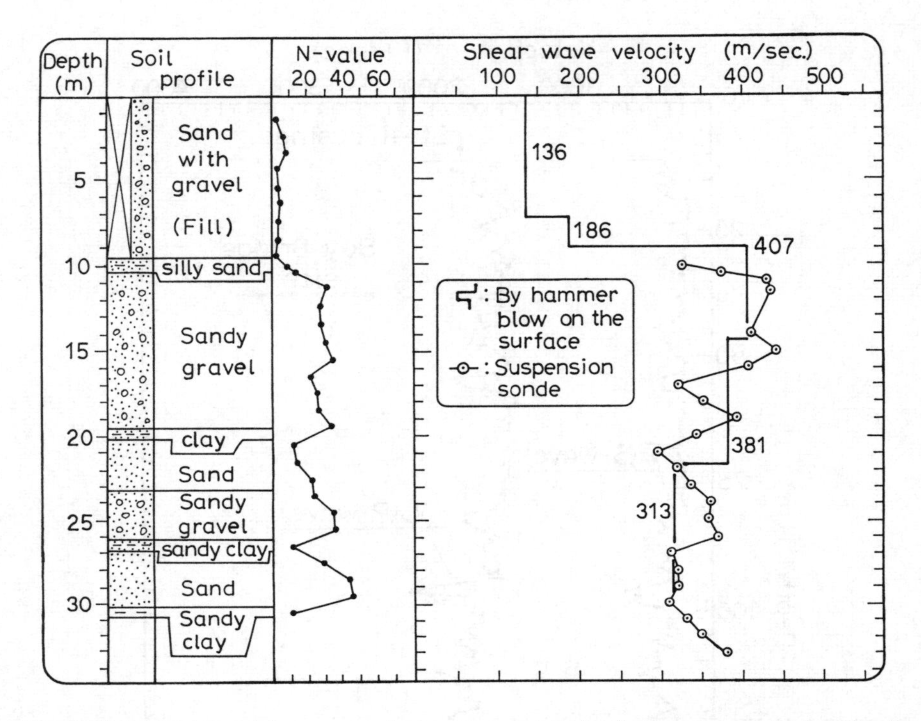

Fig. 5.14 Comparison of shear wave velocity measurements by the downhole method and by the suspension method.

the source point is sinusoidal, the surface of the half-space would be displaced also sinusoidally as illustrated in Fig. 5.17, where L_R denotes the wave length. In Fig. 5.17, two cases of surface displacement for different wave lengths are indicated. If the vertical displacement of the input motion is represented by

$$v_0(t) = v_0 \sin \omega t, \tag{5.8}$$

then the vertical displacement at any other point away from the source may be expressed as

$$v(t) = v \sin(\omega t - \varphi) \tag{5.9}$$

where φ is an angle of phase difference and v denotes an amplitude. Since the time lag between the source and a point at a distance x is equal to x/V_R, eqn (5.9) is

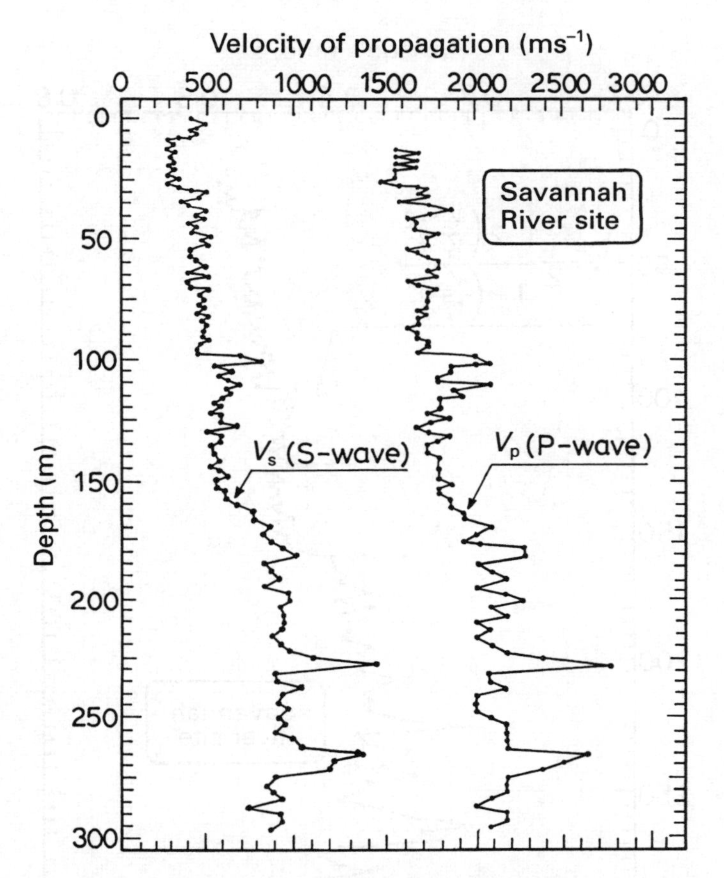

Fig. 5.15 An example of at-depth velocity logging by the suspension method (Nigbor and Imai, 1994).

rewritten as (Richart *et al.* 1970, p. 113)

$$v(t) = v_a \sin \omega(t - x/V_R) = v_a \sin\left(\omega t - \frac{2\pi f x}{V_R}\right) \tag{5.10}$$

where V_R is the velocity of R-wave propagation and f is the frequency of oscillation. Thus the phase angle is expressed as

$$\varphi = \frac{2\pi f x}{V_R}. \tag{5.11}$$

When the distance x is equal to one wave length L_R, the phase angle should be equal to 2π. Then, eqn (5.11) is reduced to the well-known relation

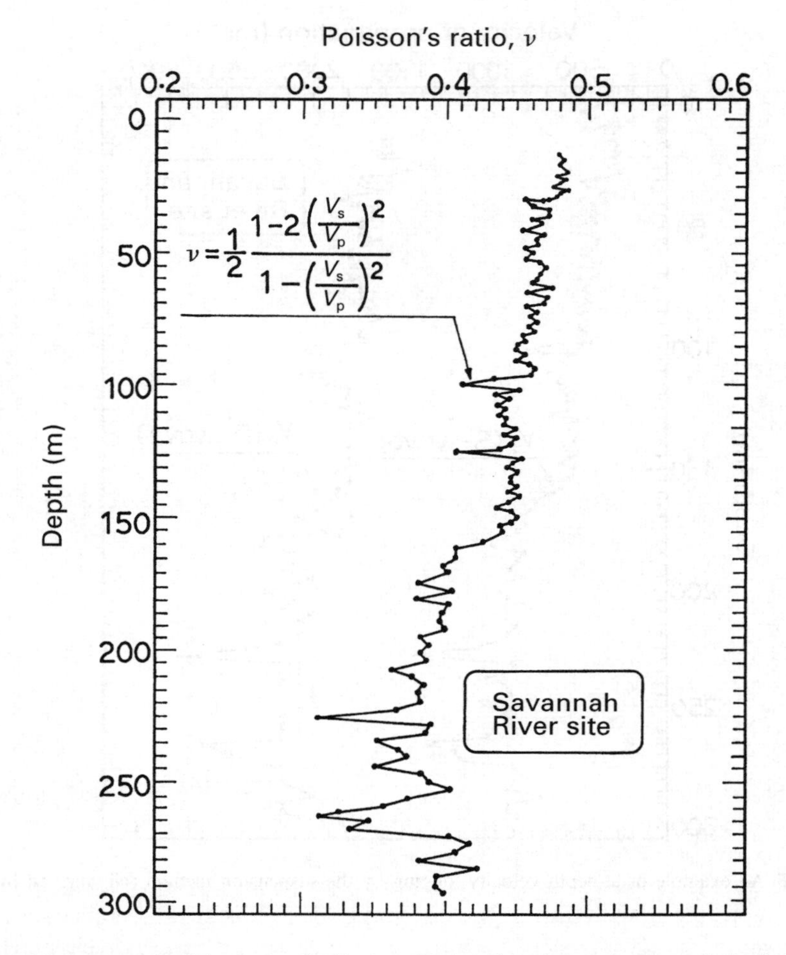

Fig. 5.16 An example of depthwise distribution of Poisson's ratio based on the velocity logging by the suspension method (Nigbor and Imai, 1994).

$$V_R = f\, L_R. \tag{5.12}$$

On the other hand, it is known from the theory of elastic waves that the velocity of R-wave propagation in a homogeneous half-space is a function of shear wave velocity V_s and compressional wave velocity V_p (or Poisson's ratio). This relation is shown in Fig. 5.18 in terms of values of V_R / V_s plotted versus Poisson's ratio ν. It can be seen that the velocity of R-wave propagation is only slightly smaller than the shear wave velocity (0.874 to 0.955 times the value of V_s), and therefore it may be taken approximately equal to V_s for practical purposes. With this fact in mind, eqn (5.12) is reduced to

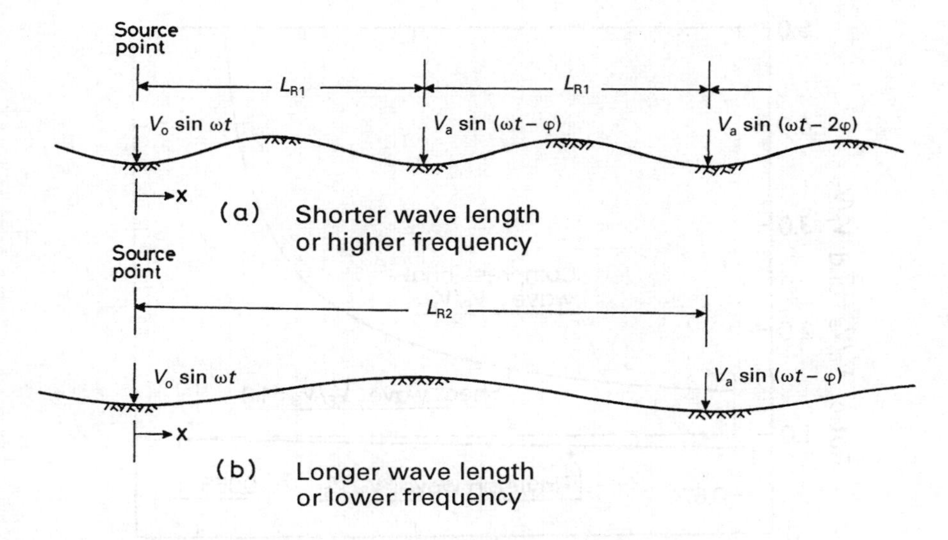

Fig. 5.17 Dependency of Rayleigh wave propagation on frequency.

$$V_s \doteqdot fL_R. \tag{5.13}$$

On the basis of the rationale as described above, the Rayleigh wave has been utilized as a non-destructive technique for subsurface exploration or for evaluating the quality of pavements in highways and airways.

In the early technique, a steady-state vibration of a known-frequency was applied vertically on the ground surface and the positions at which the vertical surface motion was in phase with that of the vibrator were detected by successively moving a vertically oriented sensor away from the source on the ground surface. The distance between any two of these successive peaks or troughs was considered to be equal to a wave length L_R of the propagating Rayleigh wave. With the known input frequency f, the velocity of the wave propagation V_R was calculated by eqn (5.12).

By changing the frequency of the oscillator, different values of the phase velocity V_R were generally obtained. Thus it became possible to obtain a curve indicating a relationship between the phase velocity V_R and frequency or wave length. Such a curve is called a *dispersion curve*.

The dispersion curve, or frequency-dependent R-wave velocity characteristics can also be derived from the theory of wave propagation through an elastic half-space. According to the theory, the R-wave is shown to propagate with a constant phase velocity which is independent of the frequency, if the half-space is composed of a homogeneous medium. The velocity of propagation in such a case has been displayed in Fig. 5.18. The pattern of displacements occurring in the homogeneous

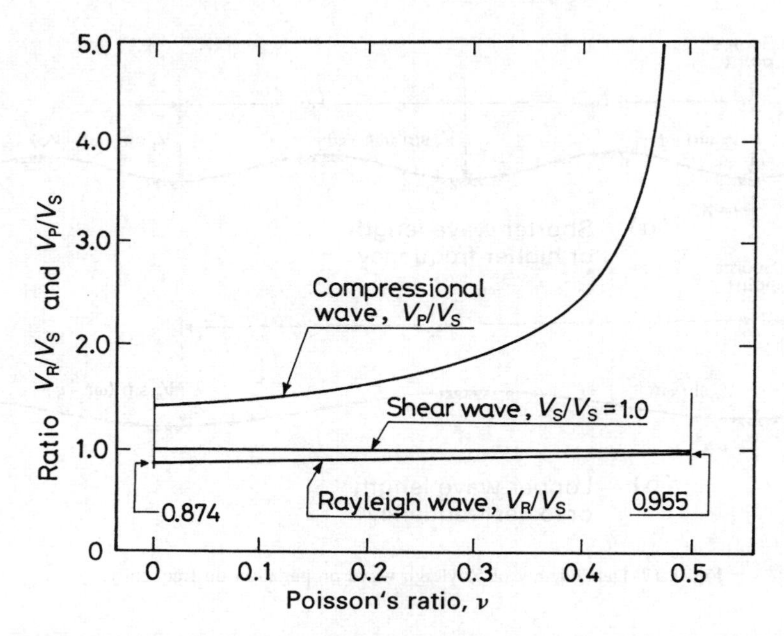

Fig. 5.18 Propagation velocities of various waves as function of Poisson's ratio (Richart, Hall, and Woods, 1970).

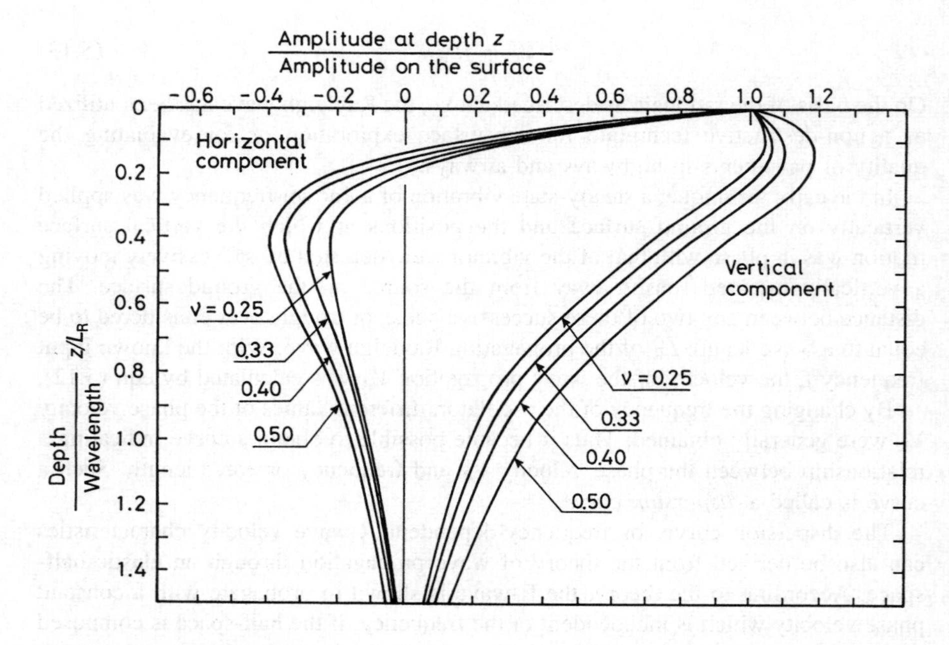

Fig. 5.19 Amplitudes of Rayleigh wave versus depth in an elastic half-space (Richart, Hall and Woods, 1970).

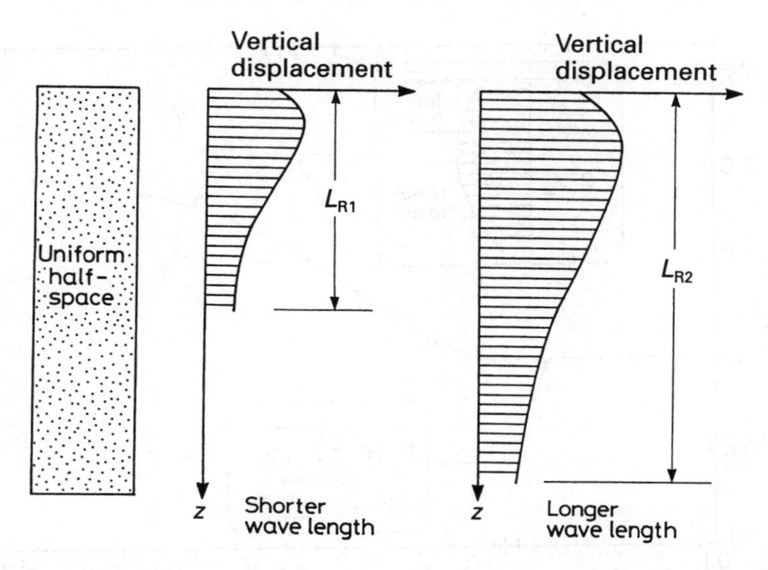

Fig. 5.20 Dependency of vertical displacement distribution on the wave length in Rayleigh wave propagation (Stokoe *et al.* 1994).

half-space during the R-wave propagation is displayed in Fig. 5.19 in which the amplitude on the surface is plotted versus the depth to wavelength ratio. Thus if the vertical displacement is plotted, for example, versus the depth z, the patterns of distribution as shown in Fig. 5.20 are obtained for the two cases shown in Fig 5.17, indicating the depth of the medium involved in the deformation becoming deeper with increasing wave length.

In an elastic half-space consisting of a single layer over a semi-infinite medium, similar analysis for R-wave propagation indicates that, with a higher frequency of input excitation, the wave propagates mainly through the upper layer with the result that the phase velocity is determined by the elastic properties of the top layer. This could be the case, as illustrated in Fig. 5.21, because of the deformation of the half-space taking place mainly in the medium within the upper layer with a shorter wave length. In contrast, if the input frequency is low, the deformation of the two-layered half-space caused by the R-wave propagation takes place involving the medium mainly in the lower half-space because of the longer wave length associated with the low frequency. Thus in the case of the two-layered system, the phase velocity of R-wave propagation is dependent on the frequency or wave length with which the wave is propagating. The feature of change in R-wave velocity of propagation in the two layered half-space is illustrated in Fig. 5.21 where the ratio of V_R/V_s is plotted versus L_R/H where H is the thickness of the top layer. As shown in the figure, the influence of the lower layer becomes more predominant with increasing wave length because of encroachment of the motions on the zone at deeper depth. This characteristic feature of the R-wave propagation can be utilized in turn for practical

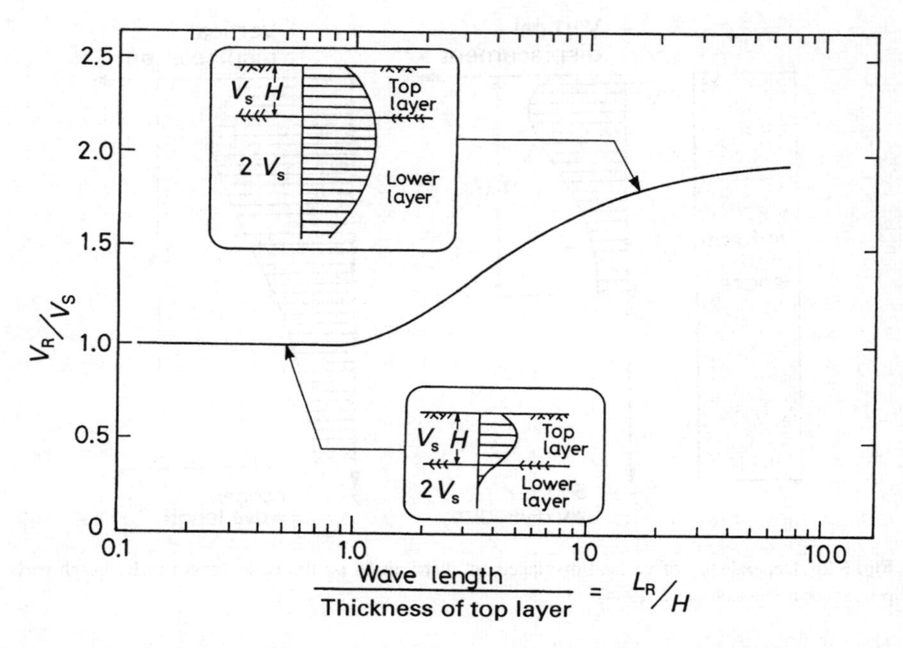

Fig. 5.21 Velocity of Rayleigh wave propagation in a two-layer half-space as a function of wave length (Stokoe *et al.* 1994).

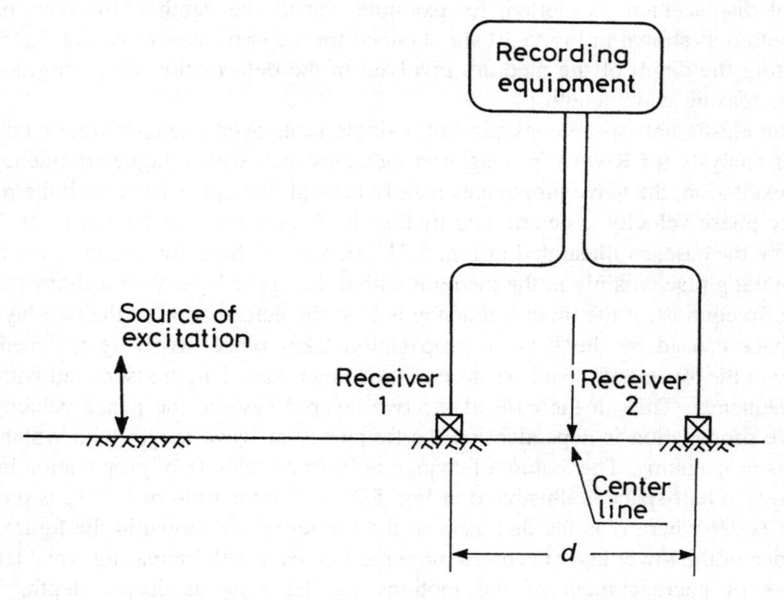

Fig. 5.22 Layout of in situ measurements by SASW.

purposes such as identifying elastic properties of each of the two layers by monitoring the velocity of R-wave generated with two sufficiently high and low frequencies of input oscillation.

The analysis above can be similarly developed for a multiple-layered half-space consisting of alternating layers of media with difference elastic properties. The frequency-dependent variation of R-wave propagation may be utilized similarly in practice for subsurface exploration of in situ soil or rock deposits. However the application of steady-state vibration on the ground surface is a somewhat time-consuming and costly operation. Thus the method of applying an impulse, swept-sinusoidal, or random-noise load on the surface was sought and as a result a new technique named SASW (*spectral analysis of surface waves*) has been developed with the help of digital electronic equipment (Stokoe and Nazarin, 1983; Nazarin and Stokoe, 1984; Stokoe *et al.* 1994).

The general scheme of testing for the SASW method is illustrated in Fig. 5.22. To start with, an imaginary centre line is selected for the receiver array. Two receivers are placed on the ground surface at an equal distance apart from the centre line. A vertical impulse is applied on the ground by means of a hammer. A group of surface waves with various frequencies is produced and propagates with different velocities. This group of waves is monitored by the receivers and captured by a recording device. After this measurement is over, the receivers are kept in their original position and the source is moved to the opposite side of the centre line, and the test is repeated. By averaging the records from these two test runs, better quality data are obtained for the site condition surrounding the point of the centre line. A pair of the above tests is performed by changing the spacing between the two receivers. With closer receiver spacing the properties of the material near the surface is sampled and with increasing distance between the receivers, the material properties at deeper deposits can be detected. In field practice, distances between the two receivers are chosen, for example, as 1, 4, 16, and 64 m, if a soil profile is to be mapped out to a depth on the order of 50 m. Several types of sources are used to generate impulse loading to cover a required range of wave length or frequency. A small hand-held hammer can be used for measurements at close receiver spacing of 1 to 4 m. A sledge hammer or a large drop weight of 20–70 kg is employed for the measurements with larger spacing of 5 to 10 m.

The signals obtained by the transducers are sent to the recording equipment. The data are then converted to the expression in the frequency domain using a Fast Fourier Transform algorithm. A spectral analyser can be used more conveniently to perform spectral analysis of the data in the frequency domain. The most important quantity to be extracted from the spectrum analysis is the cross-spectral density function which is defined as the spectrum of the output (monitored motion) multiplied by the complex conjugate of the spectrum of the input motion. The cross spectral density function is indicative of the phase difference characteristics between the two receiver points at each frequency. On the other hand, the travel time is generally given by x/V_R as indicated by eqn (5.10). Introducing $t = x/V_R$ in eqn (5.11), one obtains

$$t = \frac{\varphi}{2\pi f}. \tag{5.14}$$

The phase difference φ from the cross spectrum density function is expressed as a function of frequency. Therefore, if it is used in eqn (5.14), the travel time is also dependent on the frequency. Equation (5.14) is thus rewritten as

$$t(f) = \frac{\varphi(f)}{2\pi f}. \tag{5.15}$$

It is to be noticed here that the time necessary for the R-wave to travel a specified distance d depends generally on the frequency or wave length at which the wave is travelling. This is due to the dispersive nature of the R-wave propagation through the multiple layered half-space.

Since the distance between the two receivers d is known, the wave velocity is calculated as

$$V_R = \frac{d}{t(f)}. \tag{5.16}$$

The corresponding wave length is determined as

$$L_R = \frac{V_R}{f} = \frac{d}{ft(f)} = 2\pi \frac{d}{\varphi(f)}. \tag{5.17}$$

As a result of several steps of calculation as described above, it becomes possible to obtain a dispersion curve for an in situ deposit being investigated.

The next step is to infer the depth and actual velocity of shear wave in each layer of the deposit based on the dispersion curve. This task is called the inversion process. In the practice of the SASW method, the inversion is performed by what is termed *forward modelling* (Stokoe *et al.* 1994) in which a theoretically determined dispersion curve based on a pre-assumed model of the soil profile is compared with the dispersion curve obtained from in situ measurements. If the coincidence between these two dispersion curves is not satisfactory, the previously assumed model for the soil profile is modified and the theoretical dispersion curve is renewed to be compared again to the experimentally obtained dispersion curve. This type of iterative procedure is repeated until a good match can be obtained. The theoretical background for properly performing forward modelling is described in detail by Gucunski and Woods (1991,1992).

An example of the forward modelling procedure is given in Figs 5.23 through 5.25. In Fig. 5.23, three velocity profiles assumed for the model are shown. The dispersion curves obtained from the theoretical analyses for the three profiles are shown in Fig. 5.24, along with the dispersion curve actually obtained from a site being studied. Profile 1 is the one assumed for the first trial for matching. Profile 2 is the one modified for the second trial. In the third comparison based on Profile 3, the

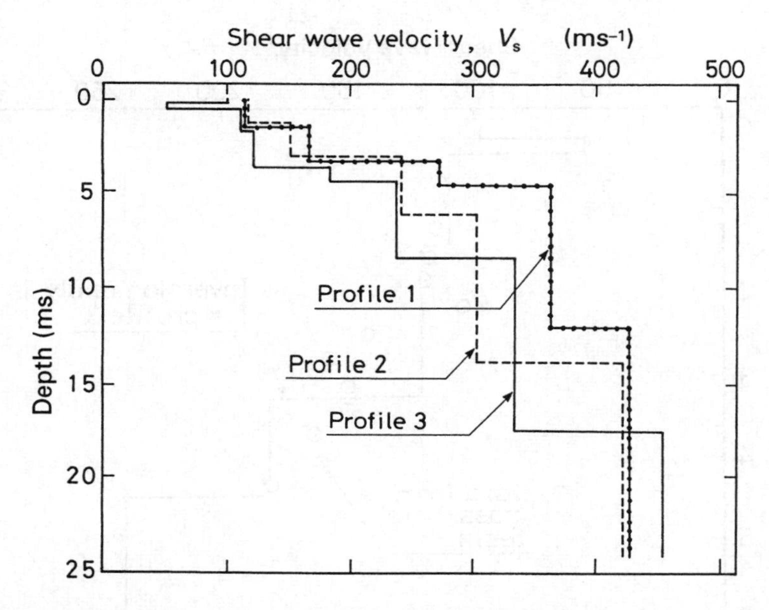

Fig. 5.23 Illustration of forward modelling procedure in SASW (Stokoe *et al.* 1994).

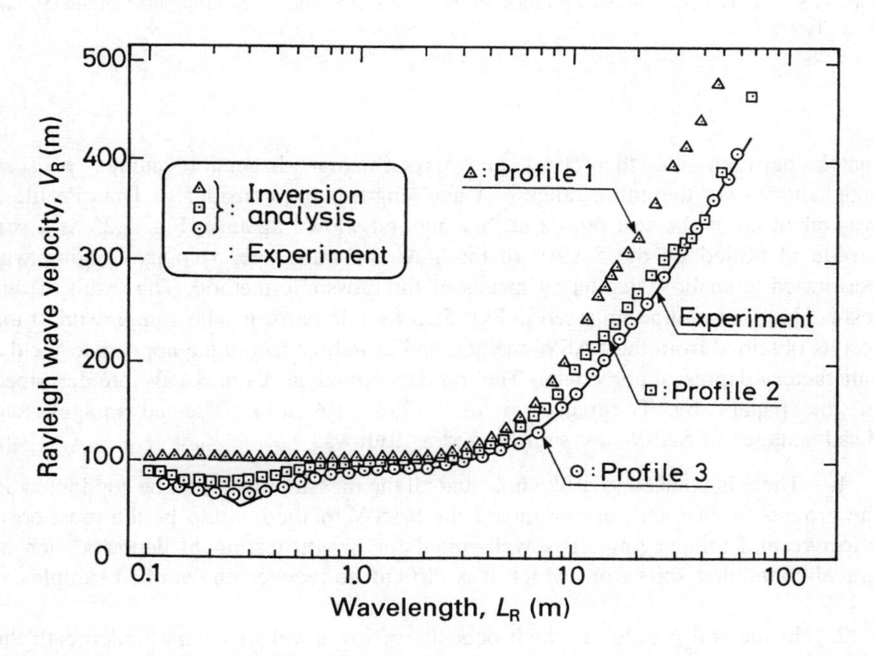

Fig. 5.24 Comparison of dispersion curves obtained by experiment and by analysis in SASW (Stokoe *et al.* 1994).

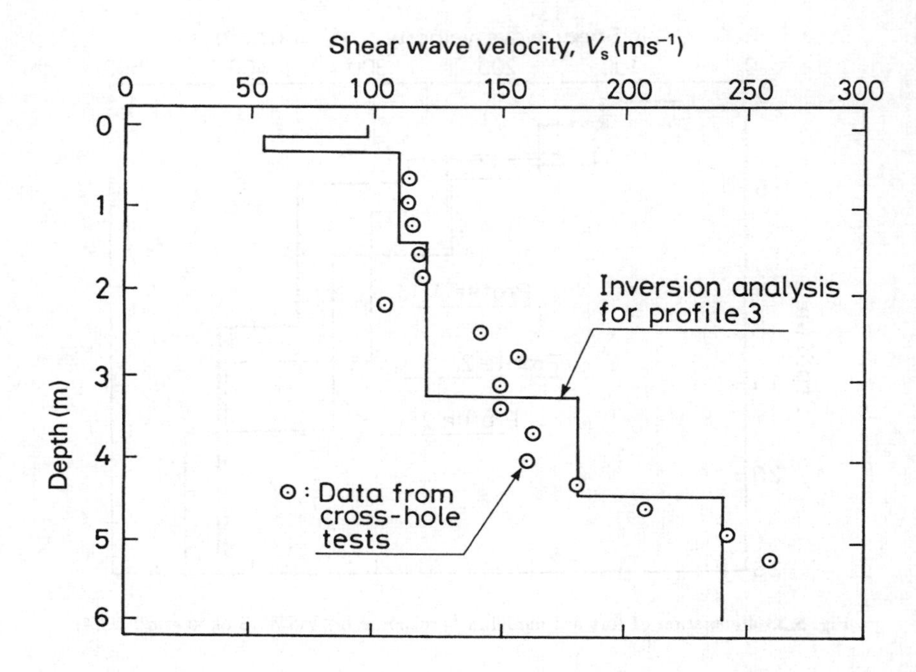

Fig. 5.25 Comparison of velocity profiles obtained by SASW and by the crosshole method (Stokoe *et al.* 1994).

match between the theoretical and experimental dispersion curves is seen satisfactory over the entire range of wave length being considered. Thus Profile 3 was taken up as the soil profile at this site and shown again in Fig. 5.25 as a soil profile identified by the SASW method. At the same site, velocity logging was performed at shallow depths by means of the crosshole method. The result of this test is also shown superimposed in Fig. 5.25 for comparison. The comparison of the results obtained from the SASW method and crosshole technique appears to yield a satisfactory degree of agreement. The more recent advances in SASW are described in the papers by Tokimatsu *et al.* (1991, 1992*a*,*b*). The advantages and disadvantages of SASW are summarized as follows.

1. There is no need to drill a hole and all the measurements can be conducted on the ground surface and, in this regard the SASW method would be the most cost-effective and time-saving. It is well-suited for in situ testing of deposits such as gravel-containing soils from which it is difficult to recover undisturbed samples.

2. In the soil profile in which deposits of lower velocity exist underneath the layer of higher velocity, care must be exercised in analyzing field data so that a proper theoretical model is established to match experimental data.

3. By SASW, soil profile data are generally picked up as a global average through some depths. Therefore, the capacity is generally poor for identifying a thin layer with a notably different shear wave velocity.

References

Anderson, D.G. and Woods, R.D. (1975). *Comparison of field and laboratory shear moduli.* Proceedings of the Conference on In-Situ Measurements of Soil Properties, ASCE, Raleigh, N.C. Vol. 1, pp. 69–92.

Auld, B. (1977). Cross-hole and down-hole V_s by mechanical impulse. *Journal of the Geotechnical Engineering Division*, ASCE, **103**, GT12, pp. 1381–93.

Ballard, R.F. and Mclean, F.G. (1975). *Seismic field methods for in situ moduli.* Proceedings of the ASCE Specialty Conference on In-Situ Measurement of Soil Properties, North Carolina State University, Vol. 1, pp. 121–50.

Grant, W.P. and Brown, F.R. (1981). *Dynamic behaviour of soils from field and laboratory tests.* Proceedings of the International Conference on Recent Advances in Geotechnical Earthquake Engineering and Soil Dynamics, St. Louis, Vol. II, pp. 591–96.

Gucunski, N. and Woods, R.D. (1991). *Use of Rayleigh modes in interpretation of SASW test.* Proceedings of the 2nd International Conference on Recent Advances in Geotechnical Earthquake Engineering and Soil Dynamics, St. Louis, Vol. 1, pp. 1399–1408.

Gucunski, N. and Woods, R.D. (1992). Numerical simulation of the SASW test. *Soil Dynamics and Earthquake Engineering*, **11**(4), pp. 213–27.

Hoar, R.J. and Stokoe, K.H. (1978). *Generation and measurement of shear waves in situ.* Dynamic Geotechnical Testing, ASTM, STP654, pp. 1381–93.

Ishihara, K., Kokusho, T., and Silver, M.L. (1989). *Recent developments in evaluating characteristics of local soils.* Proceedings of the 10th International Conference on Soil Mechanics and Foundation Engineering, Rio de Janeiro, Vol. 4, pp. 2719–34.

Kitsunezaki, C. (1982). Some basic problems of shear wave logging by means of the suspension-type sonde. *Journal of Mining College*, Akita University, Series A, **6**(2), pp. 93–108.

Nazarin, S. and Stokoe, K.H. (1984). *In situ shear wave velocity from spectral analysis of surface waves.* Proceedings of the 8th World Conference on Earthquake Engineering, San Francisco, Vol. III, pp. 31–38.

Nigbor, R.L. and Imai, Y. (1994). *The suspension P–S velocity logging method.* Geophysical Characterization of Sites, A special volume by TC10 for XIII ICSMFE, New Delhi, pp. 57–61.

Ogura, K. (1979). *Development of the suspension type S-Wave log system.* Oyo technical report, No. 1, pp. 143–59.

Ogura, K. (1988). *Expansion of applicability for suspension P–S logging.* Oyo technical report No. 10, pp. 69–98.

Richart, F.E., Hall, J.R., and Woods, R.D. (1970). *Vibration of Soils and Foundations.* Prentice-Hall.

Stokoe, K.H. and Woods, R.D. (1972). In situ shear wave velocity by cross-hole method. *Journal of the Soil Mechanics and Foundations Division*, ASCE, **98**, SM5, pp. 443–60.

Stokoe, K.H. and Richart, F.E. (1973). *In situ and laboratory shear wave velocities.* Proceedings of the 8th International Conference on Soil Mechanics and Foundation Engineering, Moscow, Vol. 1, Part 2, pp. 403–9.

Stokoe, K.H. and Nazarin, S. (1983). *Effectiveness of ground improvement from spectral analysis of surface waves.* Proceedings of the 8th European Conference on Soil Mechanics and Foundation Engineering, Helsinki, pp. 91–4.

Stokoe, K.H., Wright, S.G., Bay, J.A., and Roesset, J.M. (1994). *Characterization of geotechnical sites by SASW method*. Geotechnical Characterization of Site, Special volume of ISSMFE TC10, New Delhi.

Tokimatsu, K., Kuwabara, S., Tamura, S., and Miyadera, Y. (1991). V_s Determination from steady state Rayleigh wave method. *Soils and Foundations*, **31**(2), pp. 153–63.

6

LOW-AMPLITUDE SHEAR MODULI

6.1 Low-amplitude shear moduli from laboratory tests

A vast majority of experimental data has been accumulated to evaluate the shear modulus of various soils at very small levels of strains. This modulus is called maximum shear modulus, initial shear modulus, or low-amplitude shear modulus, and denoted by G_{max} or G_0. In the laboratory, the most widely used procedure has been the resonant column test, but because of its simplicity, the cyclic triaxial test procedure with precise monitoring of axial strains is finding overriding popularity in recent years, particularly in Japan, for measuring modulus and damping. The outcome of these investigations will be described in the following section with emphasis on laboratory studies.

6.1.1 *Shear modulus of sands*

In any type of laboratory test, the shear modulus at small strains of cohesionless soils is measured under different effective confining stresses σ'_0 for various states of packing represented by different void ratios e. In the early works by Hardin and Richart (1963), the effects of void ratio were found to be expressed in terms of a function $F(e)$ as

$$F(e) = \frac{(2.17 - e)^2}{1 + e} \quad \text{or} \quad F(e) = \frac{(2.97 - e)^2}{1 + e}. \tag{6.1}$$

Thus, it is expedient to divide the measured shear modulus G_0 by the function $F(e)$ and plot this ratio against the effective confining stress employed in the test. Typical results of cyclic tests on Toyoura sand by means of a triaxial test apparatus are arranged in this manner and presented in Fig. 6.1, where the shear modulus G_0 was determined from the stress–strain curve in the 10th cycle of load application (Kokusho, 1980). The amplitude of shear strain γ_a indicated in the figure was obtained by converting the axial strain ε_a in the triaxial test through the relation

$$\gamma_a = (1 + \nu)\varepsilon_a. \tag{6.2}$$

For saturated sand used in the test, Poisson's ratio was taken as $\nu = 0.5$. Figure 6.1 shows that the data points for each strain amplitude plotted on the log–log scale may be represented by a straight line as

$$G_0 = AF(e)(\sigma'_0)^n \tag{6.3}$$

LOW-AMPLITUDE SHEAR MODULI

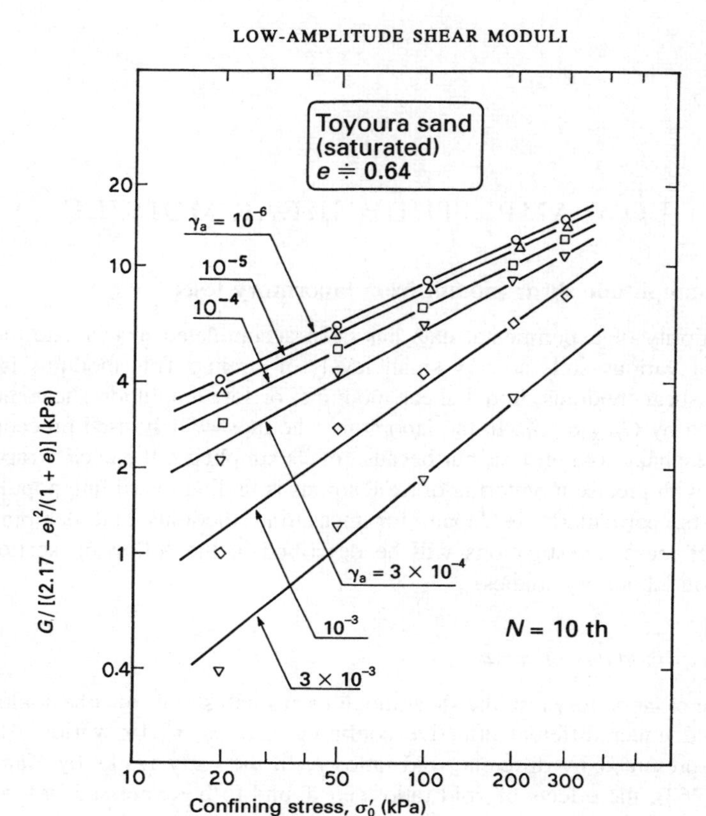

Fig. 6.1 Effects of confining stress on low-amplitude shear modulus (Kokusho, 1980).

where G_0 and σ'_0 are in terms of kPa. In the above form of empirical relation, the constants A and n take values which are dependent on the shear strain amplitude. For a sufficiently small shear strain of $\gamma_a = 10^{-5}$, the test data in Fig. 6.1 gives a typical formula as

$$G_0 = 8400 \frac{(2.17 - e)^2}{1 + e} (\sigma'_0)^{0.5} \text{ kPa.} \tag{6.4}$$

A number of similar formulae have been proposed for various sands, but most of them can be expressed in the general form of eqn (6.3). A summary of these empirical formulae was provided by Kokusho (1987) as shown in Table 6.1, where it may be seen that, in the majority of the cases studied, the value of n is taken as 0.5, whereas the value of A varies over a wide range reflecting variability of the modulus for individual sands. With respect to $F(e)$, the first function given by eqn (6.1) is used generally for round-grained clean sands having a void ratio variation ranging approximately between 0.5 and 1.2. For angular-grained clean sands or silty sands

Table 6.1 Constants in proposed empirical equations on small strain modulus:* $G_0 = A \cdot F(e)(\sigma_0')^n$ (Kokusho, 1987)

	References	A	$F(e)$	n	Soil material	Test method
Sand	Hardin–Richart (1963)	7000	$(2.17 - e)^2/(1 + e)$	0.5	Round grained Ottawa sand	Resonant column
		3300	$(2.97 - e)^2/(1 + e)$	0.5	Angular grained crushed quartz	Resonant column
	Shibata–Soelarno (1975)	42000	$0.67 - e/(1 + e)$	0.5	Three kinds of clean sand	Ultrasonic pulse
	Iwasaki *et al.* (1978)	9000	$(2.17 - e)^2/(1 + e)$	0.38	Eleven kinds of clean sand	Resonant column
	Kokusho (1980)	8400	$(2.17 - e)^2/(1 + e)$	0.5	Toyoura sand	Cyclic triaxial
	Yu–Richart (1984)	7000	$(2.17 - e)^2/(1 + e)$	0.5	Three kinds of clean sand	Resonant column
Clay	Hardin–Black (1968)	3300	$(2.97 - e)^2/(1 + e)$	0.5	Kaolinite, etc.	Resonant column
	Marcuson–Wahls (1972)	4500	$(2.97 - e)^2/(1 + e)$	0.5	Kaolinite, $I_p^{**} = 35$	Resonant column
		450	$(4.4 - e)^2/(1 + e)$	0.5	Bentonite, $I_p = 60$	Resonant column
	Zen–Umehara (1978)	$2000 \sim 4000$	$(2.97 - e)^2/(1 + e)$	0.5	Remolded clay, $I_p = 0 \sim 50$	Resonant column
	Kokusho *et al.* (1982)	141	$(7.32 - e)^2/(1 + e)$	0.6	Undisturbed clays, $I_p = 40 \sim 85$	Cyclic triaxial

$^*\sigma_0'$: kPa, G_0 : kPa, $^{**}I_p$: Plasticity Index

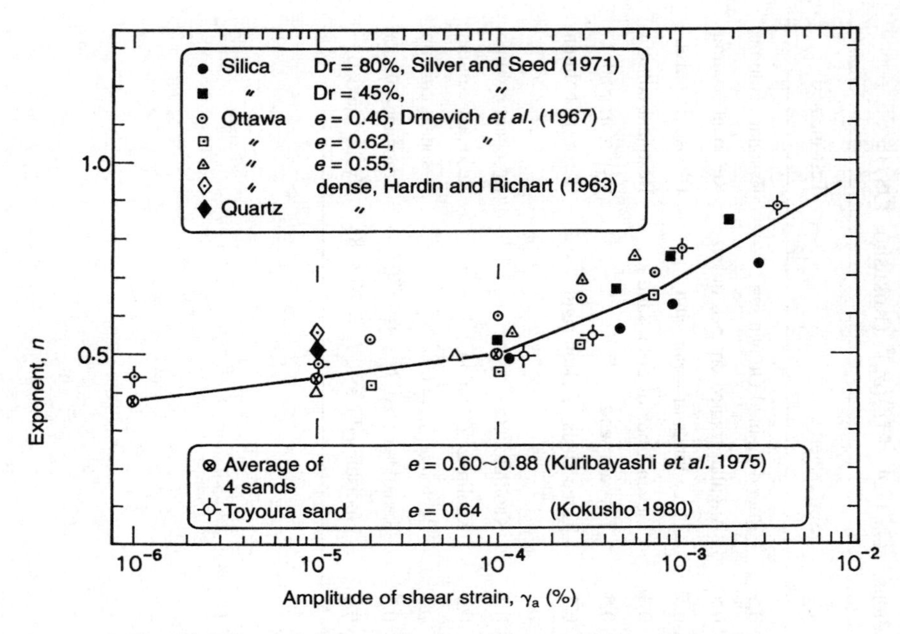

Fig. 6.2 Effects of strain amplitude on the n power of confining stress.

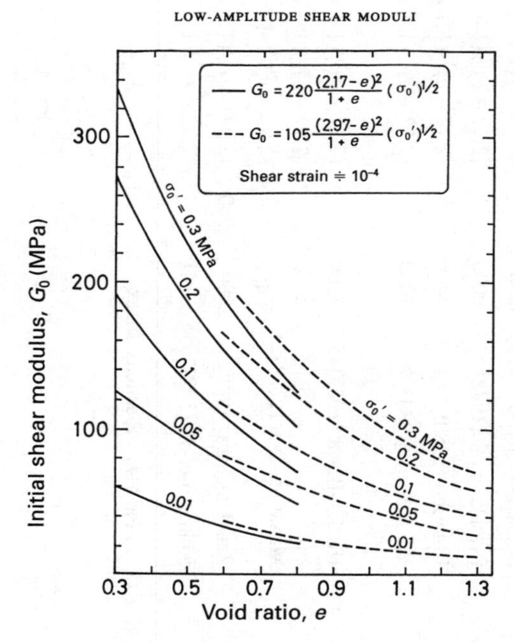

Fig. 6.3 Low-amplitude shear modulus of cohesionless soils as function of void ratio and confining stress.

whose void ratio is generally higher, the second function in eqn (6.1) is used instead.

It can be seen in Fig. 6.1 that the empirical correlation as given by eqn (6.3) changes generally with the shear strain amplitude under consideration. This change is expressed in terms of the constants A and n tending to decrease with increasing shear strain amplitude. The value of the power n as read from Fig. 6.1 is plotted versus the shear strain in Fig. 6.2, along with similar data from other sources. It may be seen that the power n tends to increase with increasing amplitude of shear strain. The lower limit of n attained at an infinitesmal strain appears to coincide with the value of 1/3 which is derived from what is called Hertz contact theory in the elasticity (See Richart *et al.* 1970, p. 145). The upper bound $n = 1.0$ at large strains is considered reasonable in the light of the well-known Coulomb failure criterion stipulating the condition on the shear strength which is proportional to the confining

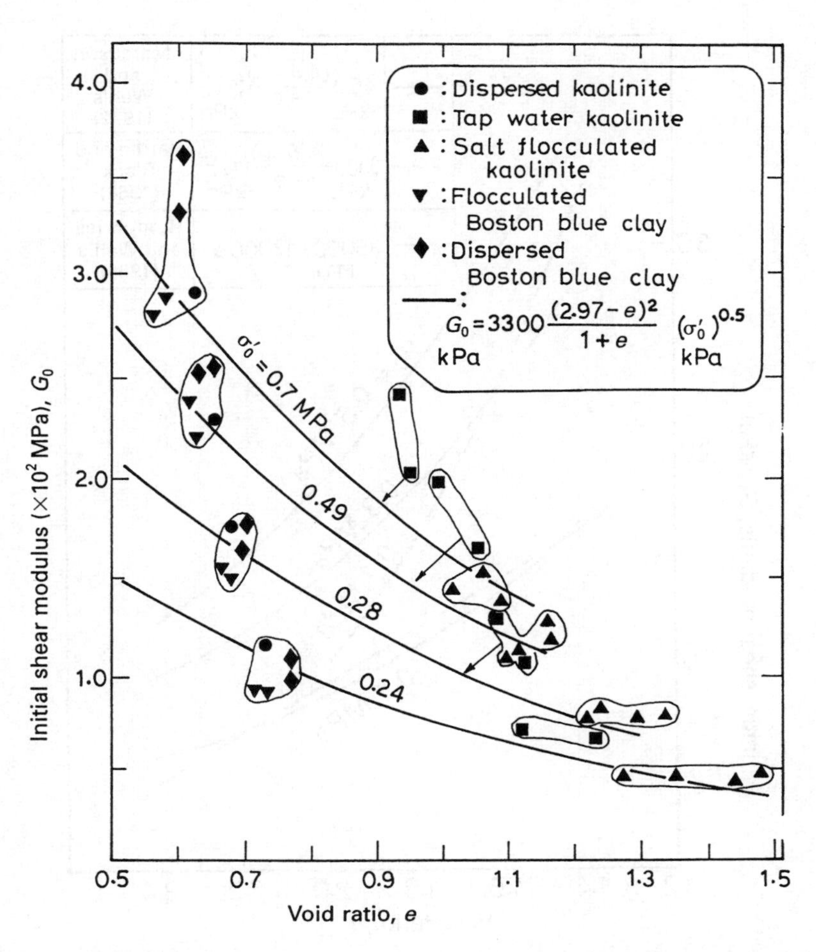

Fig. 6.4 Low-amplitude shear modulus of cohesive soils as functions of void ratio and confining stress (Hardin and Black, 1968).

stress. While the power n varies with shear strain, it has been customary to take up the strain amplitude of $\gamma_a = 10^{-5}$ for defining the low-amplitude shear modulus G_0 through the expression of eqn (6.3). The value of A tends to change depending upon the grain composition of sand and, therefore, tests on an individual material are generally needed if it is to be used in analysis for design purposes. For reference sake, the formulae proposed by Hardin and Richart (1963) are displayed in a graphical form in Fig. 6.3.

6.1.2 *Shear modulus of cohesive soils*

To investigate effects of void ratio and effective confining stress, a series of torsional vibration tests was conducted by Hardin and Black (1968) on remoulded specimens

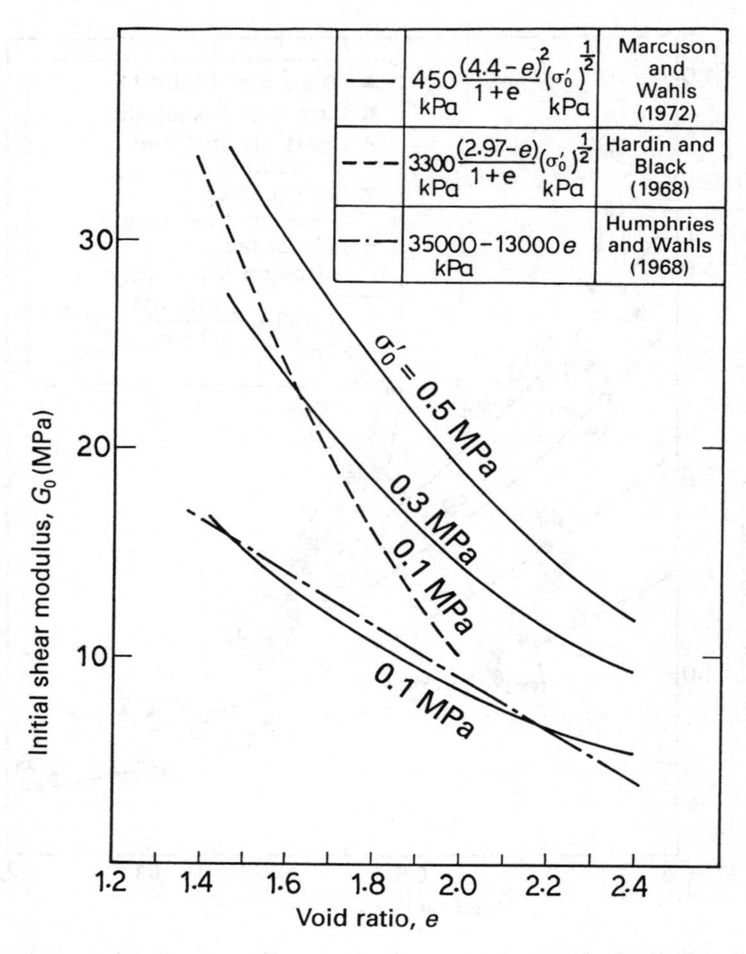

Fig. 6.5 Low-amplitude shear modulus of cohesive soils as functions of void ratio and confining stress.

of kaolinite clay with a plasticity index of $I_P = 21$. The specimens were consolidated normally to specified effective confining stress and then put to a state of resonance in torsional vibration. The shear modulus determined in this way is plotted in Fig. 6.4 against the void ratio, together with data from other sources. Also shown superimposed are the solid curves based on an empirical formula derived for angular sands. It may be seen in Fig. 6.4 that, for normally consolidated clays, the low-amplitude shear modulus at a given confining stress tends to decrease with increase in void ratio. It is of interest to note that the characteristic change in the shear modulus with void ratio and effective confining stress can be represented, with a reasonable degree of accuracy, by the empirical formula derived by Richart *et al.* (1970) for sands with angular grains. The kaolinite and Boston blue clays used in the above tests are of low plasticity having as small void ratios as those of angular sands and silty sands. Therefore the applicability of the empirical formula such as those indicated in Fig. 6.3 is limited to the clays of such low plasticity having a void ratio less than about 1.5.

Resonant column tests were conducted by Humphries and Wahls (1968) on remoulded specimens of highly compressible bentonite clay with a plasticity index of 60. The results of the tests were put in a form of an empirical equation based on the regression analysis, but if terms of minor effects are neglected, it is roughly written as

$$G_0 = 35\,000 - 13\,000\,e \quad \text{(in kPa)}. \tag{6.5}$$

In this relation, effects of confining stress are allowed for implicitly through the void ratio which in turn is a unique function of the confining stress during the normal consolidation. The formula of eqn (6.5) is displayed in Fig. 6.5.

Another series of resonant column tests was carried out by Marcuson and Wahls (1972) on the same bentonite clay and the results of the tests were expressed as

$$G_0 = 450\frac{(4.4 - e)^2}{1 + e}(\sigma'_0)^{0.5} \quad \text{(in kPa)}. \tag{6.6}$$

This relation is displayed in Fig. 6.5 in a graphical form for the confining stress of $\sigma'_0 = 0.1$, 0.3, and 0.5 MPa. It may be seen that the formula of eqn (6.5) gives about the same shear modulus as does eqn (6.6) when the confining stress is equal to 0.1 MPa. When compared with the shear modulus for low plasticity clays in Fig. 6.4, it is apparent that the high plasticity clays yield much lower values of shear modulus.

It is well known that the undrained strength of normally consolidated clay tends to increase in proportion to the confining stress where the void ratio is decreased due to consolidation. If the clay is brought to an overconsolidated state upon unloading to a certain confining stress, then the undrained strength at this state is known to become greater than that exhibited in the normally consolidated state under the same confining stress, primarily because of the reduced void ratio at the overconsolidated state. It is thus of interest to see if similar trend is observed in the shear modulus of clays tested in a dynamic loading condition. The test to this effect was conducted by

LOW-AMPLITUDE SHEAR MODULI

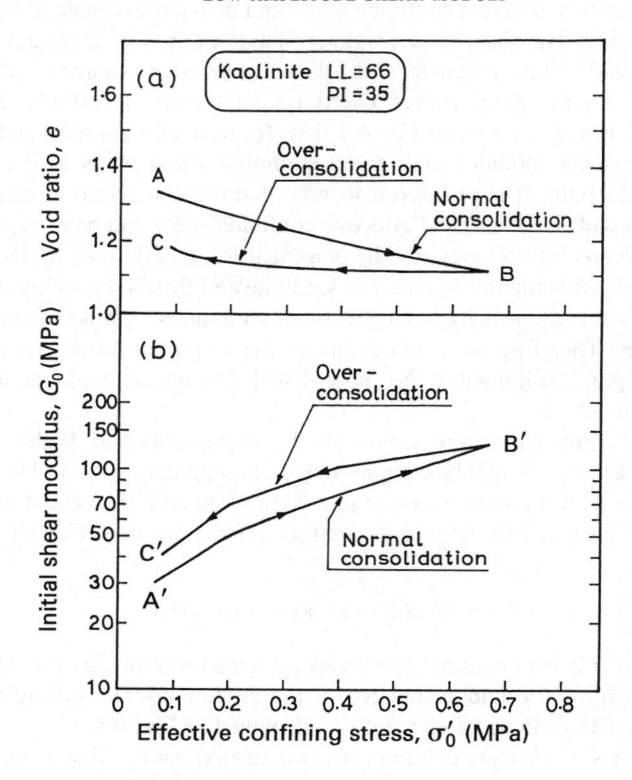

Fig. 6.6 Low-amplitude shear modulus of a low-plasticity clay as influenced by overconsolidation (Humphries and Wahls, 1968).

Humphries and Wahls (1968) on samples of artificially prepared kaolinite clay having a medium plasticity index of $I_p = 35$. As indicated in Fig. 6.6(a), the sample was consolidated isotropically from point A to B and unloaded back to the point C. At several stages in the course of loading and unloading, the clay specimen was subjected to torsional vibration. The shear modulus determined from the resonant condition is shown in Fig. 6.6(b). It can be seen that the shear modulus is about 10 to 20% greater in the overconsolidated state as compared to that obtained in the normally consolidated condition.

Similar tests were conducted also on samples of more compressible bentonite clay now having a higher plasticity index of $I_p = 60$. One of the results of these tests is demonstrated in Fig. 6.7. While the general trend is similar to the case of the kaolinite, the value of shear modulus itself is much smaller for the bentonite because of the highly compressible nature of this material as compared to the kaolinite. It is also to be noticed that the increase in shear modulus with increased OCR is more conspicuous for the kaolinite with medium plasticity than the highly plastic

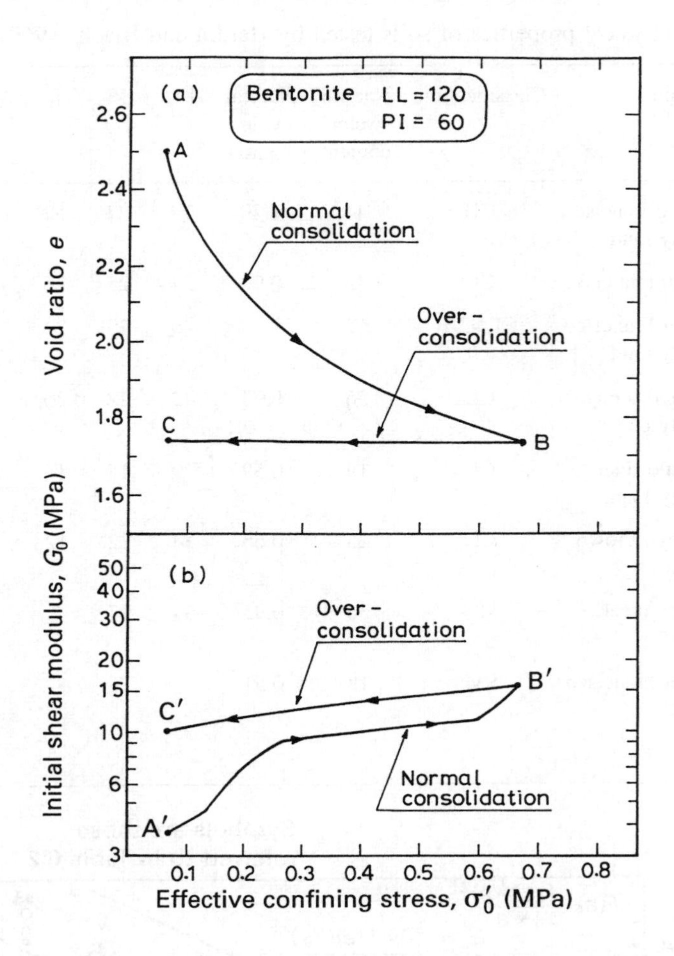

Fig. 6.7 Low-amplitude shear modulus of a high-plasticity clay as influenced by overconsolidation (Humphries and Wahls, 1968).

bentonite material. This observation leads to a hint that the greater the plasticity index of a clay, the larger the degree of stiffness gain due to overconsolidation.

Other evidence in support of the above hypothesis is provided by Hardin and Black (1968, 1969) for undisturbed specimens of many clays. The names and physical properties of the clays tested are shown in Table 6.2 in the order of decreasing plasticity index. This arrangement is also coincident with the order of decreasing water content, void ratio, and activity. The sizes of these property index values imply that the clay moves from CH to CL in the Unified Classification System. Multiple series of resonant column tests were carried out on these undisturbed samples consolidated to various confining pressures.

Table 6.2 Physical properties of soils tested by Hardin and Black (1969)

Symbol	Name	Classification	Natural water content	Natural void ratio	LL	PI	I_p	Activity
▲	San Francisco Bay mud	CH	44	1.20	79	27	52	1.16
◆	Virginia clay	CH	33	0.90	54	28	26	—
△	San Francisco Bay mud	CL–CH	52	1.42	49	25	24	1.11
▽	Rhodes creek silty clay	CL	36	0.98	42	22	20	0.68
▼	Little river silty loam	CL	19	0.59	29	15	14	0.79
◇	Floyd brown loam	CL	25	0.66	34	22	12	0.64
●	Lick creek silt	MN	36	0.95	34	27	7	0.81
○	San Francisco silt	SM	18	0.51	—	—	—	—

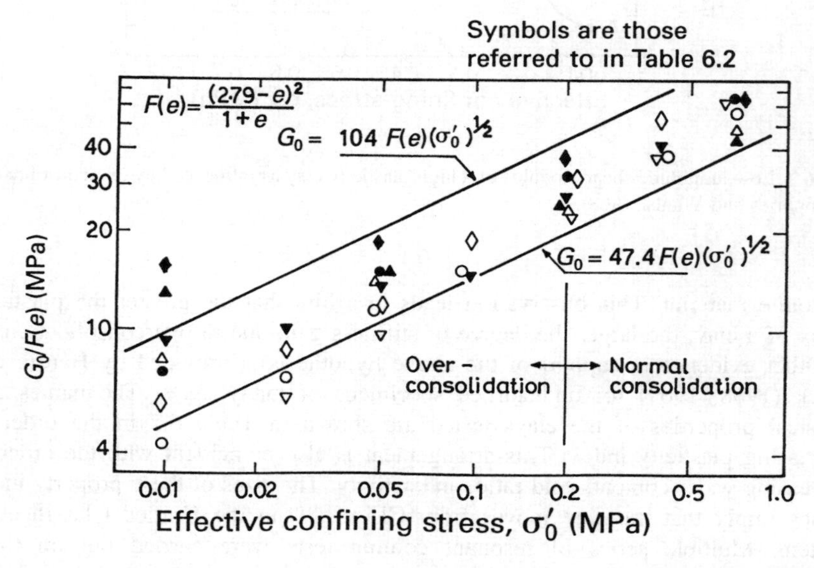

Fig. 6.8 Low-amplitude shear modulus of various plasticity clays as affected by overconsolidation (Hardin and Black, 1969).

The outcome of the tests are shown in Fig. 6.8 where the shear modulus divided by the function $F(e)$ is plotted versus the effective confining stress at which the resonant column tests were performed. In view of in situ depths of undisturbed sampling, the confining stress of 0.2 MPa was considered to roughly represent the in situ state of the overburden pressure. Therefore the shear moduli measured at a confining stress smaller than 0.2 MPa should be considered to be the values that are exhibited in an overconsolidated state of the undisturbed samples. Viewed overall, it can be seen that most of the data points in Fig. 6.8 fall within a zone bounded by two lines. If attention is drawn to the data from overconsolidated samples, it is apparent that the increase in shear modulus caused by overconsolidation is more pronounced in clays with a higher plasticity index than in soils with a low-plasticity index. The trend of increasing shear modulus at low confining stress with increasing plasticity

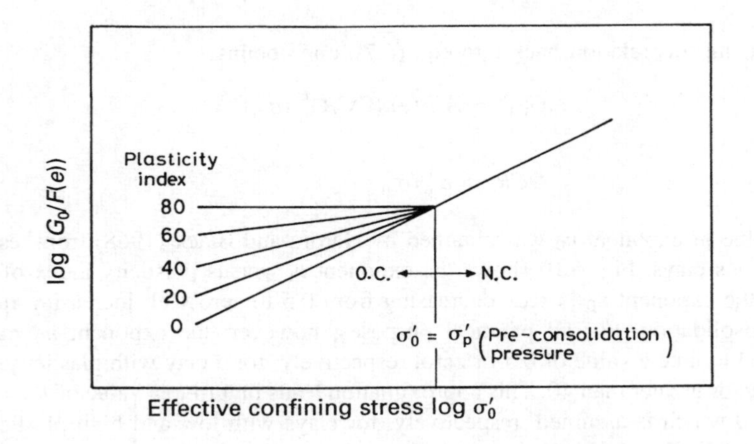

Fig. 6.9 Characteristics of shear modulus as affected by overconsolidation ratio.

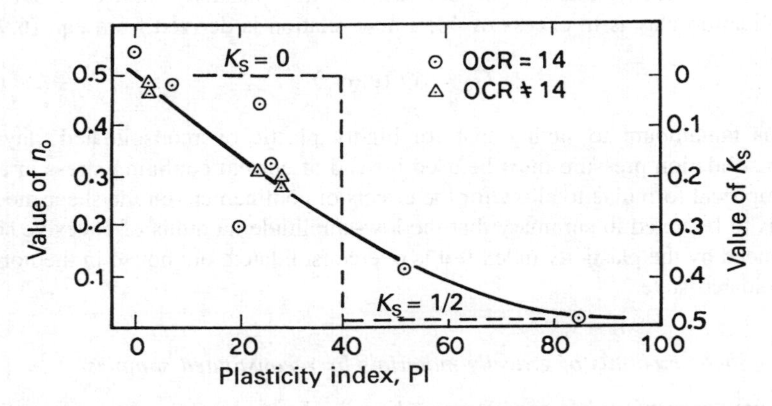

Fig. 6.10 Relation between K_s-value and plasticity index (Hardin and Black, 1969).

index is schematically illustrated in Fig. 6.9. In view of this, Hardin and Black (1968) formulated the effects of OCR as follows. Let the shear modulus in a normally consolidated clay be expressed by an empirical equation in eqn (6.3). Analogous to this expression, let the shear modulus in an overconsolidated state of the clay G_{C0} be given by

$$G_{C0} = A_{C0} F(e)(\sigma'_0)^{n_0} \tag{6.7}$$

where A_{C0} and n_0 are constants. When the current confining stress σ'_0 coincides with the preconsolidation pressure σ'_p, the shear modulus G_0 becomes equal to G_{C0}. Then one obtains

$$A_{C0} = A \cdot (\sigma'_p)^{1/2-n_0} = A \cdot (\sigma'_0)^{K_s} \tag{6.8}$$

$$K_s = \frac{1}{2} - n_0.$$

Introducing this relation back into eqn (6.7), one obtains

$$G_{C0} = A\,F(e)\,(\mathrm{OCR})^{K_s}(\sigma'_0)^{0.5} \qquad OCR = \sigma'_p/\sigma'_0. \tag{6.9}$$

The value of exponent n_0 was obtained by Hardin and Black (1968) from test data on various clays. Fig. 6.10 shows the exponent n_0 versus plasticity index of clays where the exponent n_0 is seen decreasing from 0.5 to zero with increasing ratio of overconsolidation. For all practical purposes, however, the exponent n_0 may be assumed to take a value of 0.5 or zero, respectively, for a clay with plasticity index less than or greater than 40. This approximation leads in turn to a value of $K_s = 0$ and $K_s = 1.0$ which is assumed, respectively, for clays with low and high plasticity as illustrated in Fig. 6.10. Thus, for a clay with a plasticity index less than 40 the empirical relation of the form of eqn (6.3) can be used without allowing for the effect of overconsolidation. In contrast, if the plasticity index of an over-consolidation clay is in excess of 40, a new relation is derived from eqn (6.9) as

$$G_0 = A\,F(e)(\sigma'_p)^{1/2}. \tag{6.10}$$

This is tantamount to saying that for highly plastic overconsolidated clays, the preconsolidation pressure must be used instead of current confining stress in any of the empirical formulae to allow for the effects of confinement on the shear modulus.

It is to be noted in summary that the low-amplitude modulus of cohesive soils is influenced by the plasticity index if it is overconsolidated, but not so in the normally consolidated state.

6.1.3 *Shear modulus of gravelly materials by reconstituted samples*

Dynamic properties of coarse-grained soils have recently been investigated

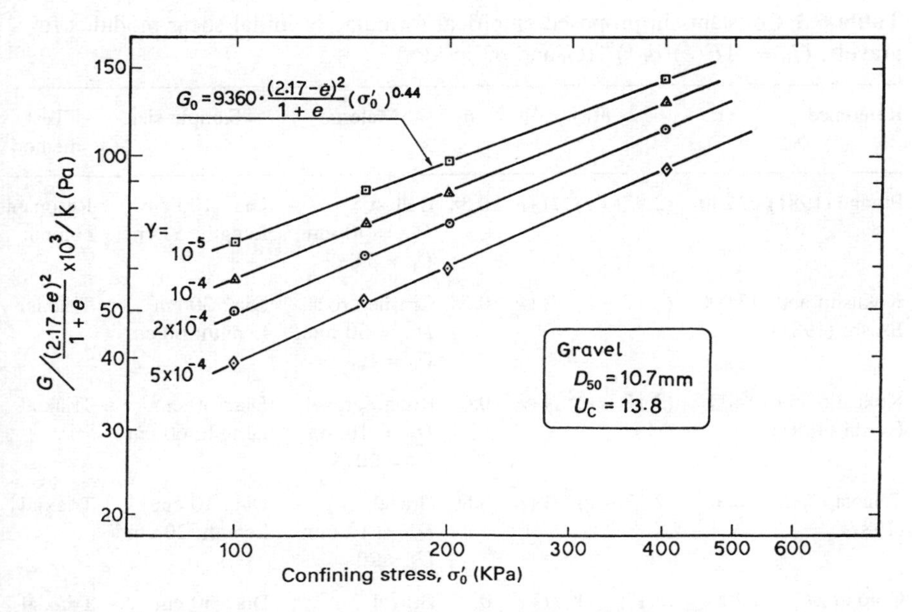

Fig. 6.11 Low-amplitude shear moduli versus effective confining stress (Nishio *et al.* 1985).

extensively in Japan in response to the needs for clarifying seismic performances of dense sand and gravelly deposits of diluvial origin. Most of the tests in the laboratory have been carried out using triaxial test apparatus in which cylindrical samples 10 to 30 cm in diameter and 20 to 50 cm in length were subjected to cyclic axial stress. Dry samples are compacted with a tamper in 5 to 10 layers to a target density and after ensuring saturation with a B value in excess of 95%, a cyclic axial load is applied in the undrained condition. Accurate monitoring of strain was made by means of a non-contact type strain gauge to infinitesimal strains as small as 10^{-5}. Since the strain level was still small in this type of test, there was no build-up of pore water pressure and samples were free from any deleterious effect due to membrane penetration. The outcome of the tests is generally expressed in terms of the initial shear modulus divided by some function of void ratio, which is plotted versus the effective confining stress σ'_0. One of such plots by Nishio *et al.* (1985) is demonstrated in Fig. 6.11, where the amplitude of axial strain is taken as a parameter. For the infinitesmal strain of 10^{-5}, the shear modulus G_0 is expressed as

$$G_0 = 9360 \frac{(2.17 - e)^2}{1 + e} (\sigma'_0)^{0.44} \tag{6.11}$$

where G_0 and σ'_0 are expressed in terms of kPa.

Similar empirical relations for reconstituted gravelly materials proposed by several groups of investigators are collected and compiled in Table 6.3, where A

Table 6.3 Constants in proposed empirical formulae of initial shear modulus for gravels, $G_0 = AF(e)(\sigma_0{}')^n$ (G_0 and σ_0' in kPa)

Reference	A	$F(e)$	n	Material	Sample size	Test method
Prange (1981)	7 230	$(2.97 - e)^2 / 1+e$	0.38	Ballast D_{50} = 40 mm, U_c = 3.0	Dia.: 100 cm Length: 60 cm	Resonant column
Kokusho and Esashi (1981)	13 000	$(2.17 - e)^2 / 1+e$	0.55	Crushed rock D_{50} = 30 mm, U_c = 10	Dia.: 30 cm Length: 60 cm	Triaxial
Kokusho and Esashi (1981)	8 400	$(2.17 - e)^2 / 1+e$	0.60	Round gravel D_{50} = 10 mm, U_c = 20	Dia.: 30 cm Length: 60 cm	Triaxial
Tanaka *et al.* (1987)	3 080	$(2.17 - e)^2 / 1+e$	0.60	Gravel D_{50} = 10 mm, U_c = 20	Dia.: 10 cm Length: 20 cm	Triaxial
Goto *et al.* (1987)	1 200	$(2.17 - e)^2 / 1+e$	0.85	Gravel D_{50} = 2 mm, U_c = 10	Dia.: 30 cm Length: 60 cm	Triaxial
Undisturbed						
Nishio *et al.* (1985)	9 360	$(2.17 - e)^2 / 1+e$	0.44	Gravel D_{50} = 10.7 mm, U_c = 13.8	Dia.: 30 cm Length: 60 cm	Triaxial

indicates a constant multiplier and n is the exponent of the effective confining stress. The grain size distribution curves of the materials used for the tests are shown in Fig. 6.12. Generally the materials are well graded, comprising more than 30% gravel with particle size greater than 2 mm. The grading of the gravelly soils tested by Tanaka *et al.* (1987) is separately displayed in Fig. 6.13. From observation of the test results in Table 6.3 it may be assumed that the exponent takes an average value of 0.5 and the constant A could vary widely depending upon the material type and its grain composition. The values of low-amplitude shear moduli as formulated in Table 6.3 are displayed in Fig. 6.14 in a graphical form. It may be seen that there are considerable variations in the shear modulus G_0 probably depending upon the type of gravel and its grain composition.

To investigate the influence of gravel content, three suites of tests were conducted by Tanaka *et al.* (1987) on materials containing gravel proportion of 0%, 25%, and 50% by weight. The grading of these three materials is shown in Fig. 6.13. In the first suite of tests, the material was placed in a container (200 cm deep 200 cm long

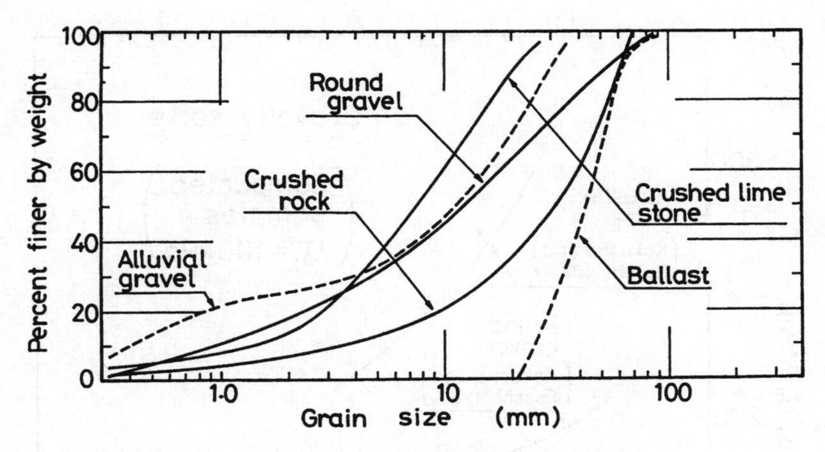

Fig. 6.12 Grain-size distribution curves of gravelly soils used in the tests.

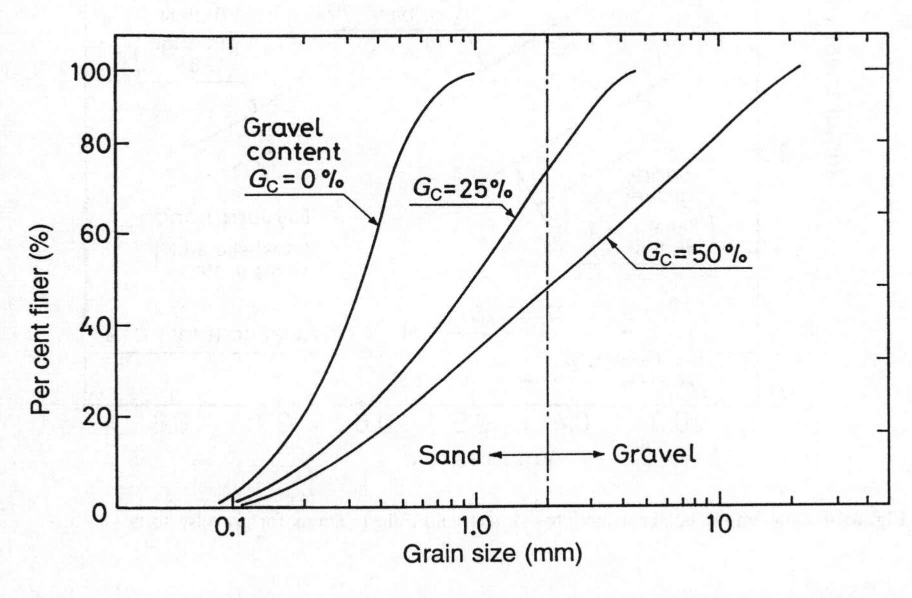

Fig. 6.13 Grain-size distribution curves of sands containing different percentage of gravel (Tanaka *et al.* 1987).

and about 100 cm wide) and after applying an overburden pressure of 100 kPa, a shear wave was generated by hitting a plate embedded in the sample. The shear modulus was determined by monitoring the velocity of the shear wave propagating across the container. The values of the shear modulus thus obtained are plotted versus void ratio by dashed lines in Fig. 6.15. Because the range of void ratio variation was different for each of the three materials tested, three curves are shown

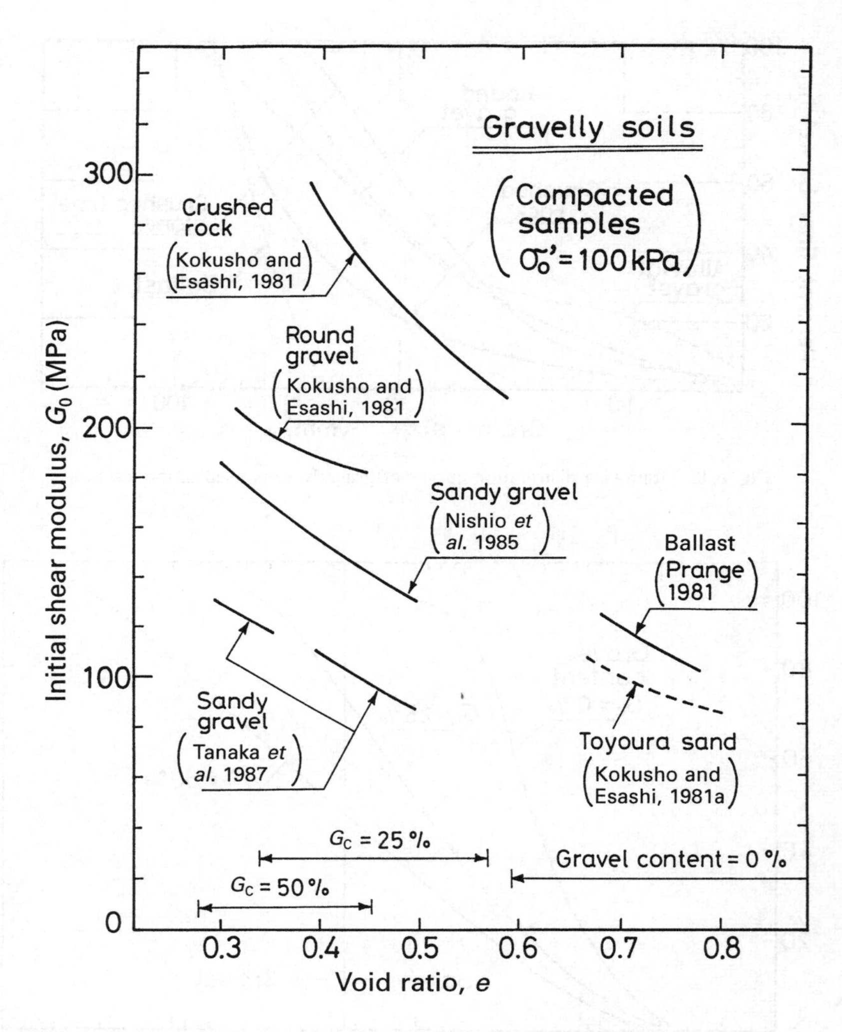

Fig. 6.14 Low-amplitude shear modulus versus void ratio relations for gravelly soils.

separately, but they are located in the diagram, though discontinuously, so as jointly to cover a wide range in void ratio. It is of interest to note that the shear modulus tends to decrease with increasing void ratio not only within the range of variation of each material but more widely over the range of void ratio covered by the mixed materials with varying gravel contents. The second suites of tests by Tanaka *et al.* (1987) consisted of measuring the velocity of ultrasonic wave propagation through the cylindrical sample in the triaxial cell. The shear modulus thus obtained is also displayed in Fig. 6.15 in terms of the plot versus the void ratio. The same trend of decreasing shear modulus with increasing void ratio is shown by the figure. The

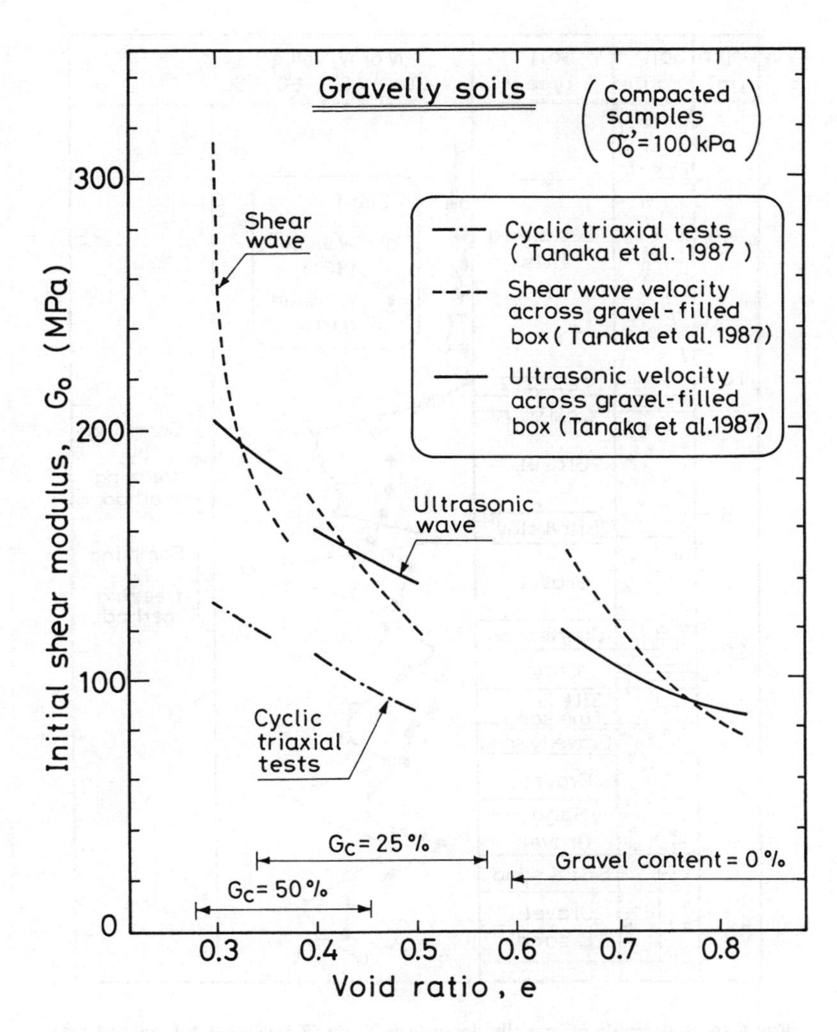

Fig. 6.15 Low-amplitude shear modulus versus void ratio relations of gravelly soils obtained by different methods.

third series of tests was conducted by means of the cyclic triaxial tests on compacted materials with gravel proportions of 25% and 50%. The outcome of the tests, proposed in the form of empirical formulae in Table 6.3, are displayed in the graphical form in Fig. 6.15 where the decreasing tendency of shear modulus with increasing void ratio is also shown. The results of the three types of tests displayed in Fig. 6.15 indicate that the shear moduli by the cyclic triaxial tests give the lowest value, as compared to those obtained from measurements of shear wave and ultrasonic wave propagation. While the reason is not known exactly for such discrepancy, imperfect coupling between the sample and its end plattens might have

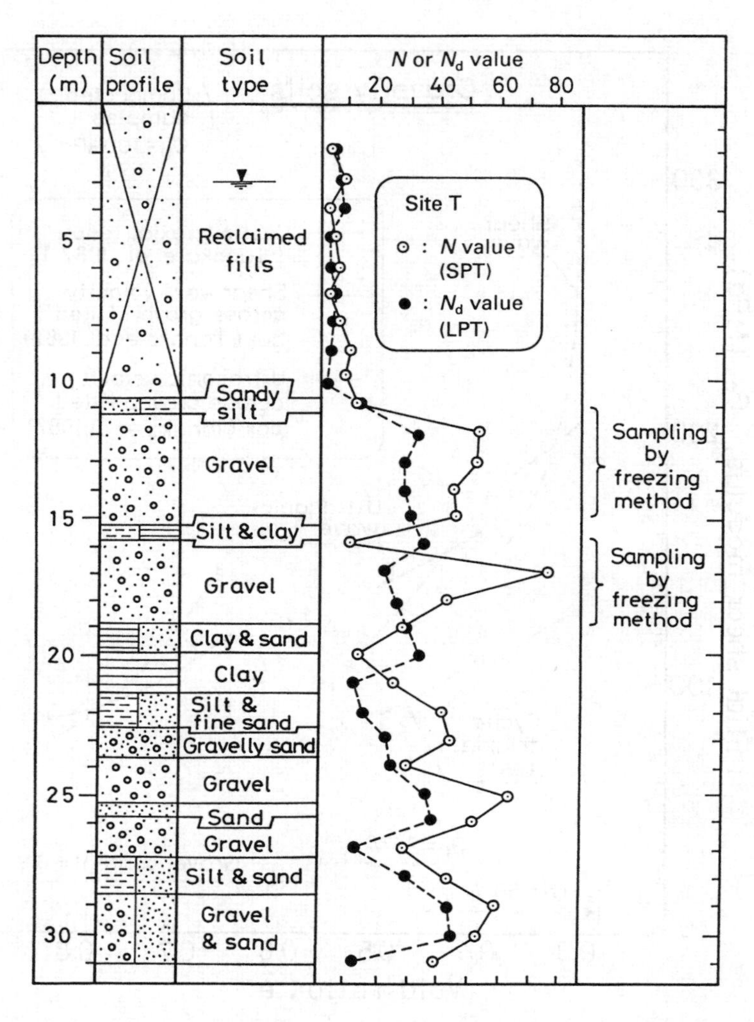

Fig. 6.16 Soil profile of gravelly deposits at T-site (Kokusho and Tanaka, 1994).

produced somewhat larger axial displacement than that occurring within the sample itself.

6.1.4 *Shear modulus of gravelly materials by intact samples*

Considerable efforts have been expended in recent years to recover high quality undisturbed samples from gravelly deposits under the ground water table. These efforts have been motivated in Japan from the needs for planning and designing facilities of nuclear energy under seismic loading conditions. To ensure a high level of safety, much higher acceleration than usual is specified in the design leading to

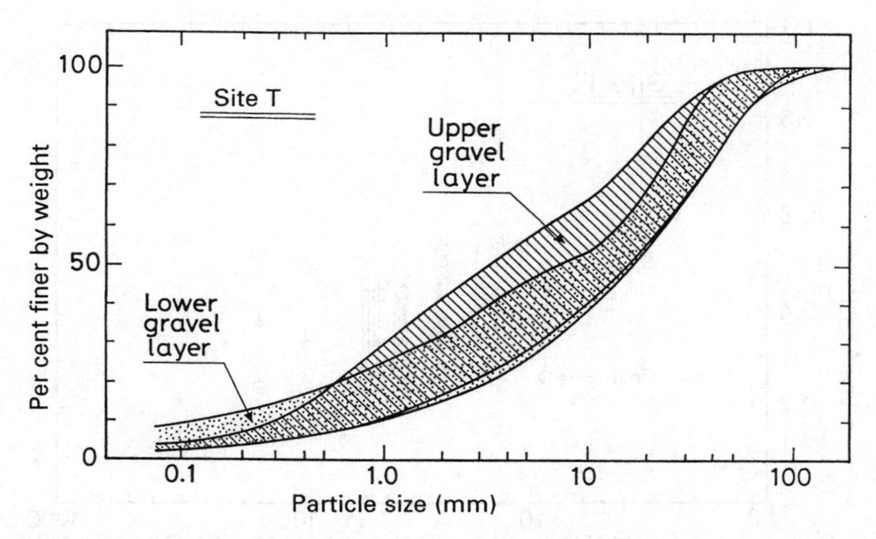

Fig. 6.17 Grain size distribution curves of two gravelly soils at T-site (Kokusho and Tanaka, 1994).

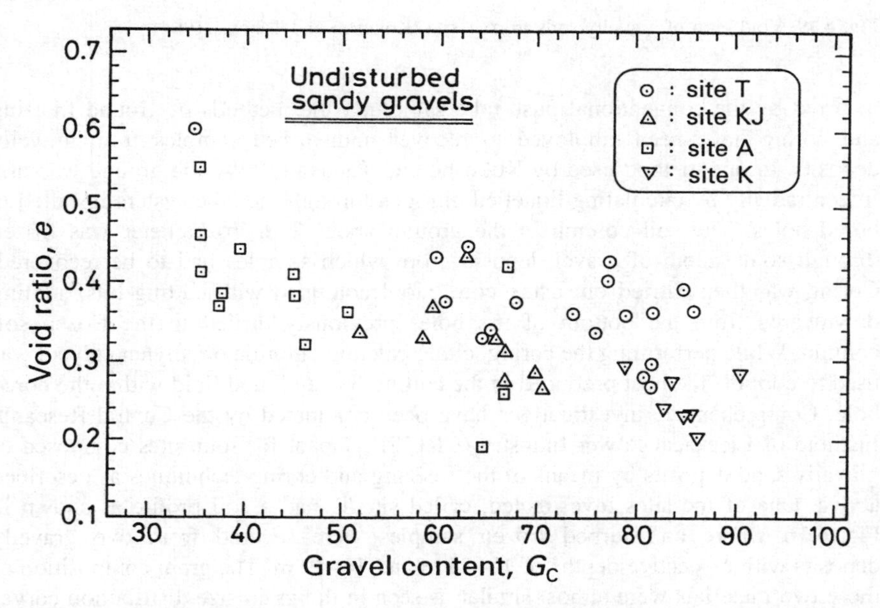

Fig. 6.18 Ranges of void ratio of undisturbed gravelly soil samples from 4 sites (Kokusho and Tanaka, 1994).

unusually great external loads to be considered in the design analyses. Thus great needs have arisen for knowing the behaviour of dense sands and gravelly sands when subjected to intense dynamic shear stress. In cognizance of the disturbance

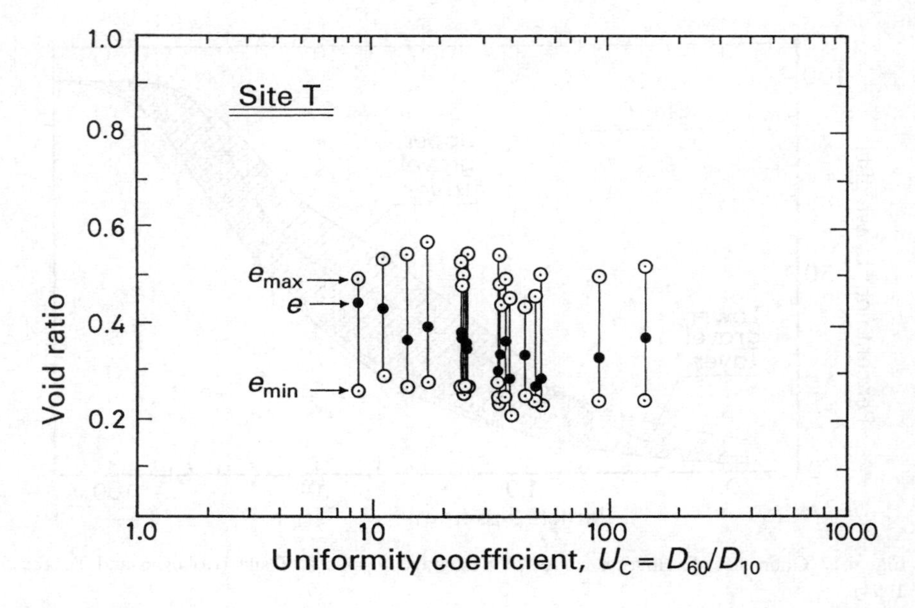

Fig. 6.19 Void ratio of gravelly soils from T-site (Kokusho and Tanaka, 1994).

incurred by the conventional push-tube sampling, the methods of ground freezing and coring have been employed to recover undisturbed samples from gravelly deposits. In one method used by Kokusho and Tanaka (1994), the ground was first frozen radially by circulating liquefied nitrogen through the tube system installed in bored holes. The soil column in the ground about 2 m in diameter was frozen through some depth of gravel deposits from which samples had to be recovered. Coring was then carried out using core barrel equipped with cutting bits, starting downwards from the bottom of the holes previously drilled in the frozen soil column. While performing the coring, either calcium chloride or ethylene glycol was used to cool off the heat produced in the cutting bits and mud fluid within the bored hole. Comprehensive investigations have been conducted by the Central Research Institute of Electrical Power Industry (CRIEPI), Japan, for four sites composed of gravelly sand deposits by means of the freezing and coring techniques as described above. One of the sites investigated, called site T, had a soil profile as shown in Fig. 6.16 where undisturbed frozen samples were secured from two gravelly deposits with respective depths of 11–15 m and 16–19 m. The grain composition of these two materials were almost similar as seen in the grain size distribution curves in Fig. 6.17. The cored undisturbed samples had a dimension 30 cm in diameter and 60 cm in length. The void ratio ranged between 0.28 and 0.60, as shown in Fig. 6.18. It may be seen that the void ratio has a tendency to decrease with increasing proportion of gravel content. The in situ void ratios determined from the undisturbed samples from site T are alternatively displayed in Fig. 6.19 in relation to the maximum and minimum void ratios, e_{max} and e_{min}, respectively. It may be seen that the relative density of most samples lies in the range of 60 to 90%.

The low-amplitude shear modulus G_0 obtained as a result of cyclic triaxial tests is displayed in a form of the plot versus the logarithm of the effective confining stress in Fig. 6.20 where data from two gravel deposits at site T are indicated by circles. Each sample was tested in sequence with stepwise increased confining stress. Assuming the dependency on the effective confining stress to be represented in a general form of eqn (6.3), the value of n in the exponent can be determined by reading off the slope of straight lines drawn in Fig. 6.20. The exponent is shown to take a large value of 0.84 and 0.93 for the deposits at site T. The low-amplitude shear modulus of undisturbed samples of gravelly materials from other sites is also shown in Fig. 6.20 where it may be seen that the exponent n takes a value between 0.56 and 0.93 with an average of 0.75 which is substantially larger compared to the corresponding value for reconstituted gravelly samples shown in Fig. 6.11. It is also seen in Fig. 6.20 that the average value $n = 0.87$ for the materials from site T is definitely larger compared to the average value $n = 0.72$ and 0.70 for the sandy gravels respectively from site KJ and site A.

To examine the effects of void ratio, those data from 5 sites tested under the confining stresses representative of those in the field are assembled and demonstrated in Fig. 6.21 in terms of the initial shear modulus G_0 plotted versus the void ratio. Reflecting the potential variability existing in undisturbed natural deposits, there is much scatter in the test data in comparison with those from reconstituted samples. However, it is apparent in Fig. 6.21 that the modulus tends to decrease with an increase in void ratio for the materials from the same site.

At site A, sampling by means of a tube-pushing technique was also conducted nearby together with the undisturbed sampling by the method of ground freezing. The cyclic triaxial tests were also performed on allegedly undisturbed samples by means of the tube sampling. The results of the tests are also presented in Fig. 6.21 with black rectangles where it can be seen that the low-amplitude modulus of the samples by the push-tube technique is generally smaller than that obtained from the undisturbed samples by means of the ground freezing technique.

6.2 Time dependency of low-amplitude shear modulus

It has been recognized by Hardin and Black (1968) and Humphries and Wahls (1968) that the low-amplitude shear moduli of remoulded clays tend to increase with the duration of time under a sustained confining stress after the primary consolidation is over. Several efforts have been made by Afifi and Woods (1971). Marcuson and Wahls (1972) and Afifi and Richart (1973) to investigate this aspect via the laboratory tests on artificially prepared clays using the resonant column device.

Figure 6.22 shows a typical example of a kaolinite clay exhibiting a modulus change with time under a constant confining stress (Anderson and Stokoe, 1977). It can be seen that the shear modulus increases noticeably with progression of primary consolidation but the rate of its increase becomes small after the consolidation is completed. The increase due to consolidation can be accounted for by the

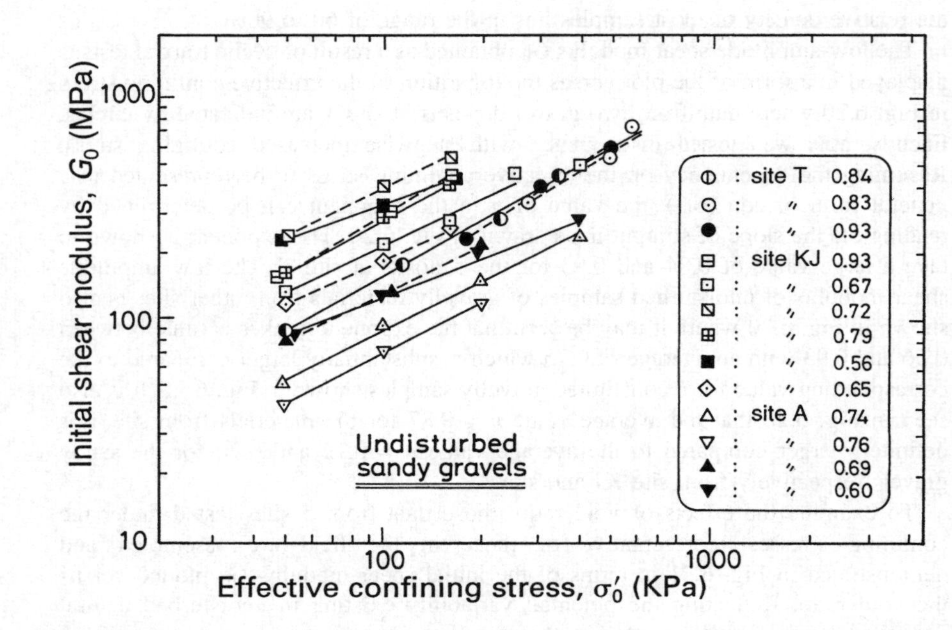

Fig. 6.20 Initial shear modulus versus confining stress for undisturbed samples of sandy gravels (Kokusho and Tanaka, 1994).

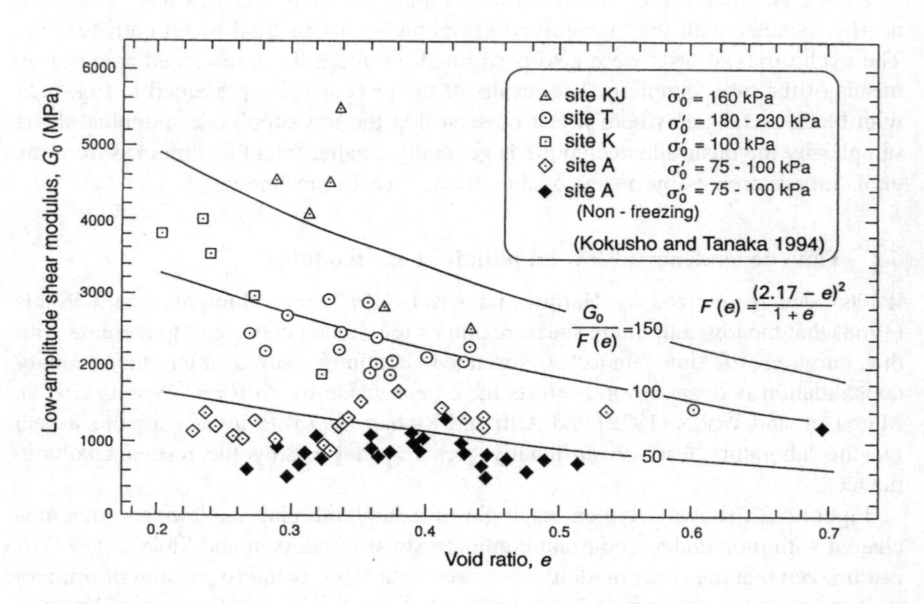

Fig. 6.21 Shear modulus versus void ratio for undisturbed specimens of gravelly sands (Kokusho and Tanaka, 1994).

corresponding decrease in void ratio, but the long-term increase in shear modulus under a sustained confining stress following the primary consolidation is deemed as a time effect resulting from strengthening of particle bonding. During this long-term response, the low-amplitude shear modulus G_0 is observed to increase almost linearly with the logarithm of time. To describe the long-term effect, Afifi and Richart (1973) proposed the use of a coefficient defined as

$$I_G = \frac{\Delta G_0}{\log_{10}(t_2/t_1)} \tag{6.12}$$

where t_1 and t_2 are arbitrarily chosen points of time after primary consolidation and ΔG_0 denotes the increase in shear modulus during the time period from t_1 to t_2. As illustrated in Fig. 6.22, I_G is equal to the value of ΔG_0 for one logarithmic cycle of time. To remove possible influence of soil type and confining stress, Afifi and Richart (1973) also suggested the use of a normalized coefficient N_G defined as

$$N_G = \frac{1}{G_{1000}} \frac{\Delta G_0}{\log_{10}(t_2/t_1)} \tag{6.13}$$

where G_{1000} denotes the shear modulus measured after 1000 minutes of application of constant confining stress following the primary consolidation. Thus the coefficient N_G indicates a percentage increase of shear modulus within one logarithmic cycle of time. For convenience sake, if ΔG_0 is taken specifically as an increase during any 10 times period of time, i.e. $t_2/t_1 = 10$, eqn (6.13) is rewritten as

$$N_G = \frac{\Delta G_0}{G_{1000}}. \tag{6.14}$$

Initially the time effect as above was investigated on artificially prepared specimens of cohesive soils, but later a similar effect was found to exist in undisturbed samples of cohesive soils and sands as well (Stokoe and Richart, 1973; Trudeau *et al.* 1974).

After examining a majority of laboratory test data on undisturbed soils, Afifi and Richart (1973) revealed the fact that the value of N_G tends to decrease with increasing grain size of the soil. Shown in Fig. 6.23 is a summarized plot in this context where it may be seen that the effect of elapsed time becomes small for sandy soil having a mean particle diameter in excess of $D_{50} = 0.05$ mm. The wide-ranging variation of N_G for fine-grained soils may be explicated as accuring from the character of cohesive soils as represented by the plasticity index I_p. A compilation of available test data in this vein was made by Kokusho (1987) as demonstrated in Fig. 6.24 where the value of N_G is plotted versus I_p. It is obvious from this figure that the modulus increase due to the time effect becomes more pronounced with increasing plasticity index of soils. It is to be noticed that the gain in shear modulus N_G in undisturbed specimens was approximately the same as that observed in remoulded specimens. This observation would lead to a hypothesis that it is the portion of shear modulus gain due to the time effect that could be lost by disturbance during sampling and sample handling. Suppose the stiffness of in situ soils consists of two parts, one produced by primary consolidation and the other due to the long-

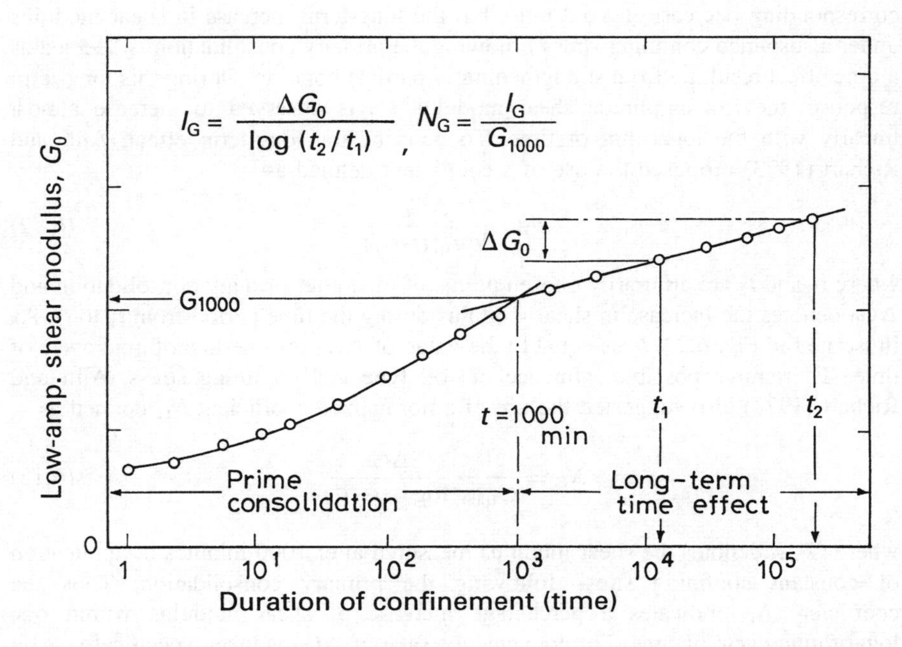

Fig. 6.22 Time effect on the low-amplitude shear modulus (Anderson and Stokoe, 1977).

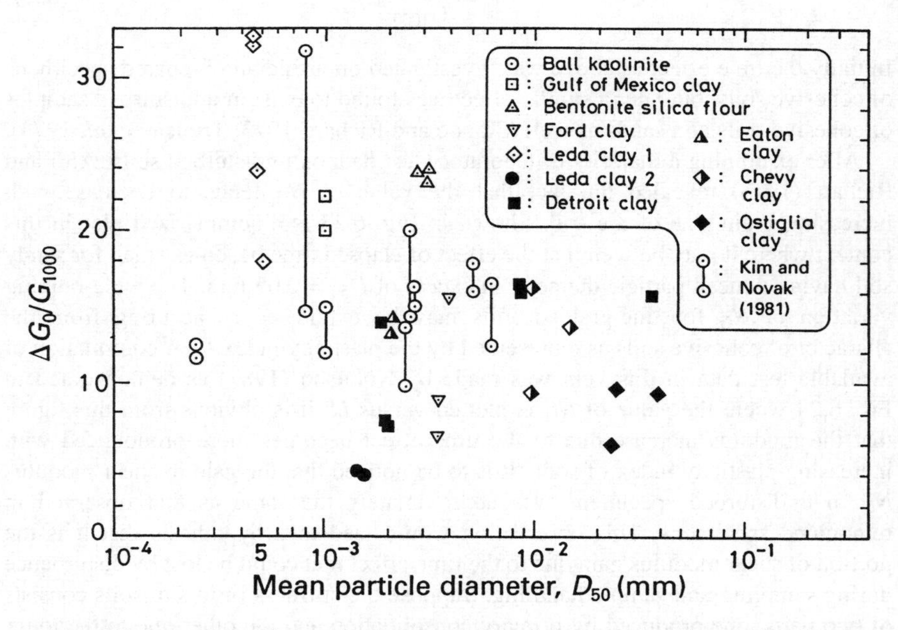

Fig. 6.23 Long-time effect on the shear modulus gain versus mean diameters of clays.

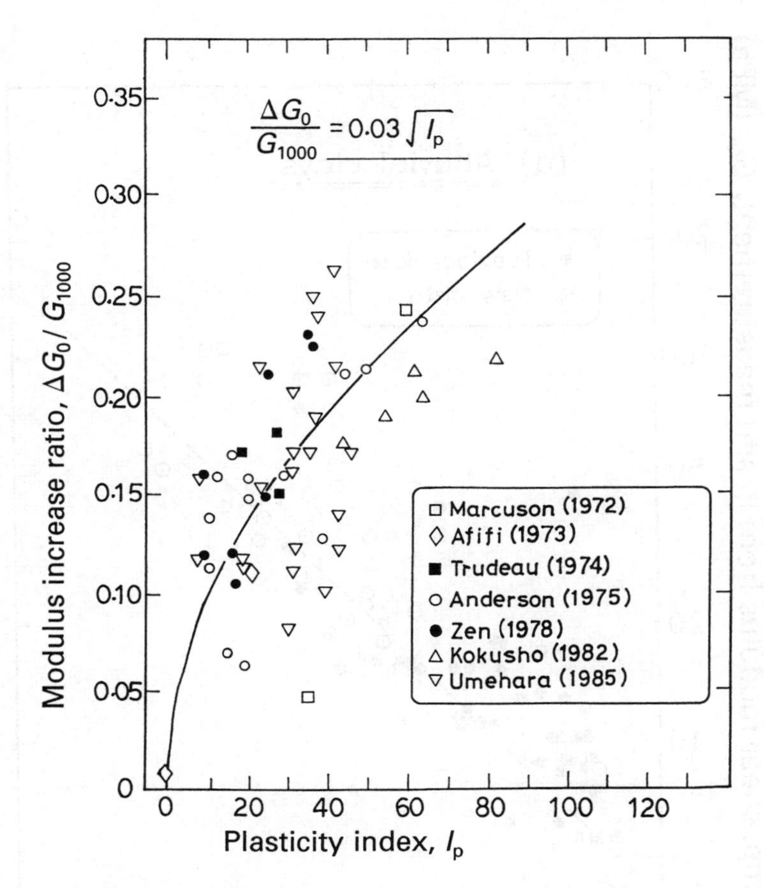

Fig. 6.24 Modulus increase ratio versus plasticity index (Kokusho, 1987).

term time effect. The above hypothesis implies that, while the stiffness due to the time effect is lost by disturbances during sampling, the portion of the stiffness produced by primary consolidation is retained because there is no change in void ratio during sampling operation.

On the basis of the above hypothesis, a procedure to incorporate the time effect was suggested by Anderson and Stokoe (1977) for assessing low-amplitude modulus G_{OF} of in situ deposits from the laboratory test results on undisturbed samples. First the low-amplitude shear modulus at the end of primary consolidation G_{OL} is measured by some laboratory tests or from an empirical relationship. The in situ shear modulus is estimated as

$$G_{OF} = G_{OL} + N_G F_A \tag{6.15}$$

where F_A is an age factor for a site. It is difficult to precisely evaluate the value of

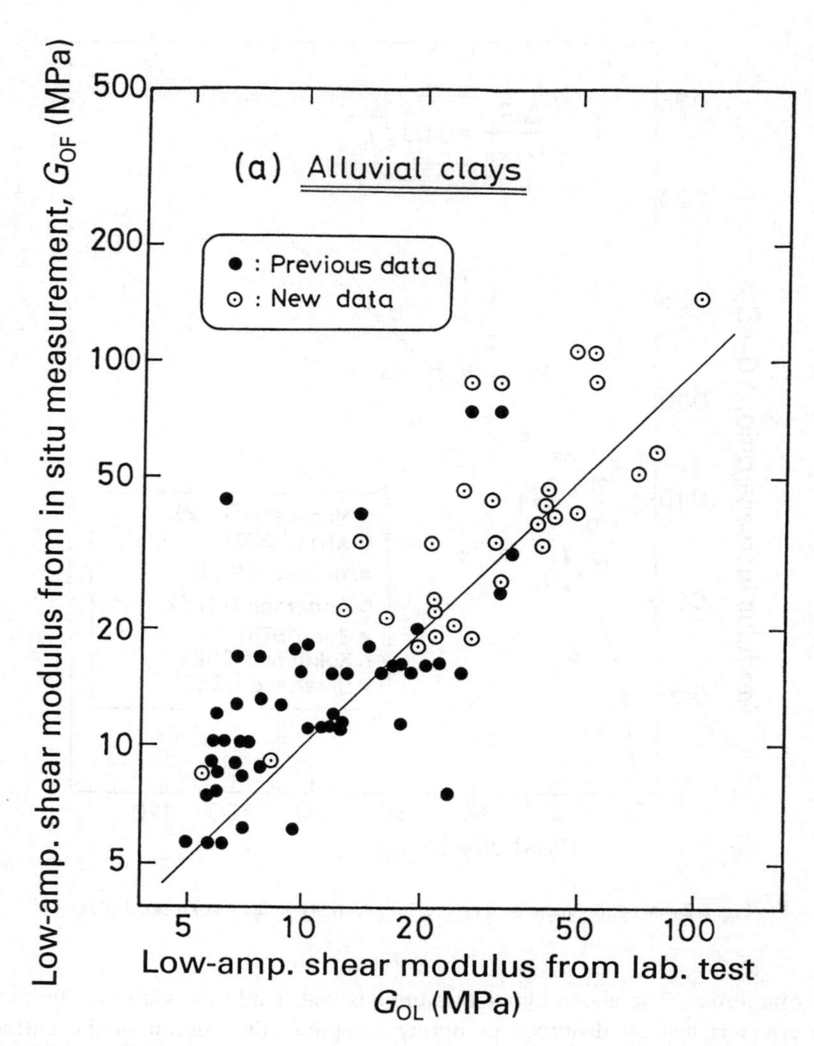

Fig. 6.25(a) Comparison of shear moduli from in situ and laboratory tests (Yokota *et al.* 1985).

F_A because it mainly depends upon the period of time from the end of primary consolidation to the present, but the data as compiled in Fig. 6.24 may be used as a reference.

6.3 Low-amplitude shear moduli from in situ tests

With the recent development of in situ techniques for soil investigations and also of the laboratory testing procedures for highly undisturbed samples, it has become

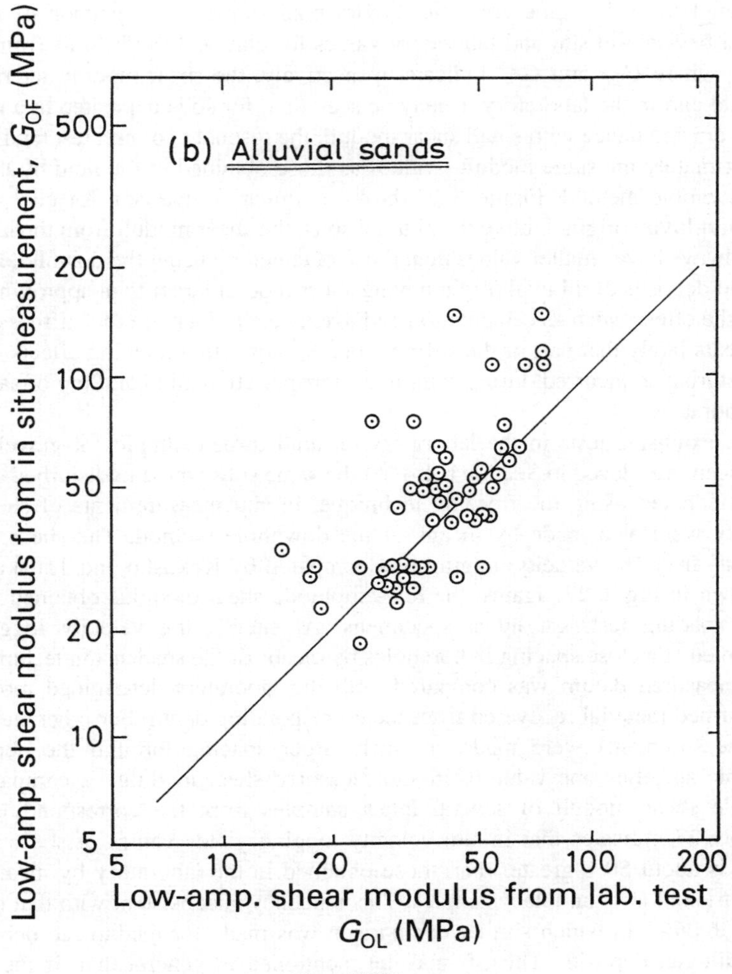

Fig. 6.25(b)

possible to obtain test data which are more reliable for the analyses and design of soil-related problems during earthquakes. In what follows, the shear moduli of in situ soil deposits will be discussed in the light of the basic rationale described in the foregoing sections.

6.3.1 *Initial shear moduli by in situ velocity loggings*

Comparison of shear moduli between in situ and laboratory values was made by Yokota *et al.* (1981) for soils of alluvial and diluvial origins. In situ shear moduli were determined by means of the downhole technique and the moduli in the laboratory were measured by the resonant column tests on undisturbed specimens recovered from the same borehole. Figure 6.25 shows a comparison of the shear moduli between in situ and laboratory values for clay and sandy soils from alluvial origin, where G_{OF} and G_{OL} indicate, respectively, the shear moduli determined in the field and in the laboratory. It may be seen that, for soils deposited in a relatively recent era and hence with small shear moduli, the resonant column test tends to yield approximately the same modulus values as those obtained in the field by the use of the downhole method. Figure 6.26 shows a similar comparison for clay and sand soils of diluvial origin. Unlike the alluvial soils, the shear moduli from the laboratory tests always have smaller values than those obtained by using the downhole method. For soil deposits of diluvial origin having shear moduli larger than approximately 50 MPa, the effects such as cementation and ageing act to increase the stiffness of soils. It appears likely that part of the stiffness due to these strengthening effects is lost by the disturbance incurred during sampling, transportation and handling of samples in the laboratory.

The extensive tests in the laboratory on undisturbed samples of gravelly sands have been introduced in Section 6.1.4. At the same sites where undisturbed sampling was conducted using the freezing technique, in situ measurements of shear wave velocity were also made by means of the downhole method. The shear modulus obtained from the velocity logging was compared by Kokusho and Tanaka (1994), as shown in Fig. 6.27, against the low-amplitude shear modulus obtained from the cyclic loading tests on intact specimens. At site T, the velocity logging was performed at a close spacing in boreholes by means of the suspension technique, and each measured datum was compared with the laboratory-determined modulus of undisturbed material recovered from the corresponding depth. For other sites, the in situ measurements were made at much larger spacing through the depth of a borehole and thus one value of in situ measured shear modulus is compared with multiple shear moduli of several intact samples from the corresponding depth. Figure 6.27 indicates that in situ velocity logging yields values of shear modulus which is about 50% greater than those obtained in the laboratory by means of the cyclic triaxial tests on intact samples. This tendency is consistent with that observed in Fig. 6.26(b) in which similar comparison was made for medium to dense sands from diluvial deposits. Thus, it may be mentioned in general that, if the sand or gravelly material is from strata of diluvial or older origin, the low-amplitude shear modulus determined from the laboratory tests on high-equality undisturbed samples would be about 50% smaller than that of the intact materials in the field deposit.

From the above observation, it is apparent that the difference in shear modulus between the laboratory-determined and in situ measured values depends upon the stiffness itself of soil deposits. This aspect was addressed by Yasuda and Yamaguchi (1985) who compiled a profusion of test data to provide diagrams in which the ratio between the laboratory-determined and in situ measured shear moduli is plotted versus the shear modulus determined from the in situ velocity logging. One of such diagrams collecting reported data outside Japan is presented in Fig. 6.28 for cohesive soils. The arrows in the figure indicate the correction for the laboratory-obtained data by allowing for the increase in shear modulus due to the elapsed time of 20 years after the completion of primary consolidation. Without regard to the time

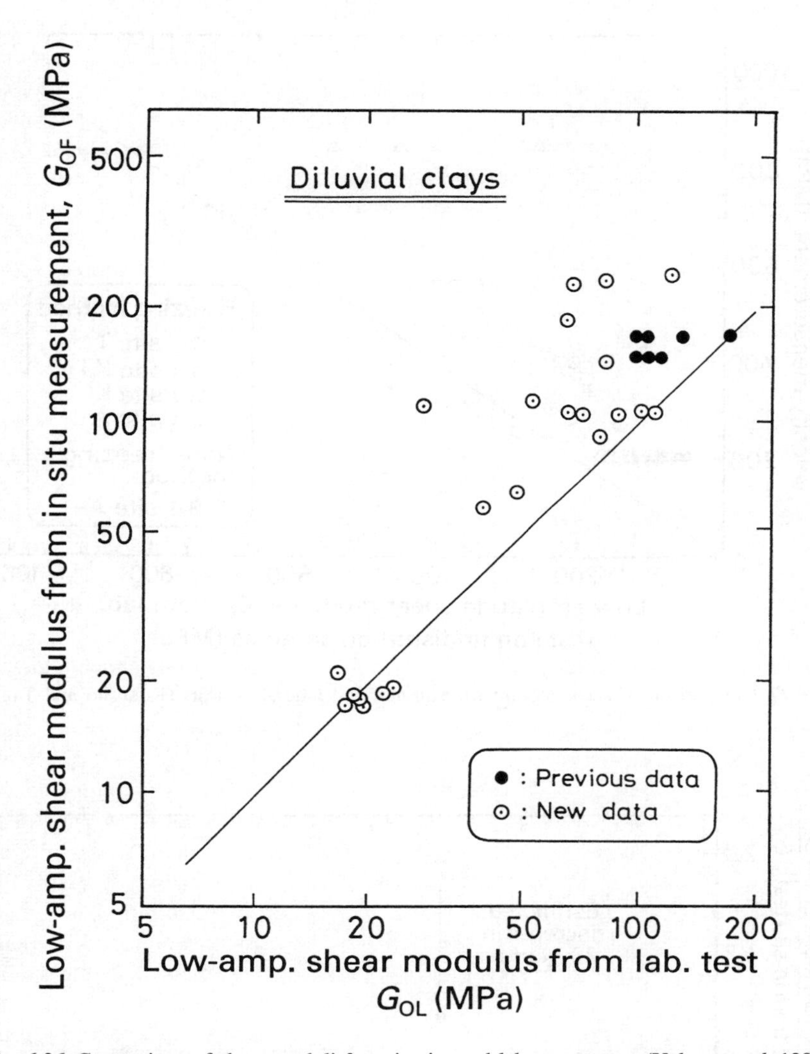

Fig. 6.26 Comparison of shear moduli from in situ and laboratory tests (Yokota *et al.* 1985).

effect, there is a tendency of the laboratory-determined shear modulus, consistent with the concept described in Section 6.2, to decrease with increasing stiffness of the soils. It is also noted in Fig. 6.28 that even if corrected for the sustained time effect, the above tendency appears still existent. A majority of test data obtained in Japan on sands was compiled as well by Yasuda and Yamaguchi (1985) as shown in Fig. 6.29. Most of the data were obtained from the tests on undisturbed samples retrieved by the tube sampling technique and as such there were presumably some effects of sample disturbance. The data from deposits of recent eras such as alluvium

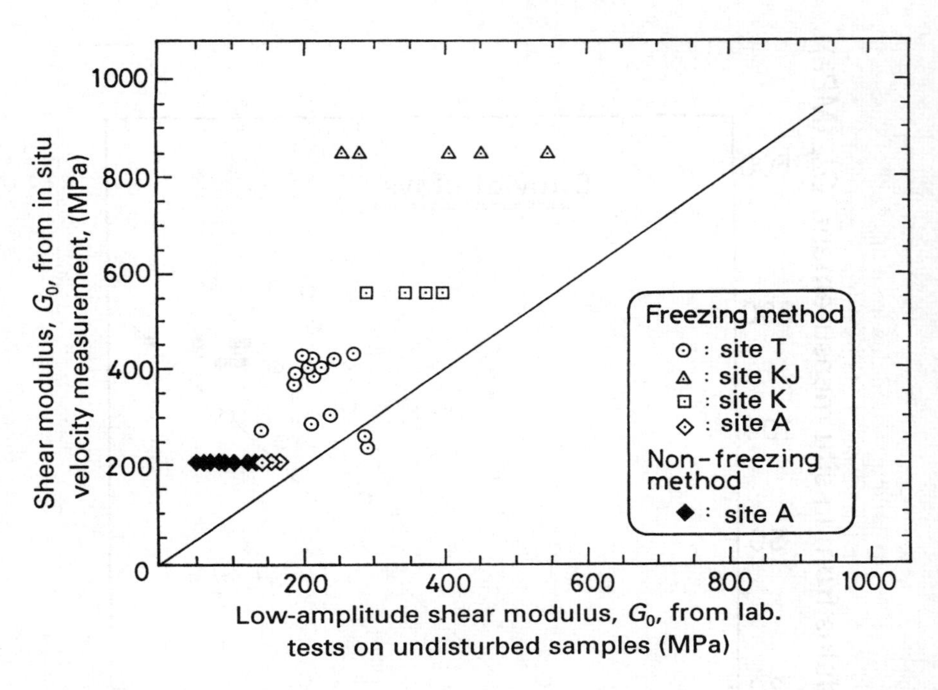

Fig. 6.27 Comparison of shear moduli from in situ and laboratory tests (Kokusho and Tanaka, 1994).

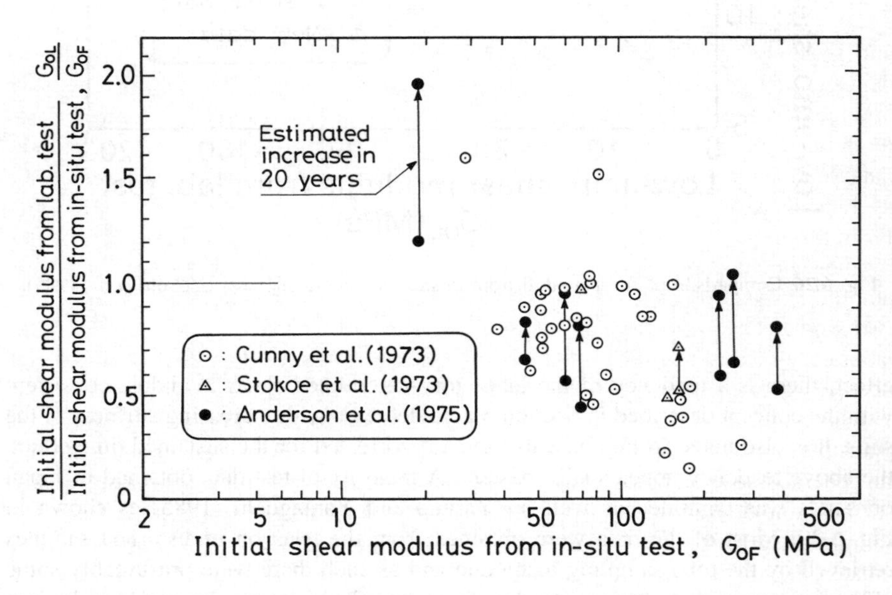

Fig. 6.28 Comparison of shear moduli from in situ and laboratory tests (Yasuda and Yamaguchi, 1985).

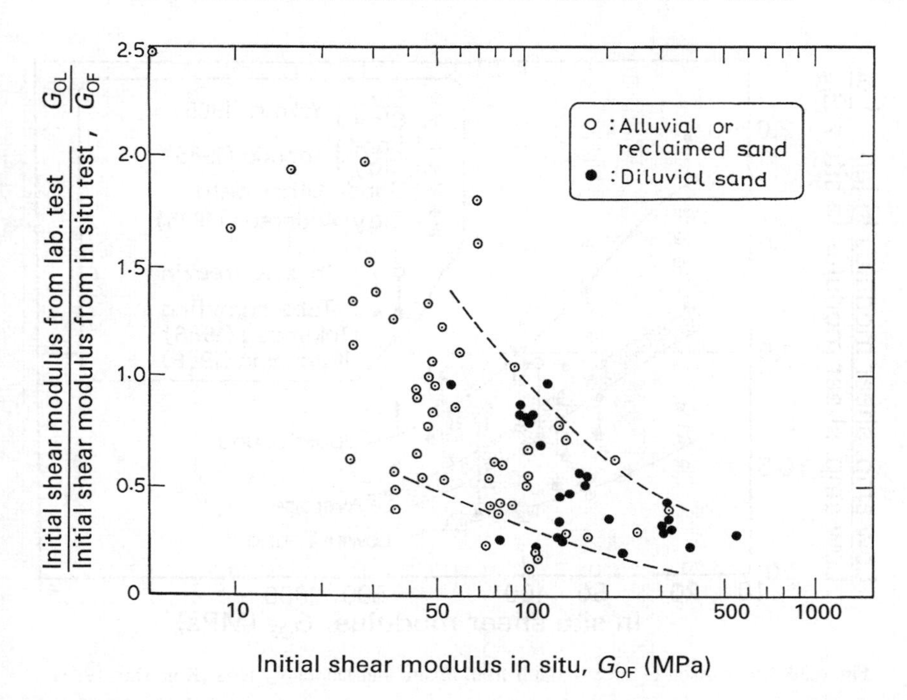

Fig. 6.29 Comparison of shear moduli from in situ and laboratory tests (Yasuda and Yamaguchi, 1985).

and reclaimed fills are indicated by open circles and the data pertaining to old deposits such as diluvium are shown by solid circles. The test data for sand deposits arranged in Fig. 6.29 show a general tendency of the laboratory-obtained shear modulus to decrease as the stiffness of the sand becomes greater. Looking over the data summarized above plus some other new data, Kokusho (1987) presented the average curve together with upper and lower bounding curves as shown in Fig. 6.30. From this summarized chart, it may be argued that, if the shear modulus determined by the velocity logging is in the range of 30–50 MPa (150–160 m/sec), the low-amplitude shear modulus obtained from the laboratory test could give a value which is reasonably close to the field modulus determined by the velocity logging. If the laboratory-determined shear modulus is found to be in excess of 30–50 MPa, it may need to be determined by in situ measurements.

There are several reasons for the above deviation to occur. One of the most plausible causes would be the disturbance of samples during sampling and handling. In general, loose samples tend to become more dense thereby exhibiting greater stiffness and dense samples to loosen showing a reduction in stiffness. The effect of sample disturbance may be illustrated as schematically shown in Fig. 6.31. The greater the disturbance, the larger the difference in the ratio of shear modulus G_{OL}/G_{OF} in the range of stiffness which is small or large, reflecting the potentiality

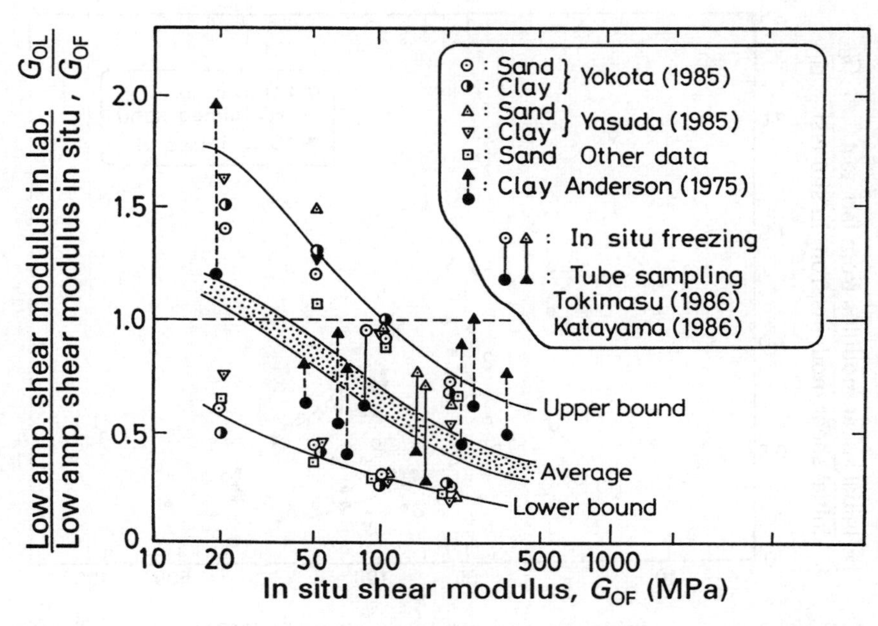

Fig. 6.30 Comparison of shear moduli from in situ and laboratory tests (Kokusho, 1987).

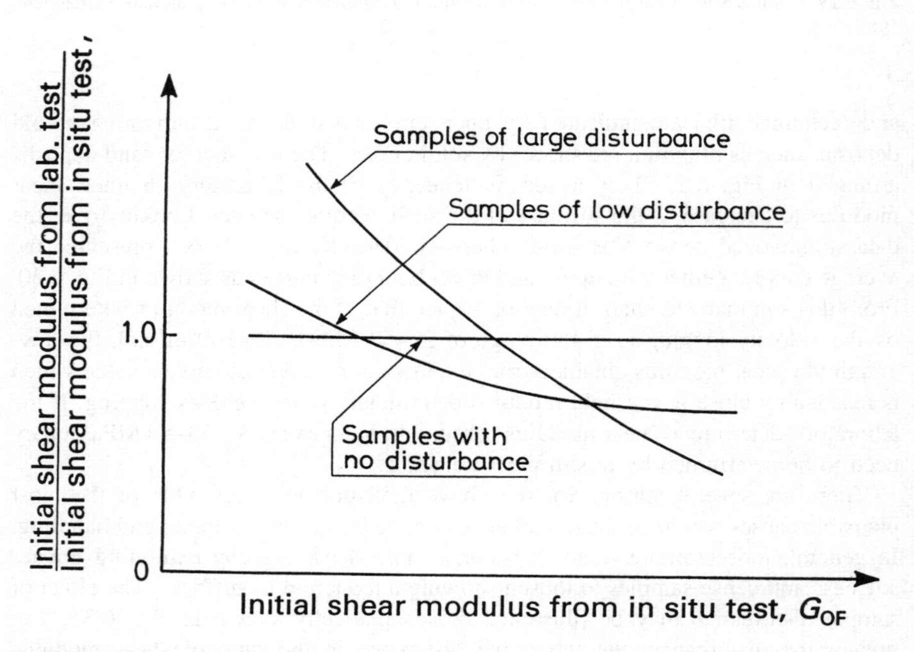

Fig. 6.31 Degree of sample disturbance influencing on the ratio of shear moduli from laboratory tests to those from in situ tests.

of volume increase or decrease due to the dilatancy. Current-practice is to use several sophisticated types of sampling device which are manufactured and used for recovering undisturbed samples from both clayey and sandy deposits. However, it appears a difficult task to retrieve perfectly undisturbed samples for all kinds of soils and to test them in the laboratory under the condition completely duplicating in situ situations. Thus the lower bounding curve shown in Fig. 6.31 may be interpreted as the one reflecting the current state of the art in sampling technique with minimum disturbance.

6.4 Estimation of shear modulus from the in situ penetration test

With a view to estimating low-amplitude in situ shear modulus, correlations were proposed by various investigators between the shear wave velocity measured in situ and the N value of the Standard Penetration Test (SPT). One of such correlations obtained by Imai and Yoshimura (1972) is shown in Fig. 6.32 where it may be seen that the velocity of shear wave propagation tends to increase with increasing N value of SPT. As mentioned in Section 12.1.1, the N value in Japanese practice is considered to be approximately 1.2 times smaller on the average than the N_{60} value in US practice. Thus $N = 0.833\ N_{60}$ is indicated in the abscissa of Fig. 6.32. Similar data compilations were performed by many investigators and output of their works is presented in a summary form in Fig. 6.33 where the shear modulus G_0 calculated from the shear wave velocity is plotted versus N value of the SPT. The straight-line relation as shown in the log–log plot in Fig. 6.33 is represented by

$$G_0 = aN^b. \tag{6.16}$$

The values of a and b are determined for each of the proposed lines in Fig. 6.33 and listed in Table 6.4. It may be seen that the exponent of N takes a value of 0.60 to 0.8 and the coefficient a takes a value between 1.0 and 1.6 in kPa.

The correlation of eqn (6.16) may be used to approximately estimate the low-

Table 6.4 a and b values in $G_0 = aN^b$

value of a (kPa)	value of b	References
1.0	0.78	Imai and Yoshimura (1970)
1.22	0.62	Ohba and Toriumi (1970)
1.39	0.72	Ohta *et al.* (1972)
1.20	0.80	Ohsaki and Iwasaki (1973)
1.58	0.67	Hara *et al.* (1974)

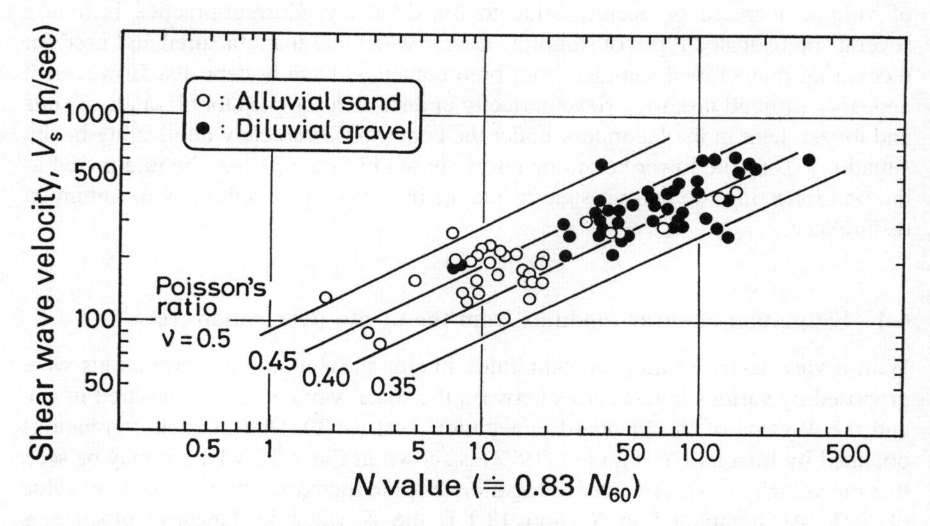

Fig. 6.32 Relation between shear wave velocity and N value of SPT (Imai and Yoshimura, 1972).

amplitude shear modulus G_0 from the blow count N value of the SPT. More comprehensive discussions on this correlation are described in papers by Sykora (1987 *a,b*).

6.5 Poisson's ratio for saturated soils

In the linear theory of elasticity, Poisson's ratio ν is expressed in terms of the shear modulus G_0 and the volumetric modulus K as

$$\nu = \frac{1}{2} \cdot \frac{3K - 2G_0}{3K + G_0}. \tag{6.17}$$

It is assumed that, while the shear modulus is not affected by the drainage condition, the volumetric modulus is influenced by whether the deformation of soils takes place in a drained or undrained condition. Let a compressional stress σ be applied undrained to a saturated soil element. This stress is divided into two parts: one component $\overline{\sigma}$ transmitted through soil skeleton and the other u carried by water in the pores. Thus one obtains

$$\sigma = \overline{\sigma} + u. \tag{6.18}$$

Let it be assumed first that the soil skeleton and the bubble structure of pore fluid are deforming independently without mutual interaction. If the volume of the skeleton V is compressed by an amount V_b due to the effective stress $\overline{\sigma}$, the following relation is obtained:

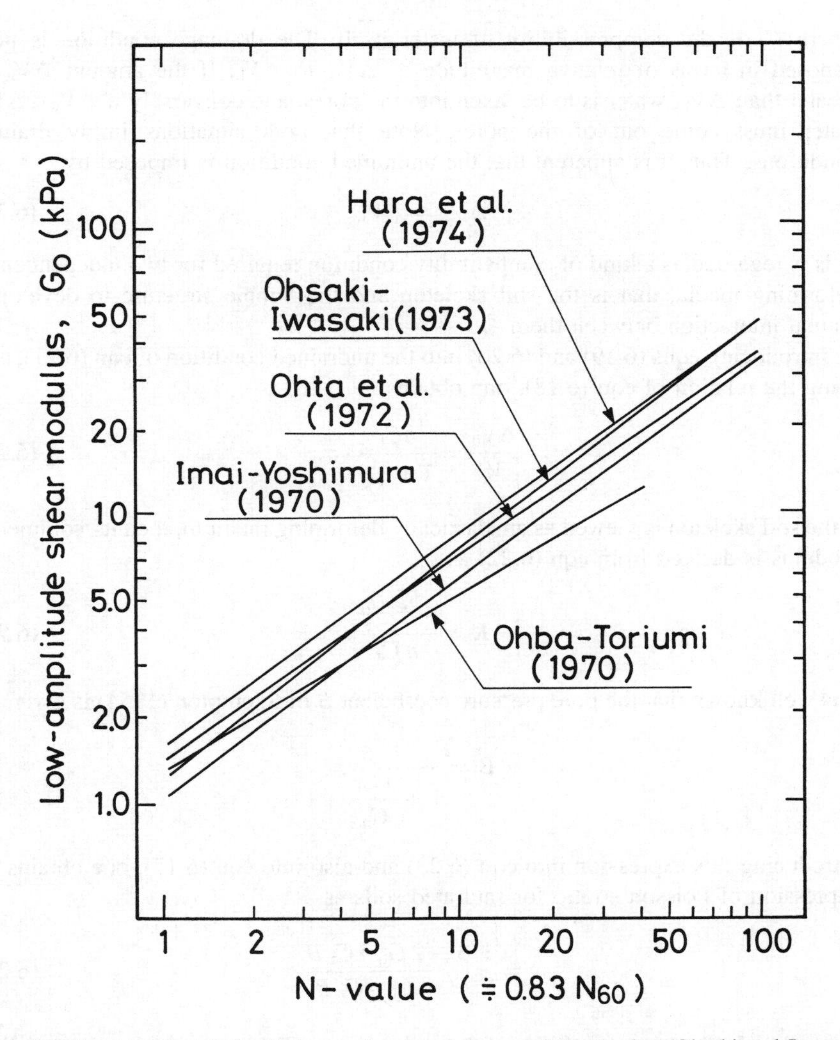

Fig. 6.33 Empirical correlation between shear modulus and SPT *N*-value (Ohsaki and Iwasaki, 1973).

$$\frac{\Delta V_{\mathrm{b}}}{V} = C_{\mathrm{b}} \cdot \overline{\sigma} \tag{6.19}$$

where C_b is the compressibility of the soil skeleton. The volume of water contained in the bubble structure is given by nV where n is porosity. Then if the pore water is compressed by an amount ΔV_w due to the pore pressure u, one obtains:

$$\frac{\Delta V_{\mathrm{w}}}{n V} = C_{\ell} \cdot u \tag{6.20}$$

where C_ℓ is the compressibility of water itself. The drainage condition is now imposed in terms of relative magnitude of ΔV_w to ΔV_b. If the amount ΔV_w is greater than ΔV_b, water is to be taken into the pores, and conversely if $\Delta V_w < \Delta V_b$, water must come out of the pores. Note that both situations imply drained conditions. Thus it is apparent that the undrained condition is imposed by

$$\Delta V_b = \Delta V_w. \tag{6.21}$$

This is regarded as a kind of compatibility condition required for two independently deforming media, that is the soil skeleton and the bubble structure to develop a mutual interaction between them.

Introducing eqns (6.19) and (6.20) into the undrained condition of eqn (6.21), and using the relation of eqn (6.18), one obtains

$$\frac{\Delta V_b}{V} = \frac{nC_\ell}{1 + \frac{nC_\ell}{C_b}} \cdot \sigma. \tag{6.22}$$

If the soil skeleton is viewed as an elastically deforming medium, then its volumetric modulus is derived from eqn (6.22) as

$$K = \frac{1 + \frac{nC_\ell}{C_b}}{nC_\ell}. \tag{6.23}$$

It is well known that the pore pressure coefficient B of Skempton (1954) is given by

$$\mathrm{B} = \frac{1}{1 + \dfrac{nC_\ell}{C_b}}.$$

Introducing this expression into eqn (6.23) and also into eqn (6.17), one obtains an expression of Poisson's ratio for saturated soils as

$$\nu = \frac{1}{2}\,\frac{3 - 2G_0 n C_\ell B}{3 + G_0 n C_\ell B}. \tag{6.24}$$

For soft soils, the compressibility of the soil skeleton C_b is known to be very small in comparison to that of water C_ℓ, and therefore $B \approx 1.0$. For such soils the value of $G_0 n C_\ell$ also becomes small. Introducing these approximations into eqn (6.24), a simple formula of Poisson's ratio for saturated soft soils is obtained as

$$\nu = \frac{1}{2}(1 - nG_0 C_\ell). \tag{6.25}$$

This approximate formula was derived alternatively by Ishihara (1970, 1971) from the linear theory of a porous elastic medium. By assuming $C_\ell = 4.8 \times 10^{-5}$/MPa and n = 0.4–0.6, the relation of eqn (6.25) is numerically shown in Fig. 6.34.

The Poisson's ratio is related to the shear wave velocity V_s and longitudinal wave velocity V_p by the following formulae:

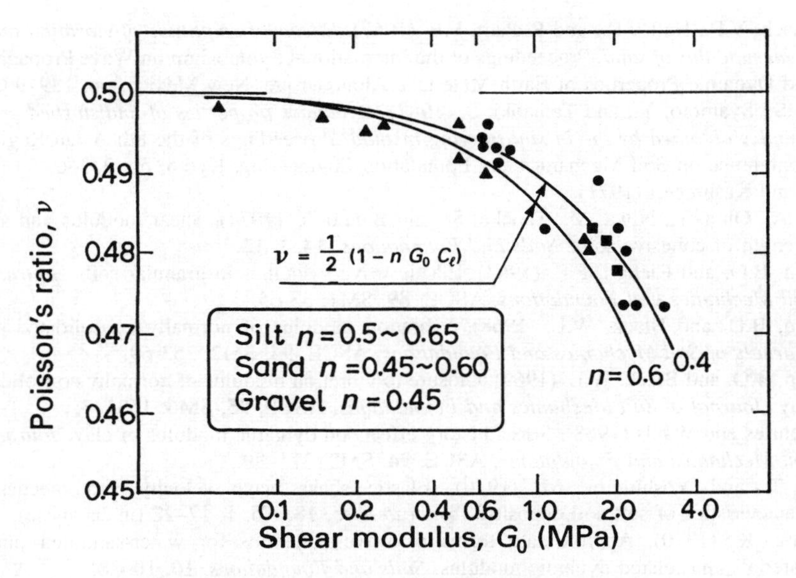

Fig. 6.34 Poisson's ratio as a function of shear modulus.

$$\nu = \frac{1}{2}\frac{(V_p/V_s)^2 - 2}{(V_p/V_s)^2 - 1} \tag{6.26}$$

$$G_0 = \rho V_s{}^2.$$

The values of Poisson's ratio obtained from a majority of in situ data on V_s and V_p using the relation of eqn (6.26) are also shown superimposed on the diagram of Fig. 6.34 for comparison sake. Poisson's ratio is shown to be approximately equal to 0.5 for soft saturated soils but to decrease with increasing stiffness of soils.

References

Afifi, S.S. and Woods, R.D. (1971). Long-term pressure effects on shear modulus of soils. *Journal of Soil Mechanics and Foundations*, ASCE, **97**, SM10, 1445–60.

Afifi, S.S. and Richart, F.E., Jr. (1973). Stress-history effects on shear modulus of soils. *Soils and Foundations*, **13**, 77–95.

Anderson, D.G. and Woods, R. D. (1975). *Comparison of field and laboratory shear moduli.* Proceedings of the Conference on in situ Measurement of Soil Properties, Raleigh, North California, pp. 69–92.

Anderson, D.G. and Stokoe, K.H. (1977). Shear modulus: a time-dependent soil property. *Dynamic Geotechnical Testing*, ASTM, SPT654, 66–90.

Cunny, R. W. and Fry, Z. B. (1973). Vibratory in situ and laboratory soil moduli compared. *Journal of Soil Mechanics and Foundations*, ASCE **99**, SM2 1022–76.

Drnevich, V.P., Hall, J.R., and Richart, F.E. (1967). *Effects of amplitude of vibration on the shear modulus of sand*. Proceedings of the International Symposium on Wave Propagation and Dynamic Properties of Earth Materials, Albuquerque, New Mexico, pp. 189–99.

Goto, S., Syamoto, Y., and Tamaoki, S. (1987). *Dynamics properties of undisturbed gravel samples obtained by the in situ freezing method*. Proceedings of the 8th Asian Regional Conference on Soil Mechanics and Foundation Engineering, Kyoto, pp. 233–6.

Gray and Kashmeeri (1971).

Hara, A., Ohta, T., Niwa, M., Tanaka, S., and Banno, T. (1974). Shear modulus and shear strength of cohesive soils. *Soils and Foundations*, **14**, 1–12.

Hardin, B.O. and Richart, F.E. (1963). Elastic wave velocities in granular soils. *Journal of Soil Mechanics and Foundations*, ASCE, **89**, SM1, 33-65.

Hardin, B.O. and Black, W.L. (1968). Vibration modulus of normally consolidated clay. *Journals of Soil Mechanics and Foundations*, ASCE, **94**, SM2, 353-69.

Hardin, B.O. and Black, W.L. (1969). Closure to vibration modulus of normally consolidated clay. *Journal of Soil Mechanics and Foundations*, ASCE, **95**, SM6, 1531–7.

Humphries and Wahls (1968). Stress history effects on dynamic modulus of clay. *Journal of Soil Mechanics and Foundations*, ASCE, **94**, SM2, 371–89.

Imai, T. and Yoshimura, M. (1970). Elastic shear wave velocity and mechanical characteristics of soft soil deposits. *Tsuchi to Kiso*, **18**, No. 1, 17–22 (in Japanese).

Ishihara, K. (1970). Approximate forms of wave equations for water-saturated porous materials and related dynamic modulus. *Soils and Foundations*, **10**, 10–38.

Ishihara, K. (1971). *On the longitudinal wave velocity and Poisson's ratio in saturated soils*. Proceedings of the 4th Asian Regional Conference on Soil Mechanics and Foundation Engineering, Bangkok, Vol. 1, pp. 197–201.

Iwasaki, T., Tatsuoka, F., Tokida, K., and Yasuda, S. (1978). *A practical method for assessing soil liquefaction potential based on case studies in various sites in Japan*. Proceedings of the 2nd International Conference on Microzonation for Safer Construction—Research and Application. Vol. 2, pp. 885–96.

Kayatama, L., Fukui, F., Goto, M., Makihara, Y., and Tokimatsu, K. (1986). *Comparison of dynamic deformation characteristics of dense sand between undisturbed and disturbed samples*. Proceedings of the 21st Annual Conference of the JSSMFE, pp. 583–4 (in Japanese).

Kim, T.C. and Novak, M. (1981). Dynamic properties of some cohesive soils of Ontario. *Canadian Geotechnical Journal*, **18**, 371–89.

Kokusho, T. (1980). Cyclic triaxial test of dynamic soil properties for wide strain range. *Soils and Foundations*, **20**, 45–60.

Kokusho, T. and Esashi, Y. (1981). *Cyclic triaxial test on sands and coarse materials*. Proceedings of the 10th International Conference on Soil Mechanics and Foundation Engineering, Stockholm, Vol. 1.

Kokusho, T., Yoshida, Y., and Esashi, Y. (1982). Dynamic soil properties of soft clay for wide strain range. *Soils and Foundations*, **22**, 1–18.

Kokusho, T. (1987). *In situ dynamic soil properties and their evaluation*. Proceedings of the 8th Asian Regional Conference on Soil Mechanics and Foundation Engineering, Kyoto, Vol. 2, pp. 215–435.

Kokusho, T. and Tanaka, Y. (1994), *Dynamic properties of gravel layers investigated by in situ freezing sampling*. Proceedings of the ASCE Specialty Conference on Ground Failures under Seismic Conditions, Atlanta. pp. 121–40.

Kuribayashi, E., Iwasaki, T., Tatsuoka, F., and Horiuchi, S. (1975). *Effects of particle characteristics on dynamic deformational properties of soils*. Proceedings of the 5th Asian Regional Conference on Soil Mechanics and Foundation Engineering, Bangalore, India, pp. 361–7.

Marcuson, W.F. and Wahls, H.E. (1972). Time effects on dynamics shear modulus of clays. *Journal of Soil Mechanics and Foundations*, ASCE, **98**, SM12, 1359–73.

Nishio, N., Tamaoki, K., and Machida, Y. (1985). *Dynamic deformation characteristics of crushed gravel by means of large-size triaxial test apparatus*. Proceedings of the 20th Annual Convention, Japanese Society of Soil Mechanics and Foundation Engineering, pp. 603–4.

Ohba, S. and Toriumi, I. (1970). *Research on vibrational characteristics of soil deposits in Osaka, part 2, on velocities of wave propagation and predominant periods of soil deposits*. Abstracts, Technical Meeting of Architectural Institute of Japan (in Japanese).

Ohsaki, Y. and Iwasaki, R. (1973). On dynamic shear moduli and Poisson's ratios of soil deposits. *Soils and Foundations*, **13**, 61–73.

Ohta, T., Hara, A., Niwa, M., and Sakano, T. (1972). *Elastic shear moduli as estimated from N-value*. Proceedings of the 7th Annual Convention of Japan Society of Soil Mechanics and Foundation Engineering, pp. 265–8.

Prange, B. (1981). *Resonant column testing of railroad ballast.* Proceedings of the 10th International Conference on Soil Mechanics and Foundation Engineering, Stockholm, Vol. 1.

Richart, F.E., Hall, J.R., and Woods, R.D. (1970). *Vibrations of soils and foundations*, Prentice Hall.

Shibata, T. and Soelarno, D.S. (1975) *Stress–strain characteristics of sands under cyclic loading*. Proceedings of the Japan Society of Civil Engineering, No. 239, pp. 57-65 (in Japanese).

Silver, M.L. and Seed, H.B. (1971). *Deformation characteristics of sands under cyclic loading*. Proceedings of ASCE, SM8, pp. 1081–98.

Skempton, A.W. (1954). The pore pressure coefficient A and B, *Geotechnique*, **4**, 143–7.

Stokoe, K.H. and Richart, F.E. (1973). *In situ and laboratory shear wave velocities*. Proceedings of the 8th International Conference on Soil Mechanics and Foundation Engineering, Moscow, Vol. 1, Part 2, pp. 403–9.

Sykora, D. (1987a). *Examination of existing shear wave velocity and shear modulus correlations in soils*. US Army WES. Miscellaneous Paper GL-87-22, Vicksburg, MS.

Sykora, D. (1987b). *Creation of data base of seismic shear wave velocities for correlation analysis*, US Army WES, Miscellaneous Paper GL-87-26, Vicksburg, MS.

Tanaka, Y., Kudo, K., Yoshida, Y., and Ikemi, M. (1987). *A study on the mechanical properties of sandy gravel—dynamic properties of reconstituted sample*, Report U87019, Central Research Institute of Electric Power Industry, (in Japanese).

Tokimatsu, K. and Hosaka, Y. (1986). Effects of sample disturbance on dynamic properties of sand. *Soils and Foundations*, **26**, No. 1, 53–64.

Trudeau, P.J., Whitman, R.V., and Christian, J.T. (1974). Shear wave velocity and modulus of a marine clay. *Journal of the Boston Society of Civil Engineers*, pp. 12–25.

Umehara, Y., Zen, K., Higuchi, Y. and Ohneda, H. (1985). *Laboratory tests and in situ seismic survey on vibratory shear moduli of cohesive soils.* Proceedings of the 7th Japan Earthquake Engineering Symposium, pp. 577–84.

Yasuda, S. and Yamaguchi, I. (1985). *Dynamic shear modulus obtained in the laboratory and in situ*. Proceedings of the Symposium on Evaluation of Deformation and Strength of Sandy Grounds. Japanese Society of Soil Mechanics and Foundation Engineering, pp. 115–18 (in Japanese).

Yokota, K., Imai, T., and Konno, M. (1981). *Dynamic deformation characteristics of soils determined by laboratory tests*. Oyo Technical Report No. 3, pp. 13–37.

Yokota, K. and Konno, M. (1985). *Comparison of soil constants obtained from laboratory tests and in situ tests*. Proceedings of the Symposium on Evaluation of Deformation and Strength of Sandy Grounds, Japanese Society of Soil Mechanics and Foundation Engineering, pp. 111–14 (in Japanese).

Yu, P. and Richart, F.E. (1984). Stress ratio effects on shear modulus of dry sands. *Journal of Geotechnical Engineering*, ASCE, **110**, GT3, 331–45.

Zen, K., Umehara, Y., and Hamada, K. (1978). *Laboratory tests and in situ seismic survey on vibratory shear modulus of clayey soils with various plasticities*. Proceedings of the 5th Japanese Earthquake Engieering Symposium, pp. 721–8.

7

STRAIN DEPENDENCY OF MODULUS AND DAMPING

7.1 Strain-dependent modulus and damping from laboratory tests

It is well known that the deformation characteristics of soil are highly nonlinear and this is manifested in the shear modulus and damping ratio which vary significantly with the amplitude of shear strain under cyclic loading.

7.1.1 *Sands*

Detailed studies have been made by Iwasaki *et al.* (1978) and Kokusho (1980) to identify the strain-dependent dynamic properties of the Japanese standard sand (Toyoura sand: $D_{50} = 0.19$ mm, $U_c = 1.3$). In the tests by Kokusho (1980), specimens were prepared by pouring saturated sand in the mould and then compacting it by a hand vibrator to obtain various target densities. The tests were performed by using a triaxial test device equipped with the non-contacting type strain gauge. The results of the tests on specimens with various densities conducted under a confining stress of 100 kPa are presented in Fig. 7.1, where the shear

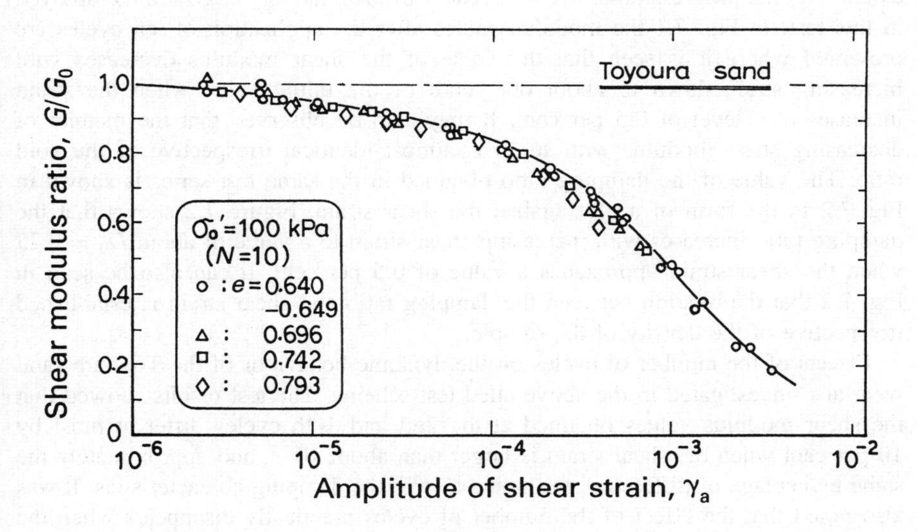

Fig. 7.1 Normalized shear modulus versus shear strain for Toyoura sand (Kokusho, 1980).

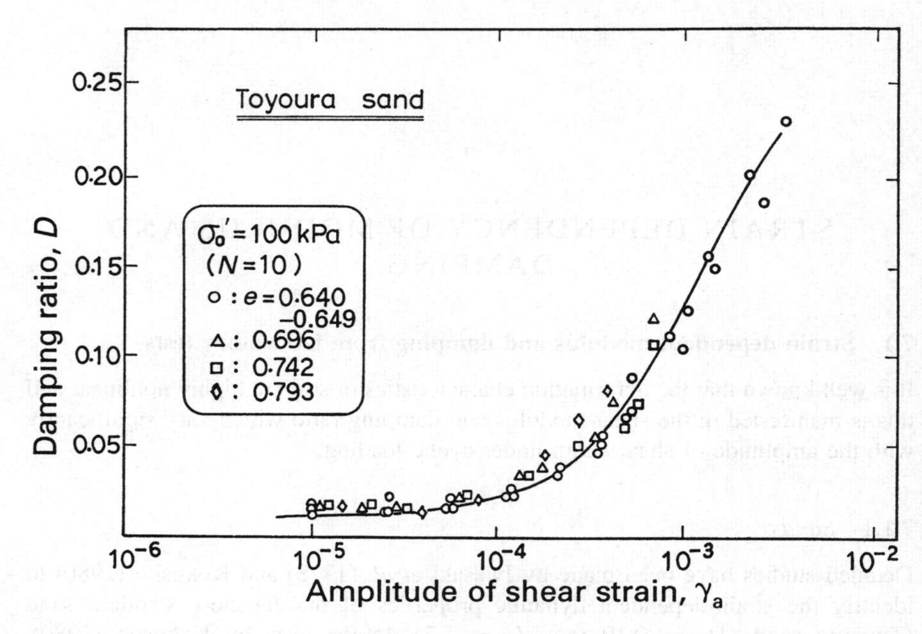

Fig. 7.2 Damping ratio versus shear strain for Toyoura sand (Kokusho, 1980).

modulus normalized to the initial modulus at a strain of 10^{-6} is plotted versus the shear strain amplitude γ_a. The value of γ_a was calculated by the relation of eqn (6.2). It was found that the shear modulus from the cyclic triaxial test changed to some extent with the progression of cycles, because of the relatively large strain employed in this test. In Fig. 7.1 the modulus values after the application of ten cycles are presented where it is seen that the value of the shear modulus decreases with increasing strain down to about one tenth of the initial value when the strain increases to a level of 0.5 per cent. It may also be observed that the manner of decreasing shear modulus with strain is almost identical irrespective of the void ratio. The value of the damping ratio obtained in the same test series is shown in Fig. 7.2 in the form of a plot against the shear strain. Figure 7.2 shows that the damping ratio increases with increasing shear strain to a value of about $D_0 = 0.25$ when the shear strain approaches a value of 0.5 per cent. It can also be seen in Fig. 7.2 that the relation between the damping ratio and shear strain is established irrespective of the density of the sample.

Effects of the number of cycles on the dynamic behaviour of the Toyoura sand were also investigated in the above cited test scheme. The test results showed that the shear modulus values obtained at the 2nd and 10th cycles differ at most by 10 per cent when the shear strain is larger than about 10^{-4}, and approximately the same percentage of difference was noted also in the damping characteristics. It was also noted that the effect of the number of cycles practically disappears when the stress application is repeated for more than 10 cycles. Consequently, for all practical

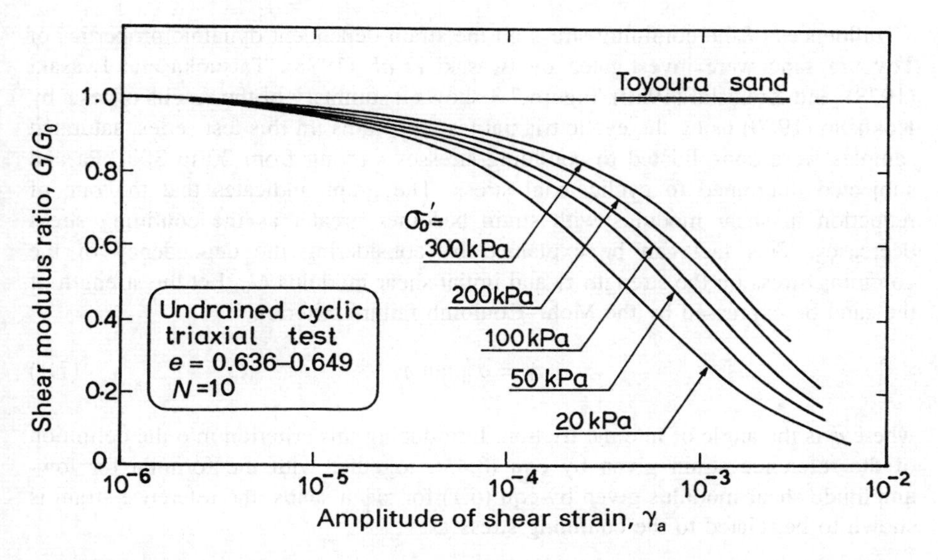

Fig. 7.3 Effects of confining stress on the strain-dependent shear modulus (Kokusho, 1980).

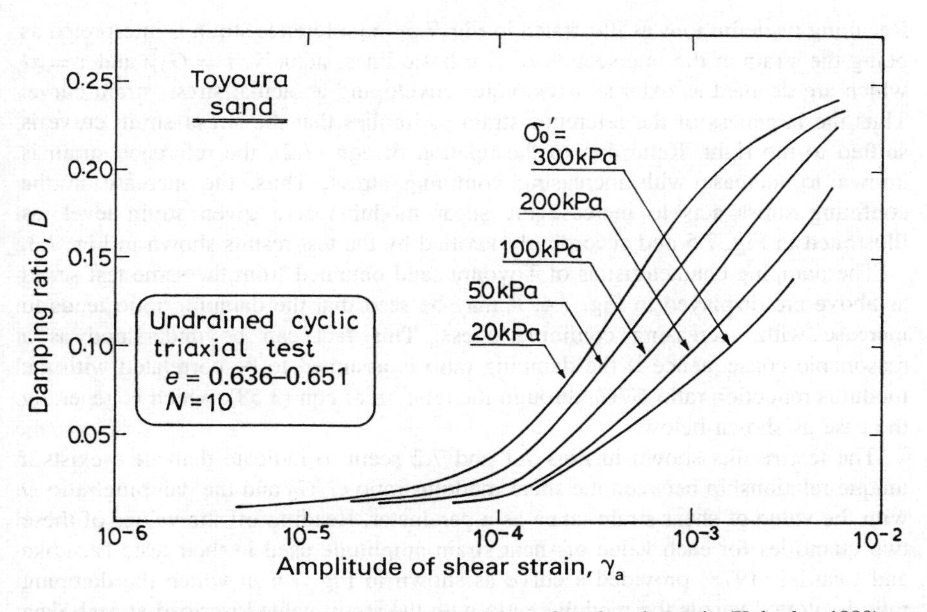

Fig. 7.4 Effects of confining stress on the strain-dependent damping ratio (Kokusho, 1980).

purposes, the change in shear modulus and damping due to the progression of cycles may be disregarded, except for the case of undrained shear with a large amplitude where pore water pressure build-up in the saturated sand is significant.

Influences of the confining stress on the strain-dependent dynamic properties of Toyoura sand were investigated by Iwasaki *et al.* (1978), Tatsuoka and Iwasaki (1978), and Kokusho (1980). Figure 7.3 shows a summary of the results of tests by Kokusho (1980) using the cyclic triaxial test apparatus. In this test series, saturated samples were consolidated to confining stresses varying from 20 to 300 kPa and subjected undrained to cyclic axial stress. The graph indicates that the rate of reduction in shear modulus with strain becomes greater as the confining stress decreases. This fact can be explained by considering the dependency of the confining stress on the strength τ_f and initial shear modulus G_0. Let the strength of the sand be expressed by the Mohr–Coulomb failure criterion,

$$\tau_f = \sigma'_0 \tan \phi \tag{7.1}$$

where ϕ is the angle of internal friction. Introducing this criterion into the definition of the reference strain given by eqn (3.43), together with the formula for low-amplitude shear modulus given by eqn (6.1) for clean sands, the reference strain is shown to be related to the confining stress as

$$\gamma_r \propto (\sigma'_0)^{0.5}. \tag{7.2}$$

Recalling its definitions as illustrated in Fig. 7.5, the reference strain is interpreted as being the strain at the intersection of two basic lines, namely, $\tau = G_0\gamma$ and $\tau = \tau_f$ which are deemed as external asymptotes enveloping an actual stress–strain curve. Thus the largeness of the reference strain γ_r implies that the stress–strain curve is shifted to the right. Returning to the relation of eqn (7.2), the reference strain is known to increase with increasing confining stress. Thus, the increase in the confining stress acts to increase the shear modulus at a given strain level, as illustrated in Fig. 7.5 and accordingly verified by the test results shown in Fig. 7.3.

The damping characteristics of Toyoura sand obtained from the same test series as above are displayed in Fig. 7.4. It may be seen that the damping ratio tends to increase with decreasing confining stress. This fact can be understood as a reasonable consequence if the damping ratio is assumed to be correlated with the modulus reduction ratio G/G_0 through the relation of eqn (3.58), which is generally the case as shown below.

The test results shown in Figs 7.1 and 7.2 seem to indicate that there exists a unique relationship between the shear modulus ratio G/G_0 and the damping ratio D with the value of shear strain taken as a parameter. Reading off the values of these two quantities for each value of shear strain amplitude used in their test, Tatsuoka and Iwasaki (1978) provided a curve as shown in Fig. 7.6 in which the damping ratio is plotted versus the modulus ratio with the strain value inscribed at each data point as a parameter. Similar plots were also made for several test data by other investigators and are shown in Fig. 7.7. It may be seen that, although there is some scatter among the individual sets of data, the damping ratio is roughly inversely proportional to the shear modulus ratio.

The outcome of the tests mentioned above has been for the behaviour of sand

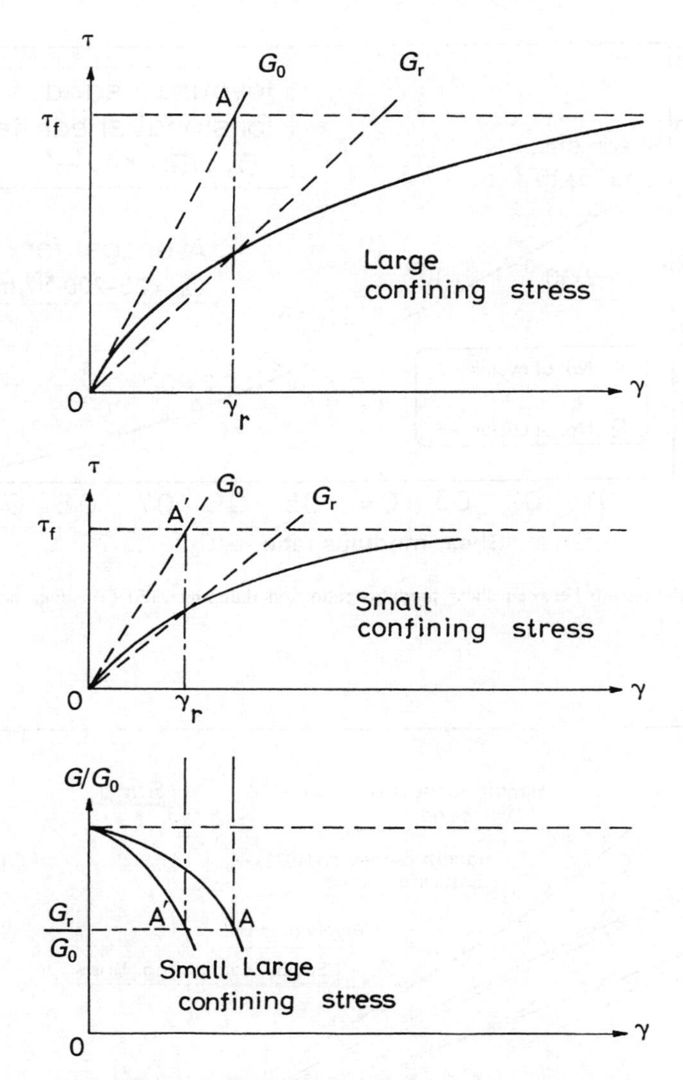

Fig. 7.5 Effects of confining stress on the reference strain.

undergoing cyclic stresses in undrained conditions. A series of cyclic triaxial tests was also conducted by Kokusho (1980) under drained conditions to investigate the effects of drainage. The results of the tests in this regard showed that there is practically no effect of drainage condition on the strain-dependent modulus and damping characteristics of Toyoura sand within the range of strain level from 10^{-6} to 5×10^{-3}. This observation may be taken as reasonable if one is reminded of the fact that the effect of dilatancy start to manifest itself when the shear strain is larger than 5×10^{-3}, as illustrated in Fig. 3.1.

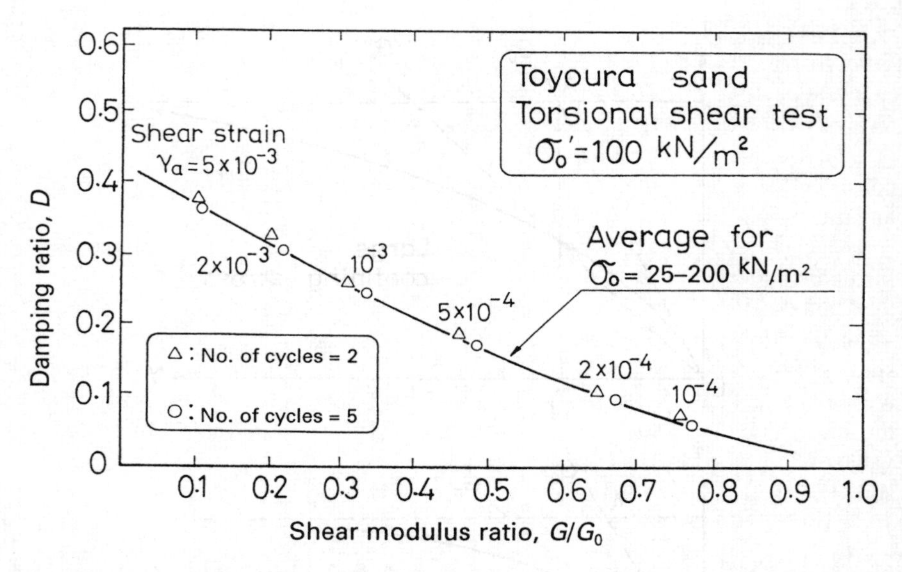

Fig. 7.6 Relationship between shear modulus ratio and damping ratio (Tatsuoka and Iwasaki, 1978).

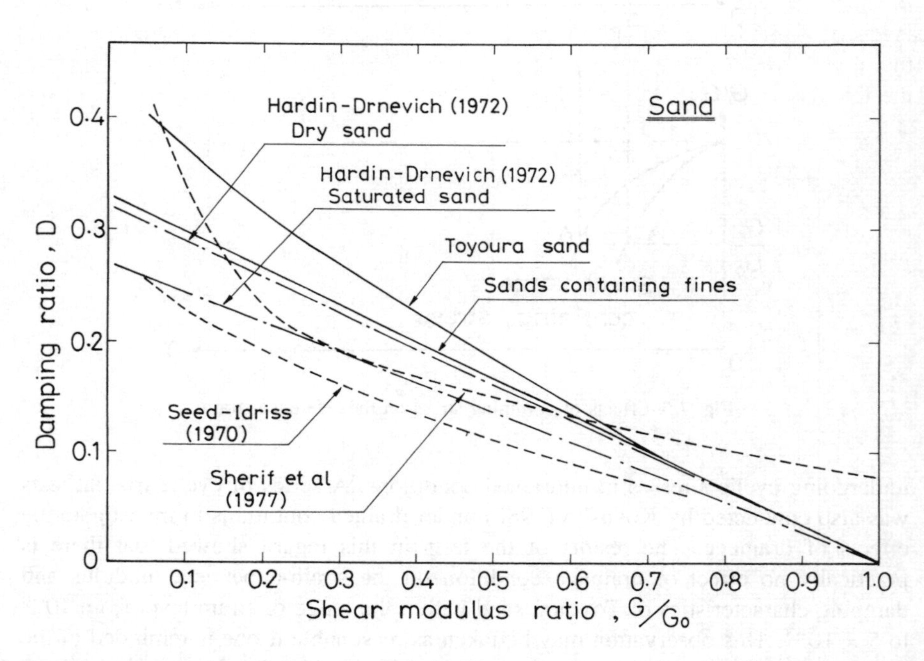

Fig. 7.7 Relationship between shear modulus ratio and damping ratio (Tatsuoka and Iwasaki, 1978).

7.1.2 *Cohesive soils*

From the early stage of research development, undisturbed samples have been used to investigate the strain dependency of deformation characteristics of cohesive soils. Extensive studies through laboratory testing by Seed and Idriss (1970), Kovacs *et al.* (1971), Hardin and Drnevich (1972 *a,b*) and Taylor and Parton (1973) showed consistently that the modulus of clays tends to decrease significantly with strain amplitude once a threshold strain is exceeded.

Extensive studies were also made by Anderson and Richart (1976) on this subject for undisturbed samples of five clays secured from naturally occurring deposits in the United States. The clays tested were of low plasticity index ranging from 20 to 45 having undrained shear strengths of about 70 to 85 kPa except one clay with a shear strength of 15 kPa. The results of the tests using the resonant column device are summarized in Fig. 7.8. One of the interesting features to be noted is the fact that the shear modulus did not start to drop until the shear strain amplitude grew to a value of 5×10^{-5}. This is in contrast to the corresponding behaviour in cohesionless soils in which the modulus reduction starts to occur from a smaller strain of about 10^{-5}, as observed in Figs 7.1 and 7.3.

Andreasson (1979, 1981) also investigated the strain dependency of the shear modulus of plastic clays obtained from three sites in the Gothenburg region, Sweden. The plasticity index of the clays was about 20 to 60. Undisturbed samples were tested in the laboratory using the resonant column device and at the same time in situ screw plate tests were carried out to determine the modulus values at large levels of strain. The results of the tests are shown in Fig. 7.8, where it may be seen the modulus reduction curve for the Gothenburg clay is similar in shape to that for the US clays reported by Anderson and Richart (1976).

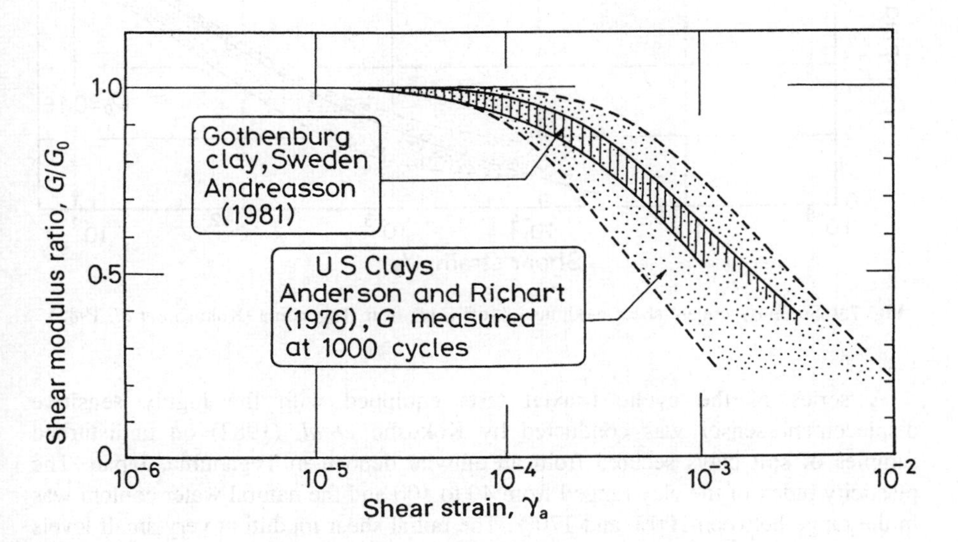

Fig. 7.8 Modulus reduction curves for clays.

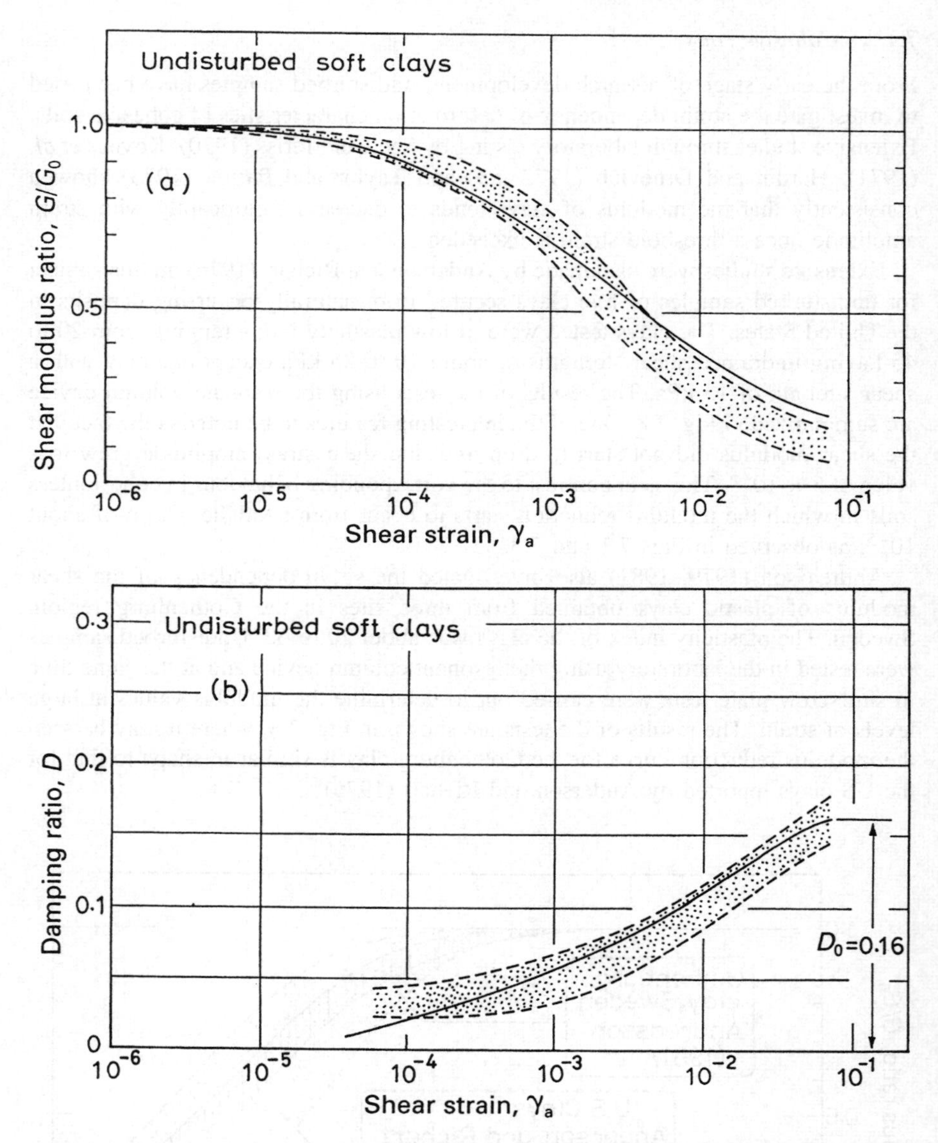

Fig. 7.9 Strain-dependent shear modulus of soft clays from Teganuma (Kokusho *et al.* 1982).

A series of the cyclic triaxial tests equipped with the highly sensitive displacement sensor was conducted by Kokusho *et al.* (1982) on undisturbed samples of soft clays secured from an alluvial deposit in Teganuma, Japan. The plasticity index of the clay ranged from 40 to 100 and the natural water content was in the range between 100% and 170%. The initial shear moduli at very small levels of strain were in the range as small as 2500 to 7500 kPa. The shear modulus value

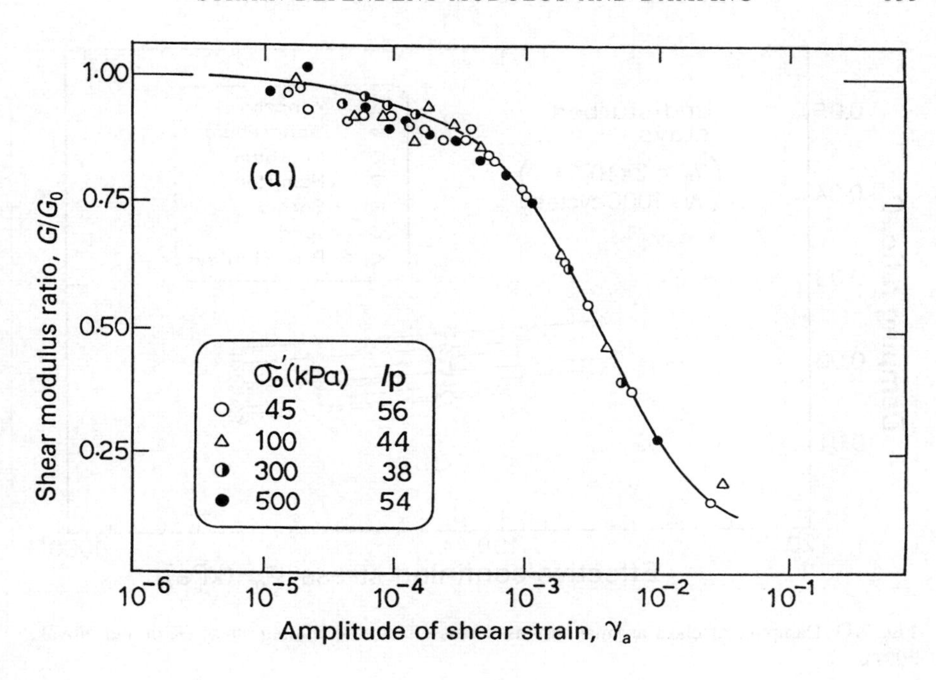

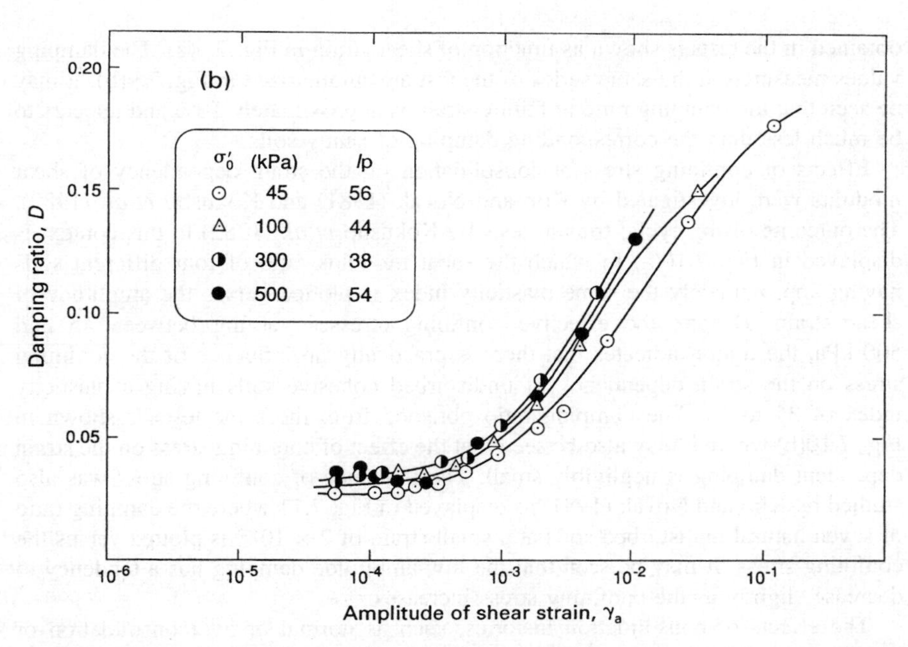

Fig. 7.10 Effects of confining stress on the strain-dependent modulus of cohesive soils (Kokusho *et al.* 1982).

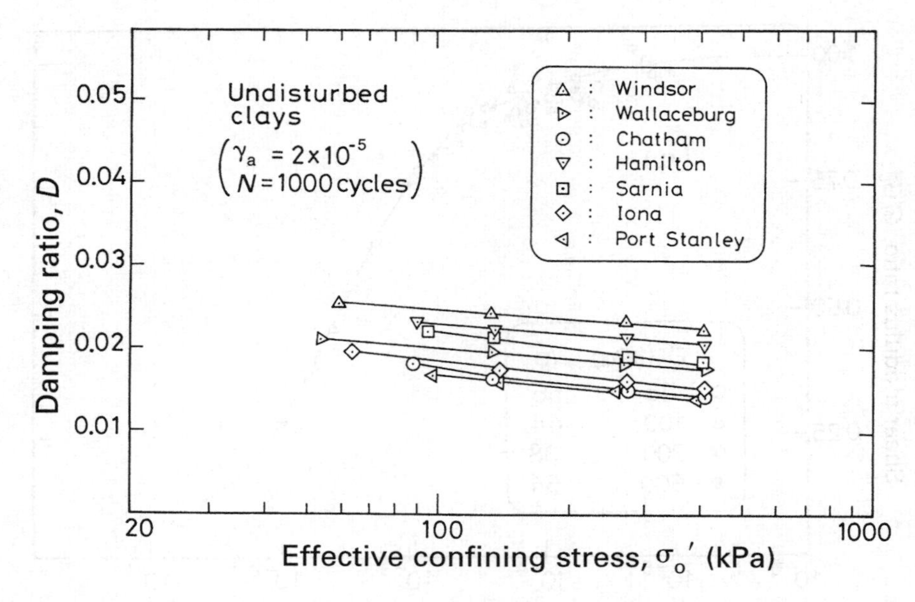

Fig. 7.11 Damping of clays at small strains versus effective confining stress (Kim and Novak, 1981).

obtained in the tests is shown as function of shear strain in Fig. 7.9(a). The damping values measured in the same series of the test are summarized in Fig. 7.9(b). It may be seen that the damping ratio at failure strain is approximately 16% and appears to be much less than the corresponding damping of sandy soils.

Effects of confining stress at consolidation on the strain dependency of shear modulus were investigated by Kim and Novak (1981) and Kokusho *et al.* (1982). The outcome of the cyclic triaxial tests by Kokusho *et al.* (1982) in this context is displayed in Fig. 7.10(a) in which the shear modulus ratio of four different soils having approximately the same plasticity index is plotted versus the amplitude of shear strain. Despite the effective confining stresses varying between 45 and 500 kPa, the figure indicates that there is practically no influence of the confining stress on the strain dependency of undisturbed cohesive soils having a plasticity index of 35 to 55. The damping ratio obtained from the same tests is shown in Fig. 7.10(b) where it may also be seen that the effect of confining stress on the strain dependent damping is negligibly small. The influence of confining stress was also studied by Kim and Novak (1981) as displayed in Fig. 7.11 where the damping ratio of seven natural undisturbed soils at a small strain of 2×10^{-5} is plotted versus the confining stress. It may be seen that the low-amplitude damping has a tendency to decrease slightly as the confining stress increases.

The effects of consolidation histories, such as normal or overconsolidation or long-term application of consolidation pressure, were investigated by Kokusho *et al.* (1982) for undisturbed natural clayey soils having medium to high plasticity index

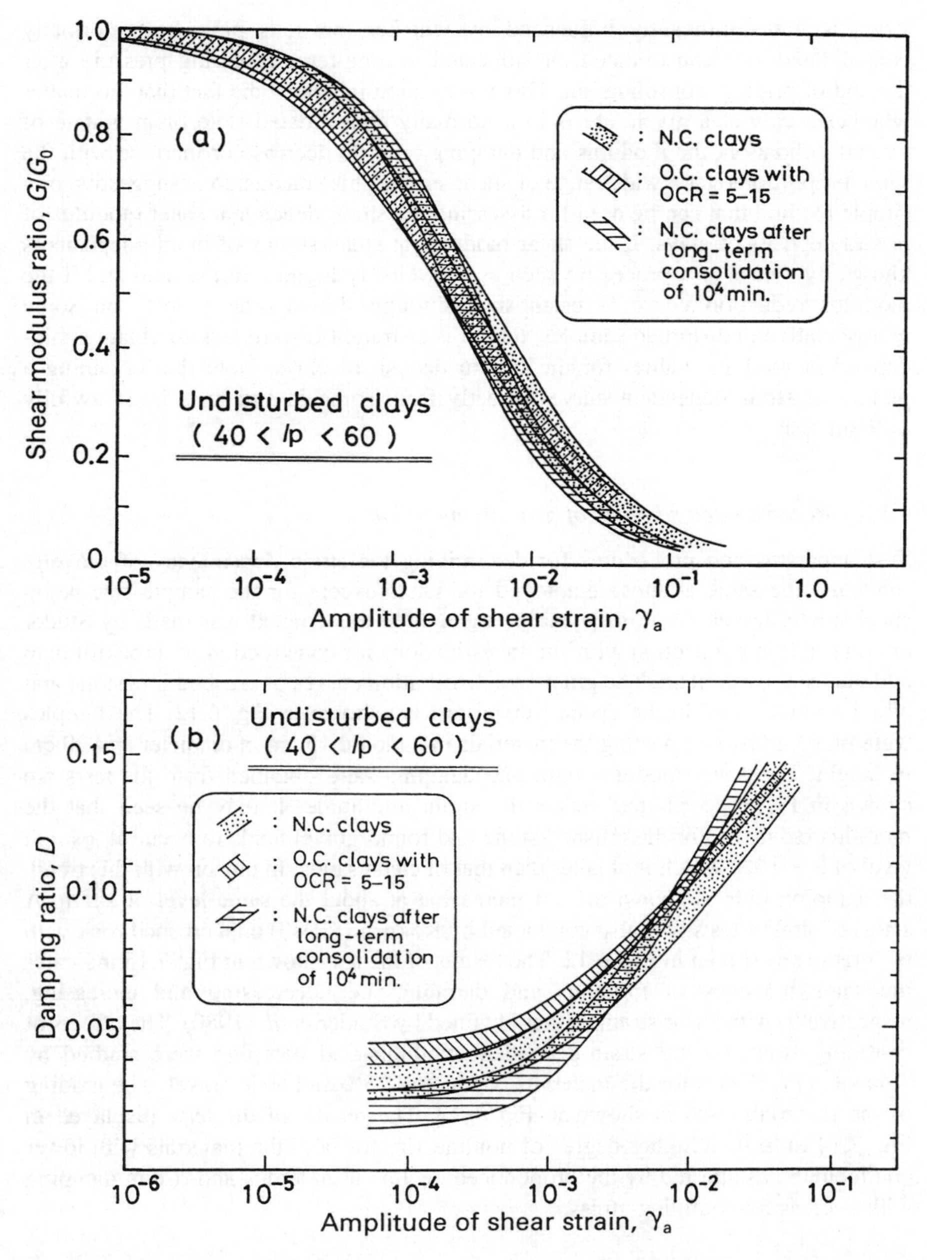

Fig. 7.12 Effects of consolidation histories on strain-dependent modulus and damping ratio (Kokusho *et al.* 1982).

of $I_p = 40–60$. The test data on the strain dependency are summarized in Fig. 7.12, where it may be seen that the changing feature of shear modulus and damping with strain is not significantly influenced by whether the soils have been normally consolidated, overconsolidated, or subjected to long-term confining pressure after the end of primary consolidation. This observation indicated the fact that, no matter whether a clay at a given site is in a normally consolidated state or in a state of overconsolidation, the modulus and damping ratio do decrease or increase with the same proportion over a wide range of shear strain. This conclusion is suggestive of a simple method that can be used for assessing the strain-dependent shear modulus of in situ deposits of clays. If the shear modulus at small strains of in situ deposits is known by means of a procedure such as the velocity logging in the field and if the modulus reduction curve is established through the laboratory tests on some representative undisturbed samples, then it is a straightforward task to obtain strain-dependent modulus values for the in situ deposit of clays. Note that obtaining a picture of strain-dependent curves directly from some in situ tests is an awfully difficult task.

7.1.3 *Reconstituted samples of gravelly materials*

Test apparatus and procedures for determining the strain dependency of gravelly sands are the same as those employed for sands except for the sample size being substantially larger. One of the earlier attempts in this context was made by Studer *et al.* (1980) in connection with site investigations for construction of a rockfill dam and nuclear power plant. The grain-size distribution curves of crushed limestone and alluvial gravel used in the cyclic triaxial test are shown in Fig. 6.12. The samples were prepared by compacting the materials in a mould 15 cm in diameter and 30 cm in height. The shear modulus ratio and damping ratio obtained from the tests are shown in Fig. 7.13 plotted versus the strain amplitude. It may be seen that the modulus reduction for the crushed stone and round gravel tends to occur at a strain level of 5×10^{-6} which is smaller than that of clean sands. In unison with this trend, the damping ratio is shown to start increasing at about the same level of strain. A series of similar tests was also conducted by Kokusho (1980) on a crushed rock with the grading as shown in Fig. 6.12. The results of the test shown in Fig. 7.13 indicates the same tendency of modulus and damping, i.e., decreasing and increasing, respectively, with shear strain as that obtained by Studer *et al.* (1980). The effects of confining stress on the strain-dependent modulus and damping were studied by Tanaka *et al.* (1987) for the materials containing 25% and 50% gravel. The grading of the materials used is shown in Fig. 6.13. The results of the tests displayed in Fig. 7.14 indicate a higher degree of nonlinearity for both the materials with lower confinement, as attested by the pronounced decline in modulus and rise in damping with decreasing confining stress.

7.1.4 *Intact samples of gravelly materials*

The outline of the extensive project involving undisturbed sampling of sandy gravels by the freezing technique and associated laboratory testing has been described in Section 6.3. As a major product of this research, data were also obtained regarding

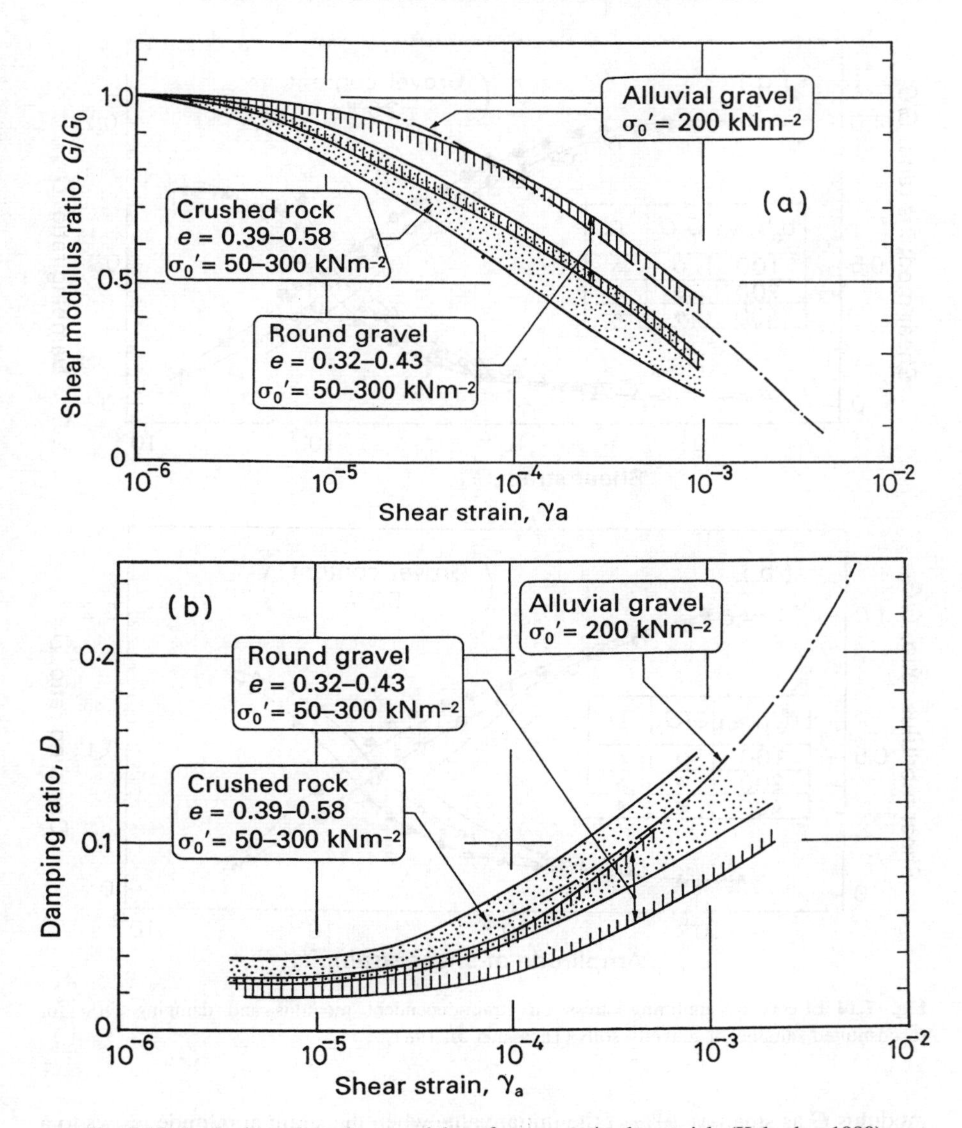

Fig. 7.13 Strain-dependent shear modulus of coarse-grained materials (Kokusho, 1980).

the nonlinear deformation characteristics of intact materials from in situ gravelly deposits. The sites of sampling and procedures for the laboratory testing are the same as described in Section 6.3. A set of the test results on undisturbed intact samples is shown in Fig. 7.15 where the shear modulus and damping of the material from site K are plotted versus the amplitude of shear strain. One sample having an initial shear modulus of $G_0 = 458$ MPa is shown to exhibit nonlinear behaviour starting from a shear strain amplitude as low as 5×10^{-6} and end up with the secant

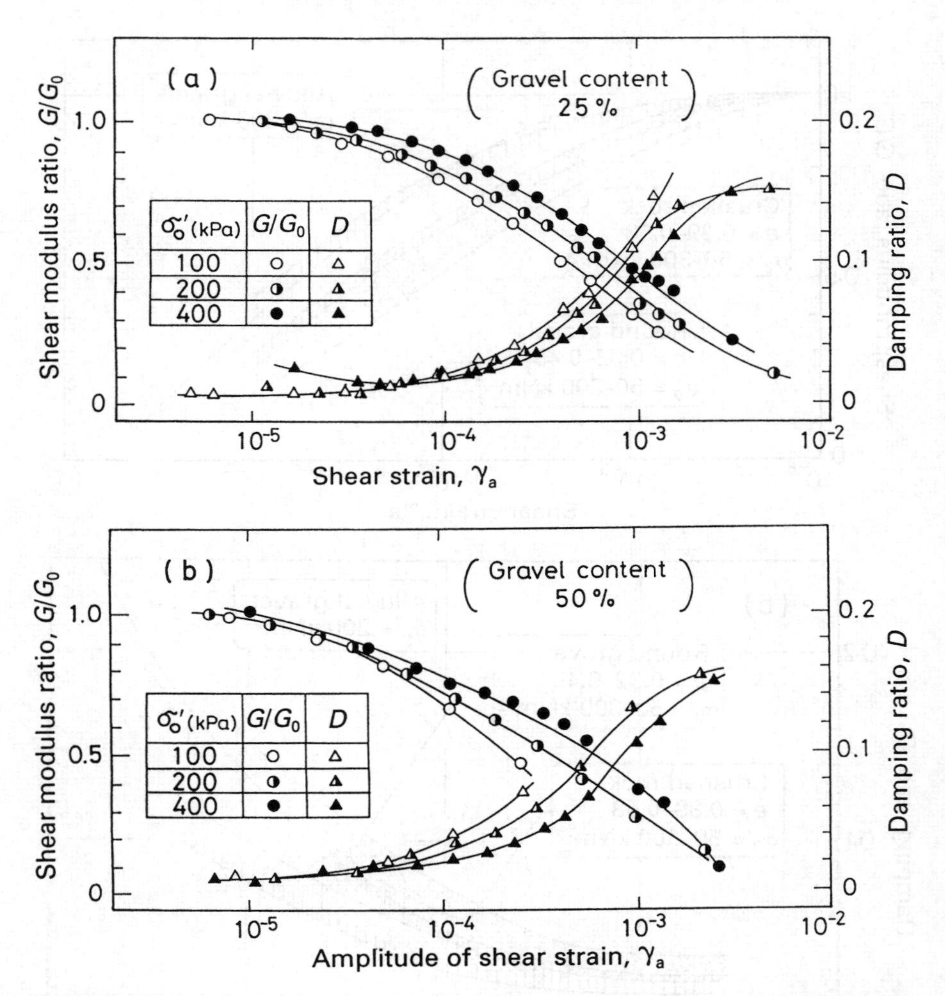

Fig. 7.14 Effects of confining stress on strain-dependent modulus and damping ratio for reconstituted samples of gravelly soils (Tanaka *et al.* 1987).

modulus G as small as 20% of the initial value when the strain amplitude grows to a value of 10^{-3}. Notable reduction in the value is observed also for another sample having an initial shear modulus of $G_0 = 288$ MPa. The damping ratio of the same undisturbed gravels shown in Fig. 7.15(b) also indicates a sharp rise with increasing amplitude of shear strain. The two intact samples as above were once disturbed and reconstituted to new samples having the same void ratio as before. The shear modulus and damping ratio obtained from the cyclic triaxial tests on these reconstituted samples are also plotted in Fig. 7.15 versus the shear strain amplitude. It is to be noted, first of all, that the initial modulus itself drops significantly to one third and to one half of the values from the intact materials. Figure 7.15 shows that

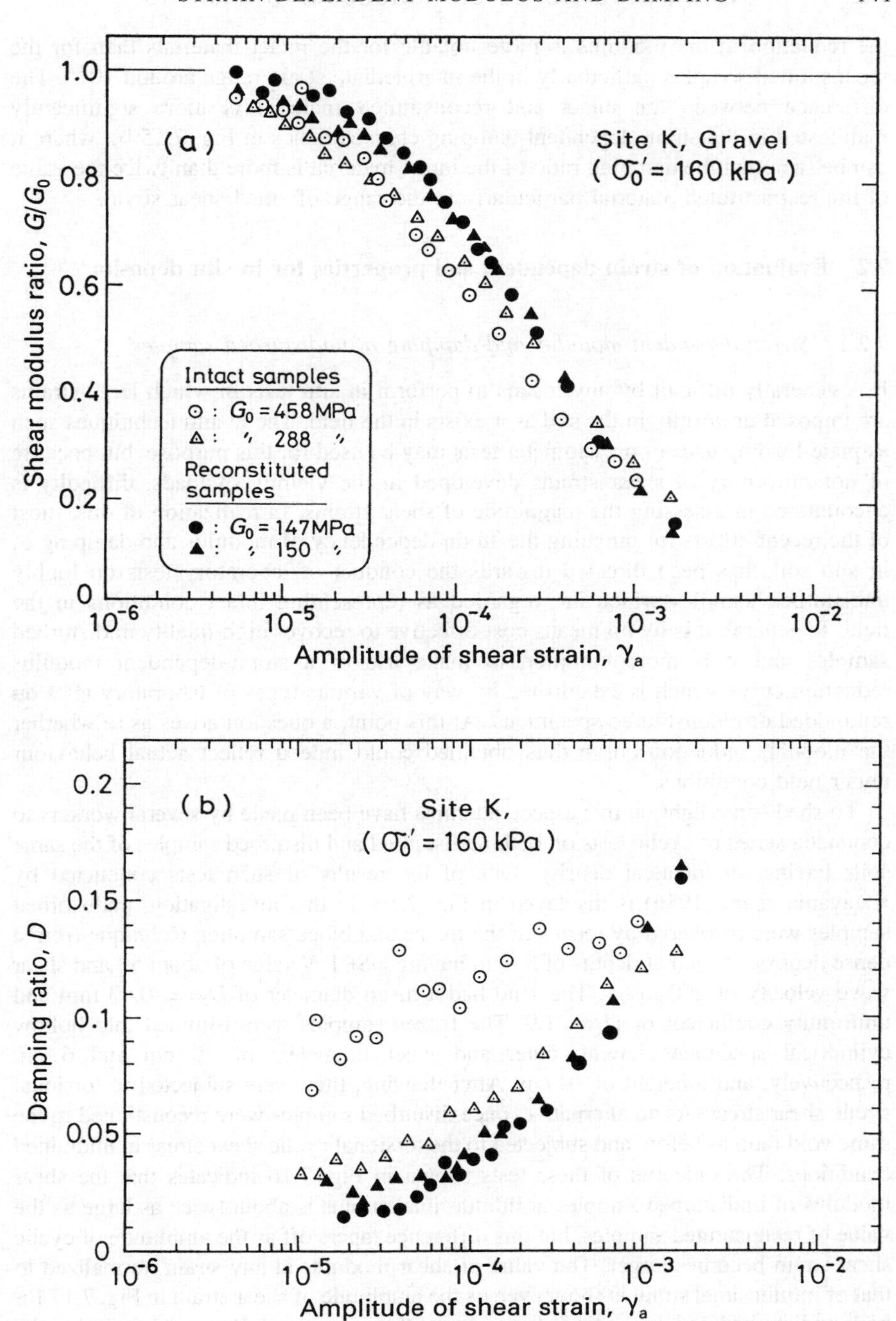

Fig. 7.15 Comparison of strain-dependent shear moduli of gravelly soils from intact samples and from reconstituted samples.

the reduction in the modulus is more notable for the intact materials than for the reconstituted samples particularly in the intermediate strain range around 10^{-4}. The difference between the intact and reconstituted material is more significantly manifested in the strain-dependent damping characteristics in Fig. 7.15(b), where it can be seen that the damping ratio of the intact material is more than twice the value of the reconstituted material particularly in the range of small shear strain.

7.2 Evaluation of strain-dependent soil properties for in situ deposits

7.2.1 *Strain-dependent modulus and damping of undisturbed samples*

It is generally difficult by any means to perform in situ tests in which large strains are imposed uniformly in the soil as it exists in the field. The in situ techniques such as plate-loading tests or pressiometer tests may be used for this purpose, but because of nonuniformity of shear strains developed in the vicinity of loads, difficulty is encountered in assessing the magnitude of shear strains. In realization of this, most of the recent efforts for pursuing the strain dependency of modulus and damping of in situ soils has been directed towards the conduct of laboratory tests on highly undisturbed samples which are regarded as representing intact conditions in the field. In general, it is by no means cost-effective to recover high-quality undisturbed samples and it is more common to make use of a strain-dependent modulus reduction curve which is established by way of various types of laboratory tests on remoulded or reconstituted specimens. At this point, a question arises as to whether the modulus reduction curve thus obtained could indeed reflect actual behaviour under field conditions.

To shed some light on this aspect, attempts have been made by several workers to conduct a series of cyclic tests on both undisturbed and disturbed samples of the same soils having an identical density. One of the results of such tests conducted by Katayama *et al.* (1986) is displayed in Fig. 7.16. In this investigation, undisturbed samples were recovered by means of the freeze and block sampling technique from a dense deposit of sand at depths of 5–9 m having a SPT N value of about 50 and shear wave velocity of 260 ms^{-1}. The sand had a mean diameter of $D_{50} = 0.43$ mm and uniformity coefficient of $U_c = 1.9$. The frozen samples were trimmed into hollow cylindrical specimens having outer and inner diameters of 10 cm and 6 cm respectively, and a height of 10 cm. After thawing, they were subjected to torsional cyclic shear stress. As an alternative, once disturbed samples were reconstituted to the same void ratio as before and subjected to the torsional cyclic shear stress in undrained conditions. The outcome of these tests shown in Fig. 7.16 indicates that the shear modulus of undisturbed samples at infinitesimal strains is about twice as large as the value of reconstituted samples, but this difference tapers off as the amplitude of cyclic shear strain becomes larger. The value of shear modulus at any strain normalized to that of infinitesimal strain is shown versus the amplitude of shear strain in Fig. 7.17 for both undisturbed and reconstituted samples of the dense sand. It may be seen that the drop in the modulus takes place more notably in the medium range of strains for undisturbed samples as compared to that of reconstituted samples.

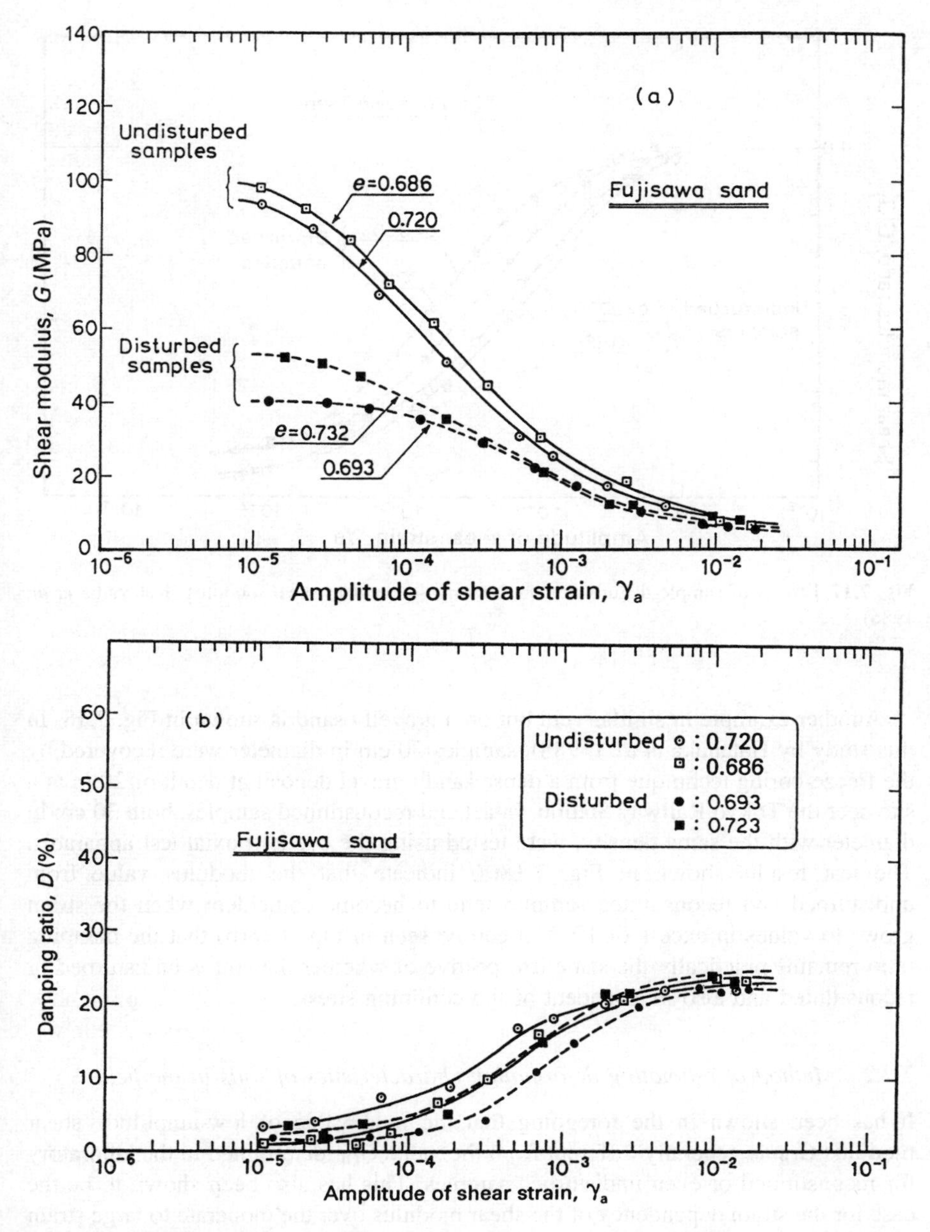

Fig. 7.16 Comparison of strain-dependent shear modulus of dense sand from undisturbed samples and from disturbed samples (Katayama *et al.* 1986).

The damping ratio obtained in this test series is displayed in Fig. 7.16(b), where it may be seen that, unlike the shear modulus, there is no noticeable difference in damping between reconstituted and undisturbed sand.

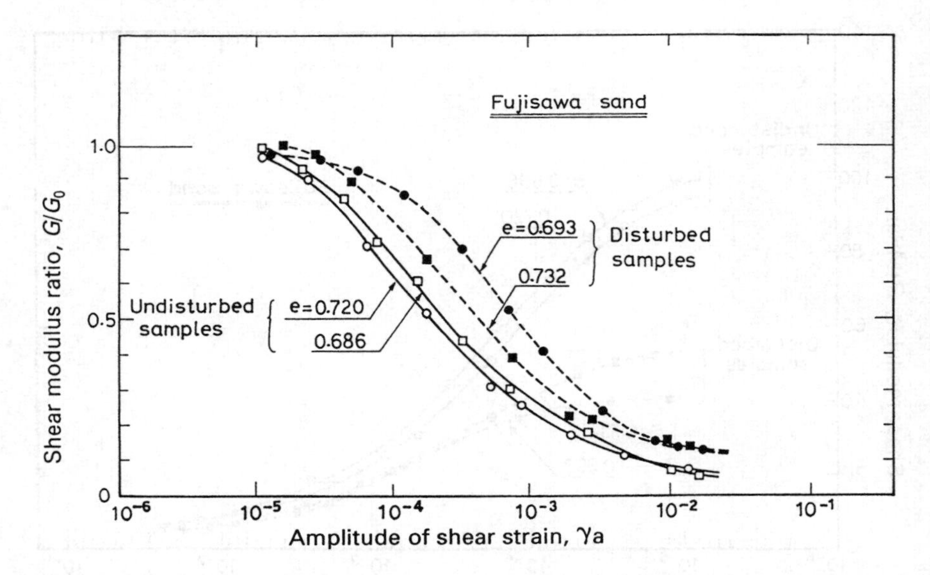

Fig. 7.17 Effects of sample disturbance on the strain-dependent shear modulus (Katayama *et al.* 1986).

Another example in similar vein but on a gravelly sand is shown in Fig. 7.18. In this study by Hatanaka *et al.* (1988), samples 30 cm in diameter were recovered by the freeze-coring technique from a dense sandy gravel deposit at depth of 20 m at a site near the Tokyo Railway Station. Intact and reconstituted samples, both 30 cm in diameter with the same density, were tested using the cyclic triaxial test apparatus. The test results shown in Fig. 7.18(a) indicate that the modulus value from undisturbed and reconstituted samples tend to become coincident when the strain grows to values in excess of 10^{-3}. It can be seen in Fig. 7.18(b) that the damping ratio remains practically the same irrespective of whether the soil is undisturbed or reconstituted and also independent of the confining stress.

7.2.2 *Method of estimating deformation characteristics of soils in the field*

It has been shown in the foregoing that the field value of low-amplitude shear modulus G_{OF} is generally different from the value G_{OL} determined in the laboratory for reconstituted or even undisturbed samples. This has also been shown to be the case for the strain dependency of the shear modulus over the moderate to large strain range. With all these circumstances in mind, it might be desirable now to formulate a procedure of practical use for estimating strain-dependent shear moduli in the field over a wide range of shear strains. This procedure may consist of the following steps:

1. The first step is to conduct measurements of shear wave velocity in the field

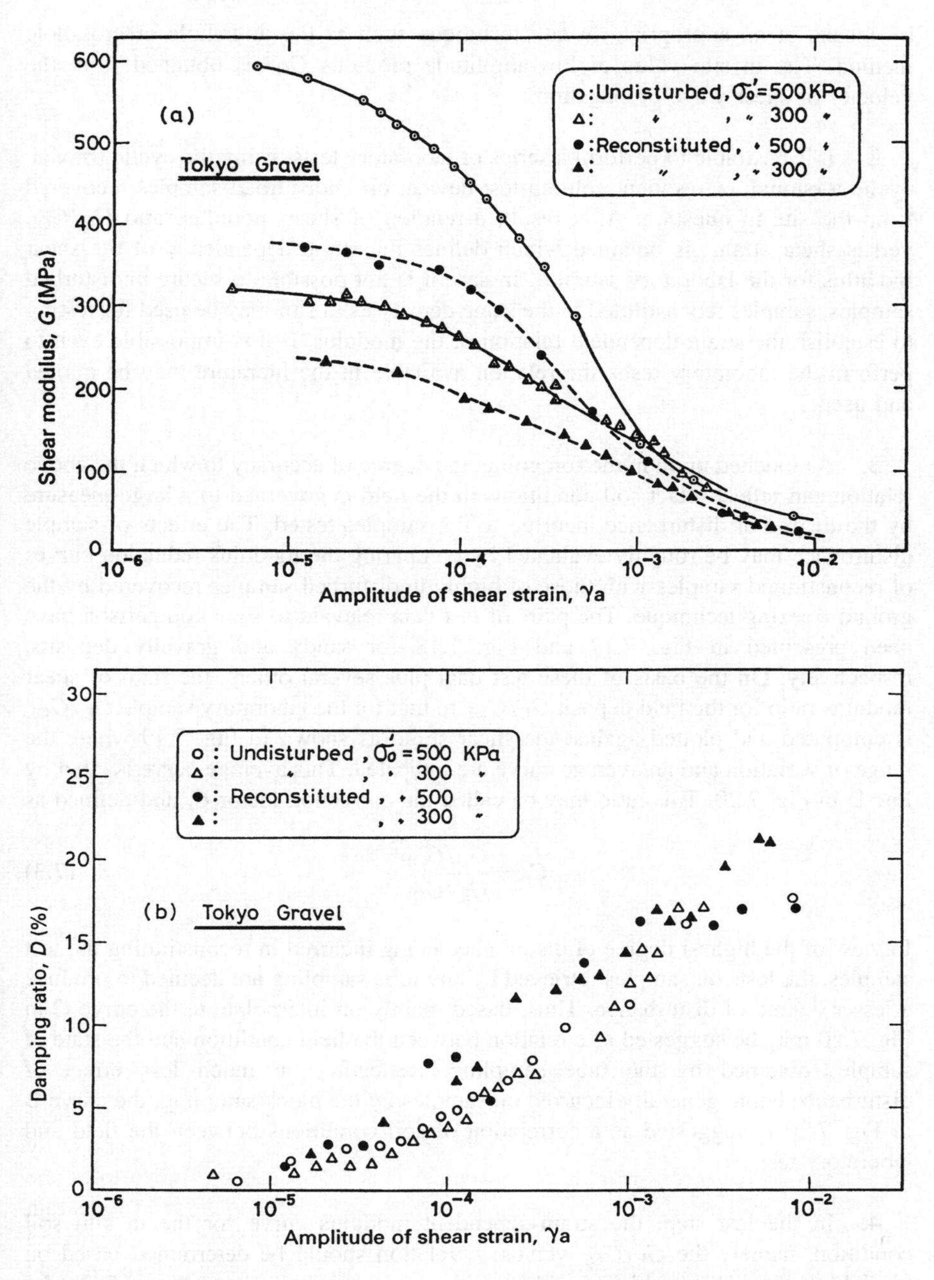

Fig. 7.18 Effects of sample disturbance on the strain-dependent shear modulus of gravelly soils (Hatanaka *et al.* 1988).

by means of an appropriate in situ technique such as the downhole or crosshole method. The in situ value of low-amplitude modulus G_{OF} is obtained from the velocity of shear wave propagation.

2. It is desirable to perform a series of laboratory tests, using the cyclic triaxial, cyclic torsional, or resonant column test device, on undisturbed samples recovered from the site in question. As a result, a relation of shear modulus ratio G_L/G_{OL} versus shear strain is obtained which defines the strain dependency of the shear modulus for the laboratory samples. In case it is not possible to secure undisturbed samples, samples reconstituted to the same density as in situ may be used for testing to establish the strain-dependent relation of the modulus. If it is impossible even to perform the laboratory tests, the relation available in the literature may be quoted and used.

3. As touched upon in the foregoing, the degree of accuracy to which the above relation can reflect intact soil conditions in the field is governed to a large measure by the degree of disturbance incurred to the samples tested. The effects of sample disturbance may be roughly evaluated by comparing the modulus reduction curves of reconstituted samples with those of highly undisturbed samples recovered by the ground freezing technique. The pairs of test data relevant to such comparison have been presented in Fig. 7.17 and Fig. 7.18 for sandy and gravelly deposits, respectively. On the basis of these test data plus several others, the ratio of shear modulus ratio for the field deposit G_F/G_{OF} to that for the laboratory sample G_L/G_{OL} is computed and plotted against the shear strain as shown in Fig. 7.19 where the range of variation and an average curve are indicated. This average curve is cited by line D in Fig. 7.20. This ratio may be called the correction factor C_r and defined as

$$C_r = \frac{G_F/G_{OF}}{G_L/G_{OL}}. \tag{7.3}$$

In view of the highest degree of disturbance being incurred in reconstituting the test samples, the tests on samples retrieved by any tube sampling are deemed to produce a lesser degree of disturbance. Thus, based mainly on interpolation, the curve C in Fig. 7.20 may be suggested as a relation between the field condition and the state of samples obtained by the tube sampling. Reflecting on much less effect of disturbance being generally incurred on samples by the block sampling, the curve B in Fig. 7.20 is suggested as a correlation of soil conditions between the field and laboratory test.

4. In the last step, the strain-dependent modulus curve for the in situ soil condition, namely the G_F/G_{OF} versus γ_a relation should be determined based on available information which could be either from literature surveys or from the laboratory tests. Depending upon the approximate level of quality of the data that can be inferred from the method of sampling, one of the four curves shown in Fig. 7.20 is adopted and used for the correction. The correction for the sample

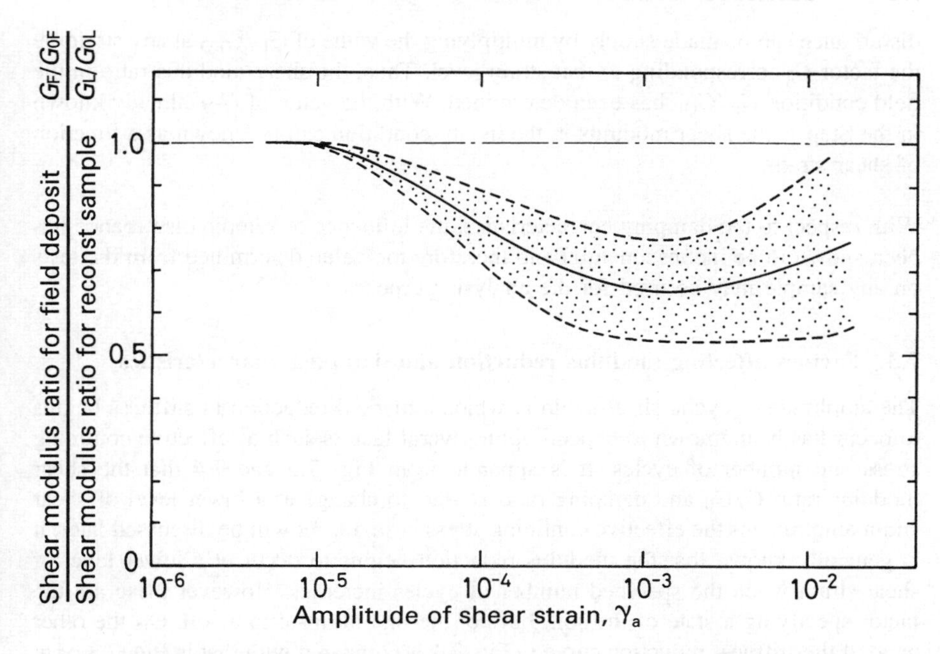

Fig. 7.19 Effects of sample disturbance on the strain-dependent shear modulus.

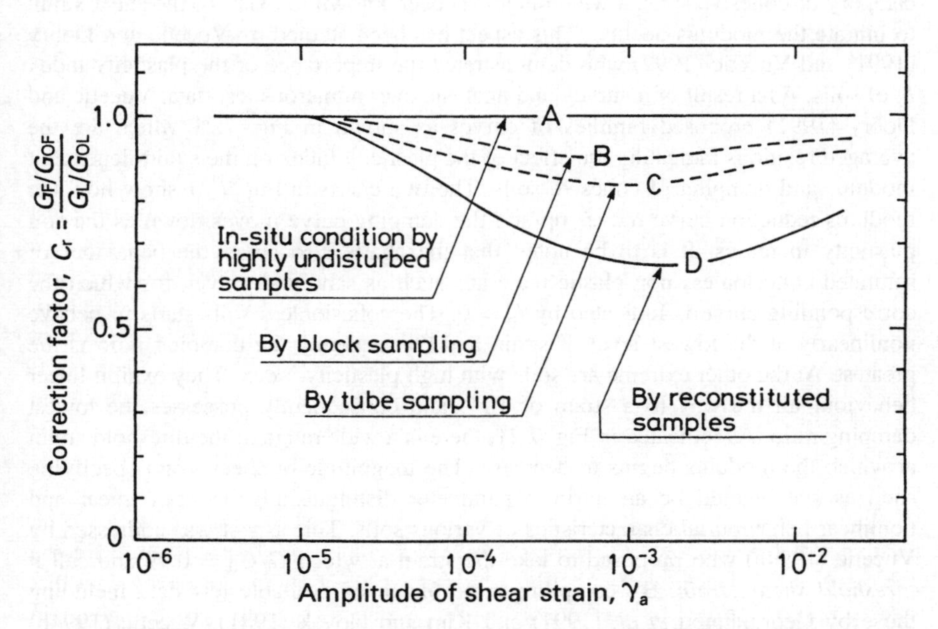

Fig. 7.20 Degree of sample disturbance by different sampling technique affecting the strain-dependent shear modulus.

disturbance can be made simply by multiplying the value of G_L/G_{LO} at any strain by the factor C_r corresponding to that strain level. Thus, the shear modulus ratio in the field condition G_F/G_{OF} has been determined. With the value of G_{OF} already known in the Step 1, the shear modulus in the in situ condition can be known as a function of shear strain.

With respect to the damping characteristics, the influence of sample disturbance has been shown to be inconsequential and therefore the value determined from the tests on any sample may be used for the analysis purposes.

7.3 Factors affecting modulus reduction and damping characteristics

The amplitude of cyclic shear strain at which a marked reduction in stiffness begins to occur has been known to depend upon several factors such as effective confining stress and number of cycles. It is apparent from Figs 7.3 and 7.4 that the shear modulus ratio G/G_0 and damping ratio D start to change at a lower level of shear strain amplitude if the effective confining stress is small. As will be discussed later, it is generally known that the modulus reduction begins to occur at a lower level of shear strain when the specified number of cycles increases. However these are the factor specifying a state of the soil, but not the one intrinsic to a soil. On the other hand, if the stiffness reduction curve in Fig. 7.1 is compared with that in Fig. 7.8 or in Fig. 7.10, it is obvious that the modulus reduction begins to occur at a smaller level of shear strain for sand as compared to that for clay. Furthermore, even within the category of cohesive soils, a wide range has been known to exist for the shear strain to initiate the modulus decline. This aspect has been studied by Vucetic and Dobry (1991) and Vucetic (1992) who demonstrated the importance of the plasticity index I_p of soils. As a result of a survey and analysis over numerous test data, Vucetic and Dobry (1991) proposed families of curves as shown in Fig. 7.21 which are the averaged relations indicating the effect of the plasticity index on the strain-dependent modulus and damping of cohesive soils. The two charts in Fig. 7.21 show how the modulus reduction curve moves up and the damping curve moves down as the soil plasticity increases. It is to be noted that the charts also cover the behaviour of saturated cohesionless non-plastic materials such as sand and gravel, for which the corresponding curve is indicated by $I_p = 0$. The cohesionless soils starts to behave nonlinearly at the lowest level of strain and consequently the damping ratio is the greatest. At the other extreme are soils with high plasticity index. They exhibit linear behaviour all the way to a strain of 10^{-4} and consequently possesses the lowest damping ratio. As indicated in Fig. 7.21, there is a wide range in the threshold strain at which the modulus begins to decrease. The magnitude of shear strain specifying such as state would be an intrinsic parameter distinguishing between linear and nonlinear behaviourial characteristics of various soils. This aspect was addressed by Vucetic (1994a) who proposed to take the strain at which $G/G_0 = 0.99$ and call it *threshold shear strain*. By compiling a profusion of available test data including those by Georgiannou *et al.* (1991) and Kim and Novak (1981), Vucetic (1994b) proposed an average curve as shown in Fig. 7.22 correlating the threshold shear strain

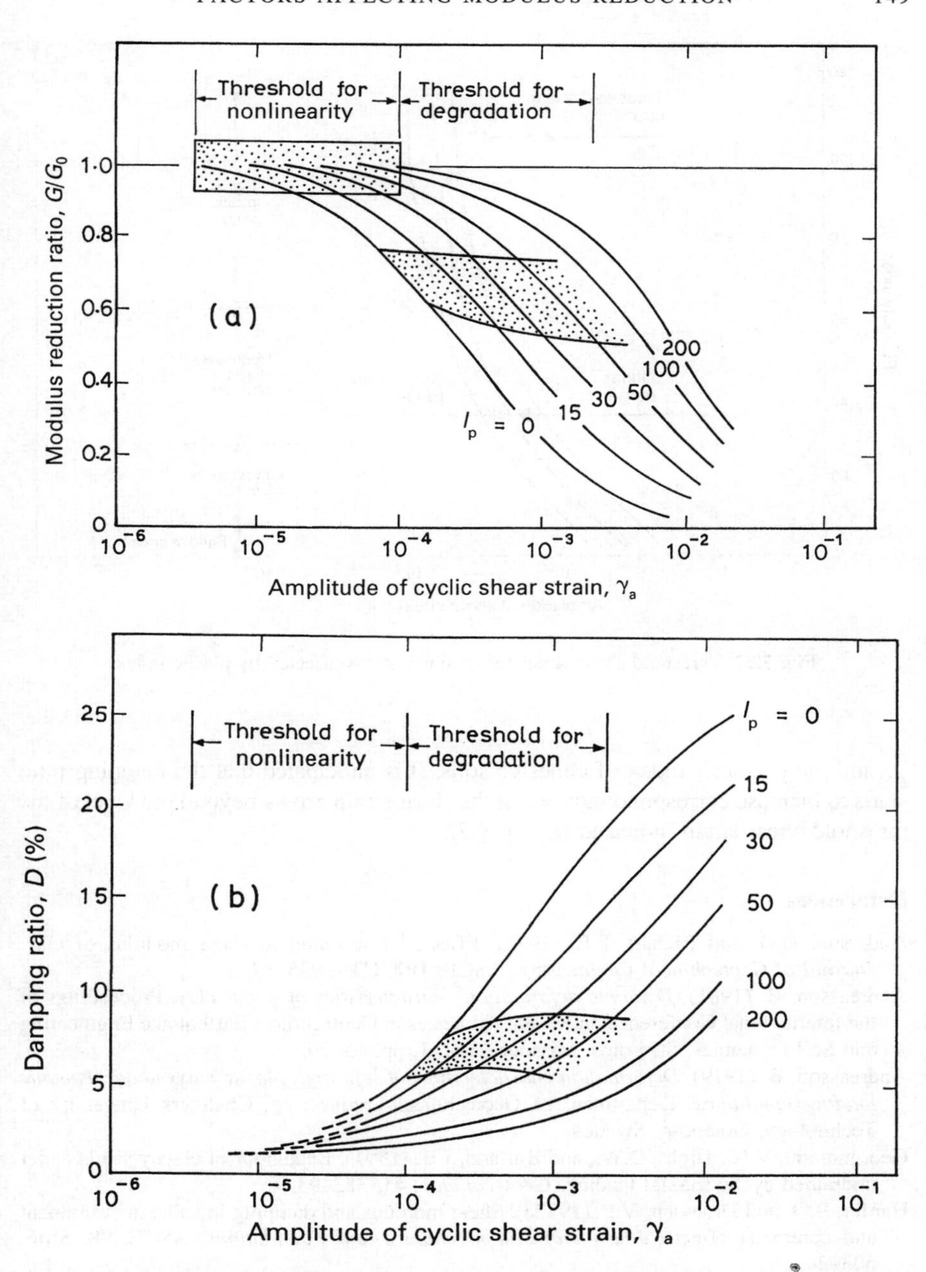

Fig. 7.21 Threshold strains of clays with different plasticity indices with respect to nonlinearity and stiffness degradation (Vucetic and Dobry, 1991).

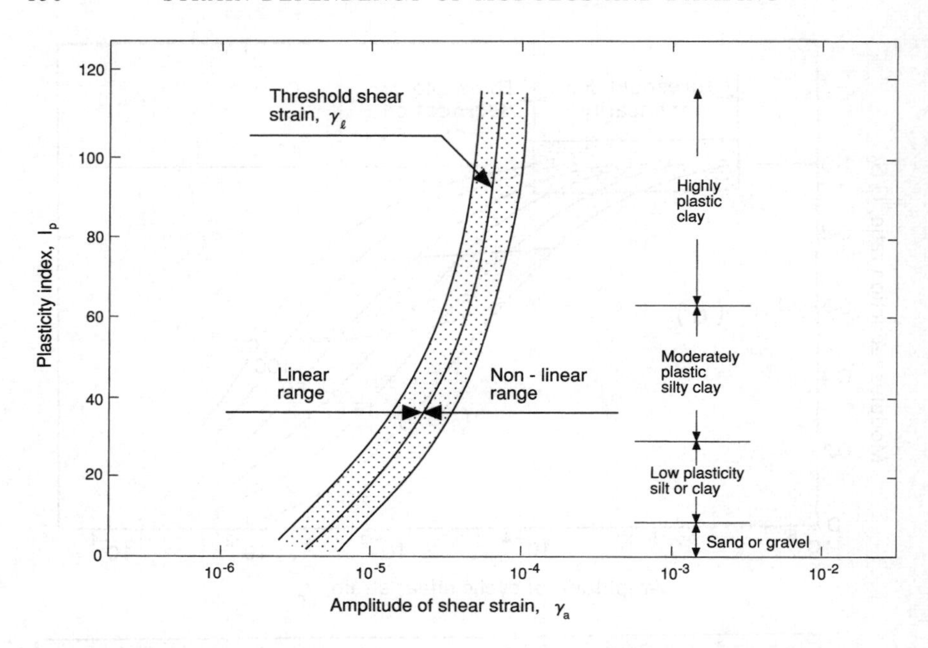

Fig. 7.22 Threshold shear strain for nonlinearity as affected by plastic index.

γ_ℓ and the plasticity index of cohesive soils. It is anticipated that the damping ratio starts to increase correspondingly when the shear strain grows beyond the level of the threshold shear strain indicated in Fig. 7.22.

References

Anderson, D.G. and Richart, F.E. (1976). Effects of straining in shear modulus of clay. *Journal of Geotechnical Engineering*, ASCE, **102**, GT9, 975–87.

Andreasson, B. (1981). *Dynamic deformation characteristics of a soft clay*. Proceedings of the International Conference on Recent Advances in Geotechnical Earthquake Engineering and Soil Dynamics, St. Louis, Missouri, Vol. 1, pp. 65–70.

Andreasson, B. (1979). *Deformation characteristics of soft, high-plastic clays under dynamic loading conditions*. Department of Geotechnical Engineering, Chalmers University of Technology, Goteborg, Sweden.

Georgiannou, V.N., Hight, D.W., and Burland, J.B. (1991). Behaviour of clayey sands under undrained cyclic triaxial loading. *Geotechnique*, **41**, 383–93.

Hardin, B.O. and Drnevich, V.P. (1972a). Shear modulus and damping in soils: measurement and parameter effects. *Journal of Soil Mechanics and Foundations*, ASCE, **98**, SM6, 603–24.

Hardin, B.O. and Drnevich, V.P. (1972b). Shear modulus and damping of soils: design equation and curves. *Journal of Soil Mechanics and Foundations*, ASCE, **98**, SM7, 667–92.

Hatanaka, M., Suzuki, Y., Kawasaki, T., and Endo, M. (1988). Cyclic undrained shear properties of high quality undisturbed Tokyo gravel. *Soils and Foundations*, **28**, 57–68.

Iwasaki, T., Tatsuoka, F., Tokida, K., and Yasuda, S. (1978). *A practical method for assessing soil liquefaction potential based on case studies at various sites in Japan*. Proceedings of the 2nd International Conference on Microzonation for Safer Construction Research and Application. Vol. 2, pp. 885–96.

Katayama, I., Fukui, F., Goto, M., Makihara, Y., and Tokimatsu, K. (1986). *Comparison of dynamic deformation characteristics of dense sand between undisturbed and disturbed samples*. Proceedings of the 21st Annual Conference of JSSMFE, pp. 583–4 (in Japanese).

Kim, T.C. and Novak, M. (1981). Dynamic properties of some cohesive soils of Ontario. *Canadian Geotechnical Journal*, **18**, 371–89.

Kokusho, T. (1980). Cyclic triaxial test of dynamic soil properties for wide strain range. *Soils and Foundations*, **20**, 45–60.

Kokusho, T., Yoshida, Y., and Esashi, Y. (1982). Dynamic properties of soft clays for wide strain range. *Soils and Foundations*, **22**, 1–18.

Kovacs, W.D., Seed, H.B., and Chan, C.K. (1971). Dynamic modulus and damping ratio for a soft clay. *Journal of Soil Mechanics and Foundations*, ASCE, **97**, SM1, 59–75.

Seed, H.B. and Idriss, I.M. (1970). *Soil moduli and damping factors for dynamic analysis*. Report No. EERC 70-10, University of California, Berkeley.

Sherif, M.A., Ishibashi, I., and Gaddah, A.H. (1977). Damping ratio for dry sands. *Journal of ASCE*, **103**, GT7, pp. 743–56.

Studer J., Zingg, N., and Prater, E.G. (1980). *Investigation on cyclic stress–strain characteristics of gravel material*. Proceedings of the 7th World Conference on Earthquake Engineering, Istanbul, Turkey, Vol. 3, pp. 355–62.

Tanaka, Y., Kudo, Y., Yoshida, Y., and Ikemi, M. (1987). *A study on the mechanical properties of sandy gravel—dynamic properties of reconstituted sample*. Central Research Institute of Electric Power Industry, Report U87019.

Tatsuoka, F. and Iwasaki, T. (1978). Hysteretic damping of sands under cyclic loading and its relation to shear modulus. *Soils and Foundations*, **18**, 25–40.

Taylor, P.W. and Parton, J.M. (1973). *Dynamic torsion testing of soils*. Proceedings of the 8th International Conference on Soil Mechanics and Foundation Engineering, Vol. 1, Part 2, Moscow, pp. 425–32.

Vucetic, M. (1992). *Soil properties and seismic response*. Proceedings of the 10th World Conference on Earthquake Engineering, Madrid, Spain, Vol. III, pp. 1199–1204.

Vucetic, M. (1994a). *Cyclic Characterization for Seismic Regions Based on PI*. Proceedings of the 13th International Conference on Soil Mechanics and Foundation Engineering, Vol. 1, pp. 329–32.

Vucetic, M. (1994b). Cyclic threshold shear strains in soils. *Journal of Geotechnical Engineering*, ASCE, **120**, pp. 2208–28.

Vucetic, M. and Dobry, R. (1991). Effect of soil plasticity on cyclic response. *Journal of Geotechnical Engineering*, **117**, 89–107.

8

EFFECT OF LOADING SPEED AND STIFFNESS DEGRADATION OF COHESIVE SOILS

The tendency of soils to dilate or to contract during drained shear and pore water pressure changes during undrained shear does not come out in the infinitesimal or small strain ranges. The effect of the dilatancy begins to manifest itself as the magnitude of shear strain increases above the level of about 10^{-3} as indicated in Section 1.3. It should be remembered that the major changes in soil properties due to cyclic loading such as degradation in stiffness of saturated soils or hardening of dry and partially saturated soils can occur as a consequence of the dilatancy effect being manifested during cyclic loading. Thus the effect of load repetition as described in Section 1.2.2 begins to crop up when the strain induced in soils grow above the level of 10^{-3}. This effect is alternatively called *cyclic degradation*.

Another important aspect of soil deformation characteristics observed in dynamic loading conditions is the influence of the speed with which loads are applied to soils. It has been well established that the resistance to deformation of soils in monotonic loading tends to increase as the speed of loading is increased and that the threshold shear strain between where the rate effect does and does not come out is on the order of about 10^{-3}. The effect of loading speed as above can also exert a profound influence on the deformation mechanism of soils under cyclic loading conditions. In the case of cyclic loading, the period at which loads are reciprocated, or more pertinently a quarter period, may be deemed to correspond to the time of loading in the case of the monotonic loading. Therefore for cyclic loading executed with shear strain amplitude in excess of 10^{-3}, the resistance of soils to deformation would be greater for a loading with the higher frequency than for the loading with the lower frequency.

In what follows, the deformation characteristics of cohesive soils undergoing cyclic loading with the shear strain above the level of 10^{-3} will be discussed in the context of the cyclic degradation and loading speed.

8.1 Classification of loading schemes

The method of load application is classified generally into three types, namely, slow static loading, rapid transient loading, and repetitive loading as illustrated in Fig. 8.1. In the static loading test, the load is applied monotonically with a time to failure on the order of several seconds to a few minutes or even longer. If the loading to failure is executed in a shorter time, the loading is called transient or rapid. The shortest time of loading ever employed in the laboratory tests is on the order of 0.001 second.

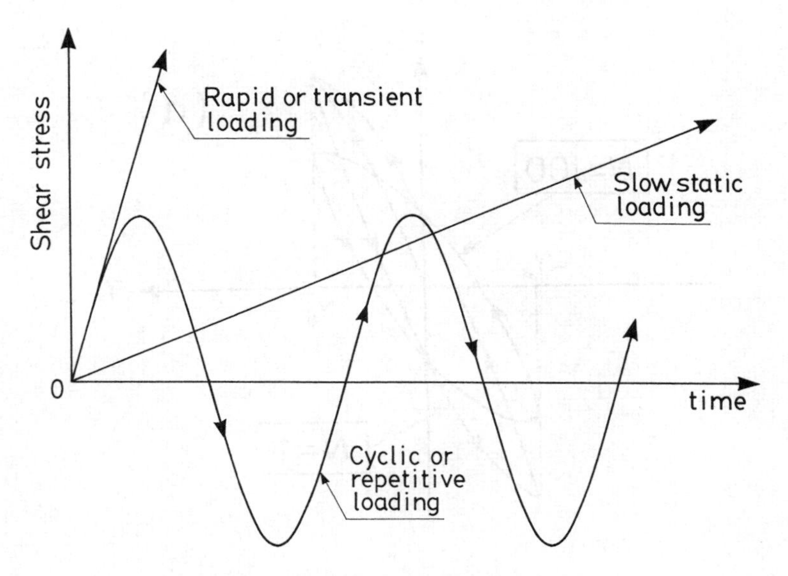

Fig. 8.1 Types of loading.

With respect to the repetitive loading, there are two kinds of load application; slow and rapid repetition of load. Tests with slowly applied repetitive loads have been conducted to clarify the effect of cyclic load on consolidation or creep characteristics of clays, but since this subject is beyond the scope of this text, this aspect is not touched upon. Tests with rapidly repeated load application have been performed to investigate the deformation response of cohesive soils subjected to medium to large shear strains. Most of the tests ever conducted with a period of 1 to 5 seconds may be deemed as the repetition of the rapid loading mentioned above. In what follows, the term repetitive loading test or cyclic loading test will be used to imply the test in which rapid loads are repetitively applied.

When a soil is subjected to cyclic loads under strain-controlled conditions with an amplitude of shear strain γ_a, the degradation of soil response will involve changes in the stress–strain curve with progression of cycles as illustrated in Fig. 8.2(a). If the amplitude is raised to γ_b, similar changes with cycles in the stress–strain relation take place as illustrated in Fig. 8.2(b). Thus if the hysteresis curves for a certain number of repetitions are put together and drawn in one diagram, a set of curves as shown in Fig. 8.3 will be obtained where each hysteresis curve is laid down along a skeleton curve or backbone curve which is also specific to the number of repetitions being considered. Thus the degradation of soil stiffness is reflected in the skeleton curve changing its shape with the progression of cyclic loading. If a skeleton curve as defined above for a certain number of cycles is extracted and redrawn, one can obtain a stress–strain curve as shown in Fig. 8.4 where it is possible to define secant moduli for a given amplitude of cyclic loading. Thus the secant modulus is a function of the number of cycles and the amplitude of shear strain as well.

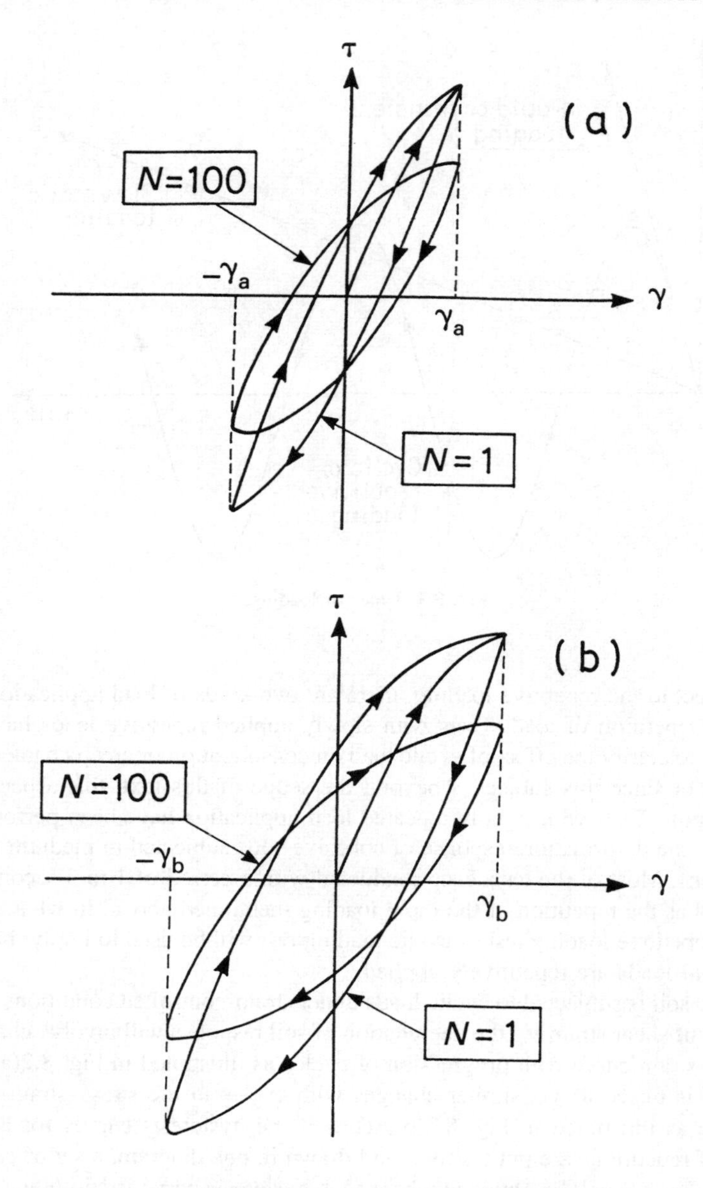

Fig. 8.2 Stress–strain curves in cyclic loading.

8.2 Deformation characteristics of soils under transient loading

It was as early as the mid-1940s that investigations were started at Harvard University regarding soil behaviour under transient loading conditions. The objectives of such studies are said to have been associated with the evaluation of stability of natural

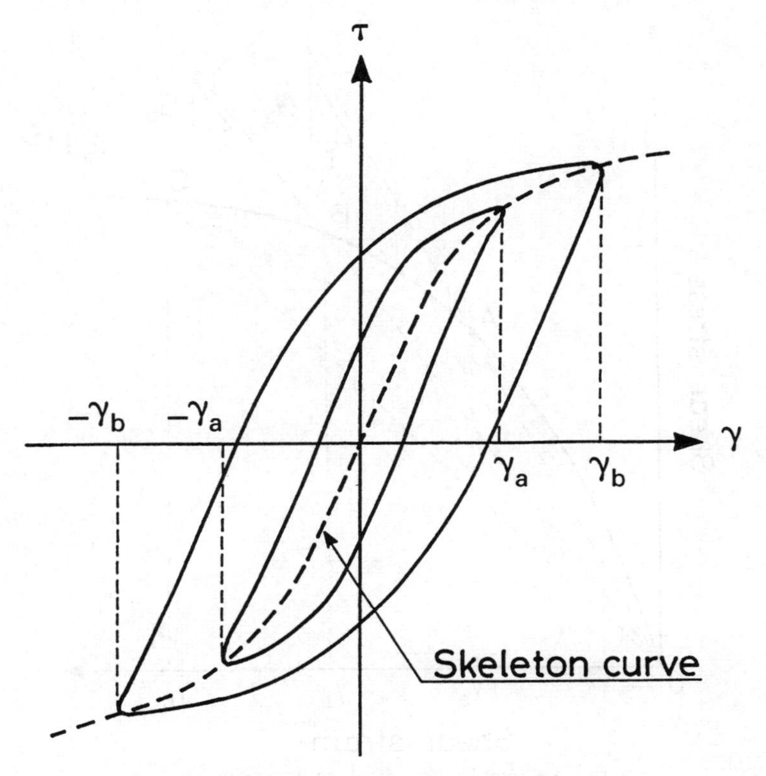

Fig. 8.3 Hysteresis curves and skeleton curve.

slopes along the Panama Canal. As a result of comprehensive laboratory tests, Casagrande and Shannon (1948) and Casagrande and Wilson (1951) clarified that the modulus as well as strength of clays were greater in dynamic loading conditions as compared to those under slow static loading conditions. One of the test results in this study is shown in Fig. 8.5 where remoulded kaolinite clay was subjected to uniaxial loads with different speeds. It is observed that the rapid loading test employing a time of 0.02 second to failure showed much greater modulus and strength value than the static test employing a longer loading time of 4 minutes. Other investigators such as Whitman (1957), Ellis and Hartman (1967), Ohsaki *et al.* (1957), Ohsaki (1964), Richardson and Whitman (1963) and Shimming *et al.* (1966) have also made some studies on the effects of loading speed and reached the same conclusion as above.

As pointed out in Fig. 8.3, the modulus value is highly dependent, be it in rapid or slow loading, on the level of shear strain at which the modulus is determined. Thus it is of interest to see if such strain dependency of modulus is different or not between rapid and slow loading conditions. For this purpose, values of secant moduli at a strain of 0.2% read off from the stress–strain curve such as those shown in Fig. 8.5 are plotted in Fig. 8.6(a) versus time of loading to failure. Viewed overall,

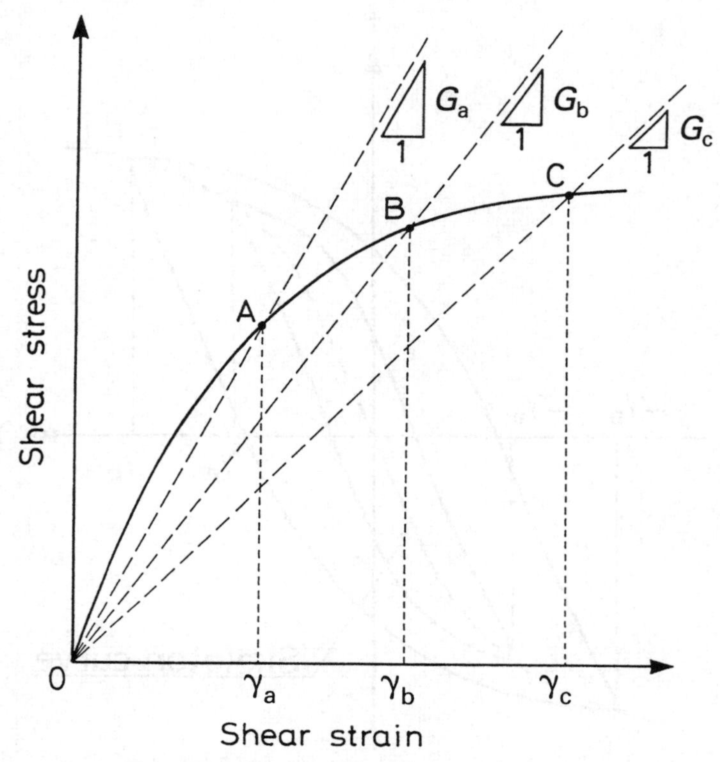

Fig. 8.4 Stiffness reduction with increasing strain.

it appears that there is practically no influence of loading speed on the modulus at the level of a strain as low as 0.2%. Similar plots are also made from the same test data, as shown in Fig. 8.6(b), by reading off the value of secant moduli at a large strain of 2%. It may now be seen that the modulus has a tendency to decrease as the time of loading becomes longer. From the observation as above, it may be mentioned that the difference in the modulus between rapid and slow loading does depend on the magnitude of shear strain, and the increasing tendency of the modulus in rapid loading over the value in slow loading becomes more pronounced as the level of strain increases. To evaluate this tendency more in a quantitative manner, the modulus obtained at a loading time of 1 minute is chosen as a value representing the modulus in the slow loading and denoted by E_{st}. The modulus obtained at a time of loading of 0.1–0.25 sec. is taken up as a value representative of that in the rapid loading and is denoted by E_d. The two times of loading to failure thus chosen are indicated by arrows in Fig. 8.6. To see the effects of time of loading on the strain dependency of the modulus, it would be convenient to compare the values of E_d and E_{st} which are determined at various levels of strain according to the procedure illustrated in Fig. 8.4. As a result of such data compilation from a majority of laboratory tests, the outcome as shown in Fig. 8.7 was obtained where the ratio

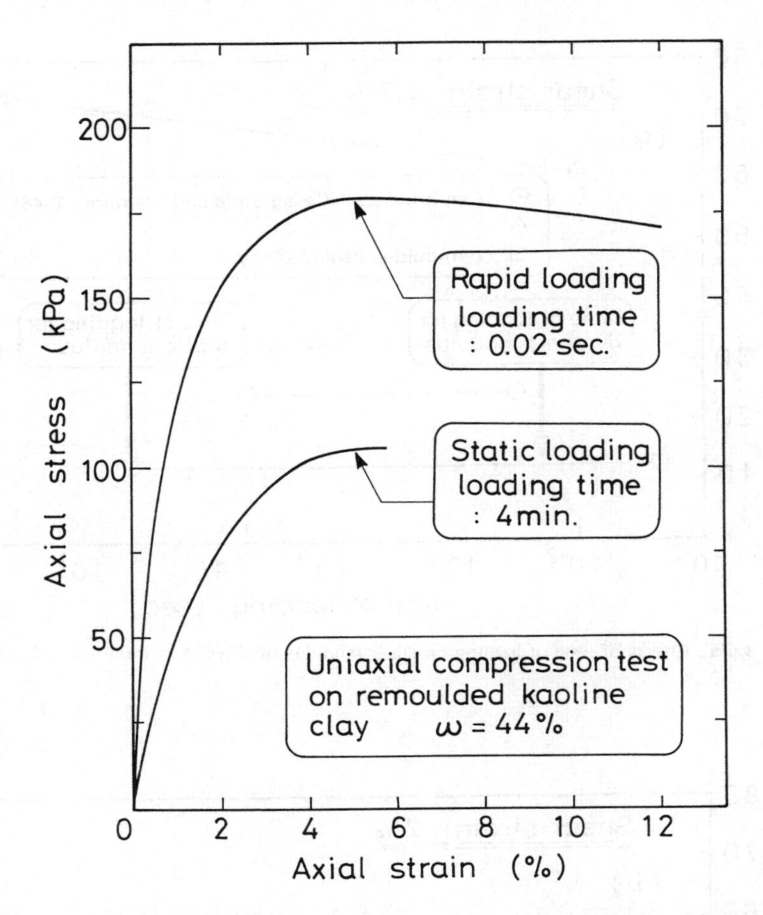

Fig. 8.5 Comparison of stress–strain curves for static slow loading and rapid loading tests.

between the dynamic modulus and static modulus E_d/E_{st} is plotted against the shear strain. The sources of each set of test data are indicated in Fig. 8.7 with the same symbols. Most of the materials tested were clays or loams having a plasticity index of $I_p = 15$–50. Water content was in the range of 20 to 50%. The majority of the data were obtained by unconfined compression tests but some by triaxial tests. Figure 8.7 shows that, although there is scatter, the ratio E_d/E_{st} tends to increase with increasing level of shear strain. More specifically, the increase in the dynamic modulus over the value of the static modulus is shown to be about 50% on average at the strain level of 1%, but the difference between these two moduli disappear when the strain becomes low on the order of 10^{-3}. For each of the above test data, the magnitude of respective strain ε was normalized to the strain at failure ε_f and the ratio $\varepsilon/\varepsilon_f$ is now used in Fig. 8.8 in the abscissa to demonstrate the effect of strain level on the dynamic to static modulus ratio E_d/E_{st}. Note that some additional data

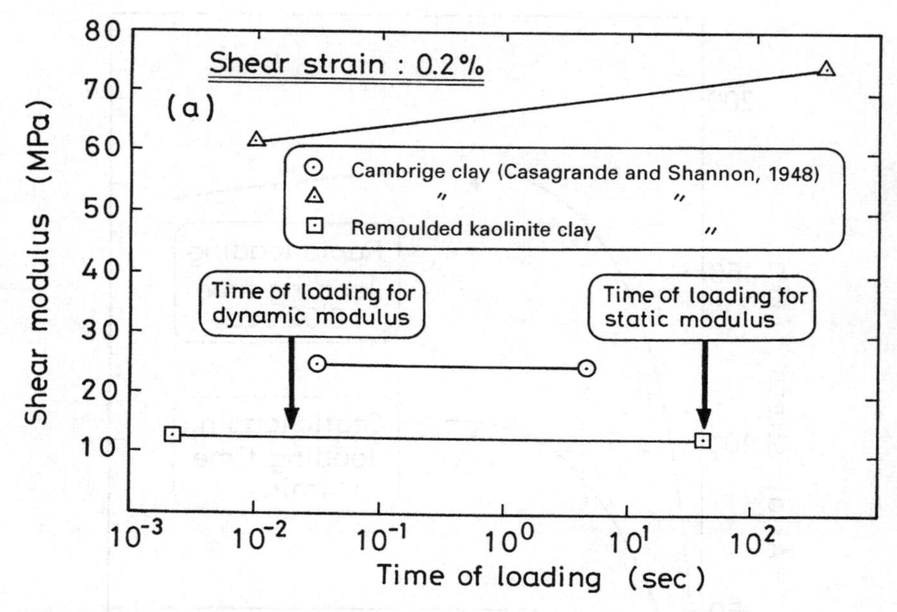

Fig. 8.6(a) Effects of time of loading on shear modulus of clays at a strain level of 0.2%.

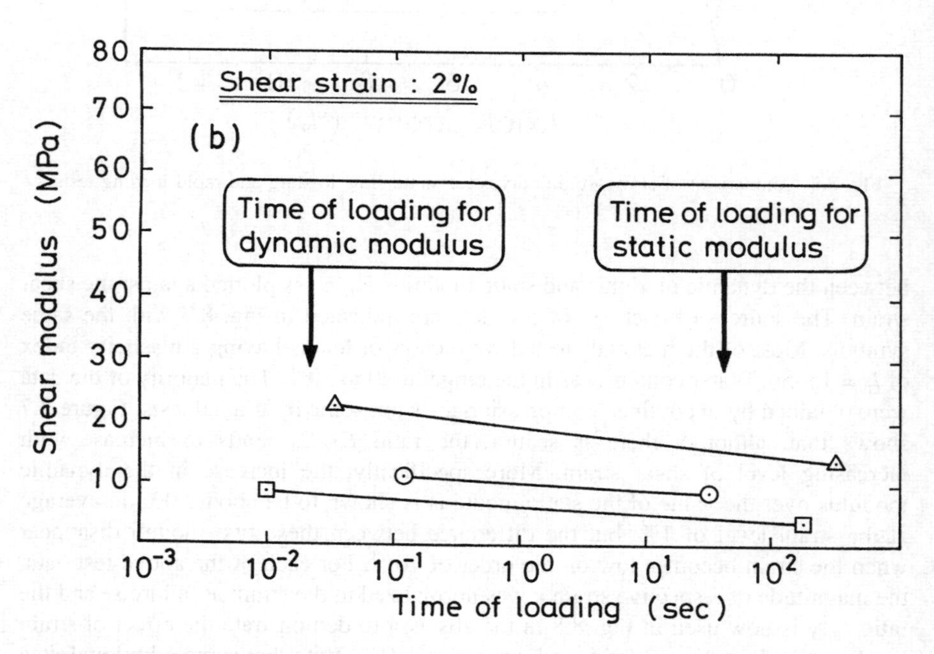

Fig. 8.6(b) Effects of time of loading on shear modulus of clays at a strain level of 2%.

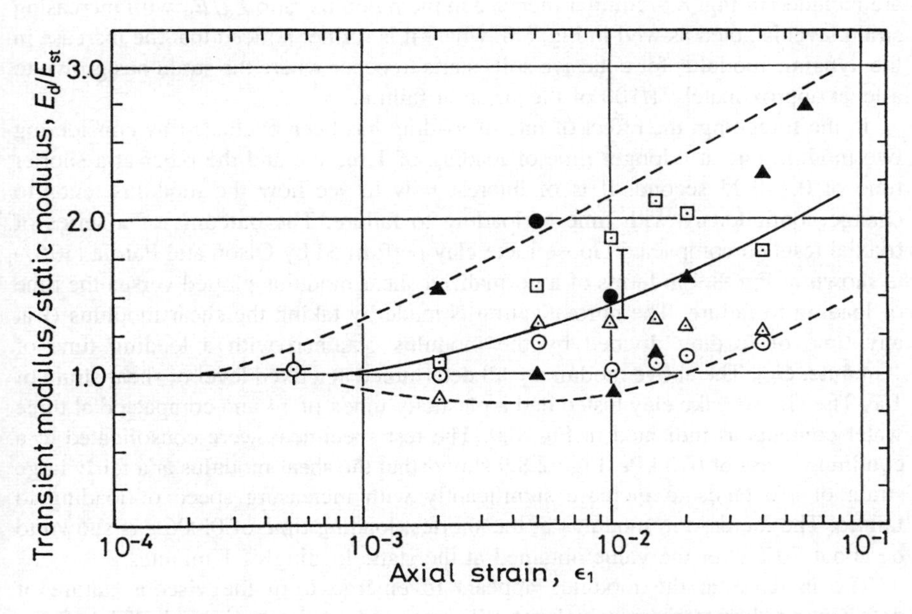

Fig. 8.7 Effects of axial strain level on the ratio of transient modulus to static modulus.

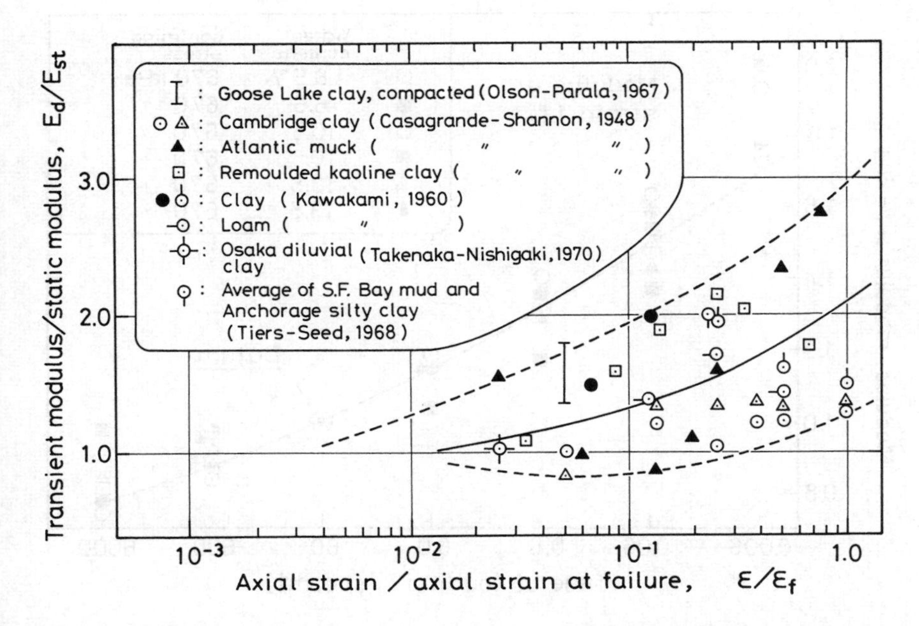

Fig. 8.8 The ratio of transient modulus to static modulus as influenced by axial strain normalized to failure strain.

are included in Fig. 8.8. Similar increase in the modulus ratio E_d/E_{st} with increasing strain level is noted as well in Fig. 8.8, where it is further noticed that the increase in the dynamic modulus for cohesive soils starts to occur where the strain has grown to a level approximately 1/100 of the strain at failure.

In the foregoing, the effect of rate of loading has been evaluated by considering two moduli, one at a longer time of loading of 1 minute and the other at a shorter time of 0.1–0.25 second. It is of interest now to see how the modulus tends to change continuously with time of loading to failure. The outcome of a series of triaxial tests on compacted Goose Lake clay performed by Olson and Parola (1967) is shown in Fig. 8.9 in terms of a normalized shear modulus plotted versus the time of loading to failure. The normalization is made by taking the shear modulus G at any time of loading divided by the modulus obtained with a loading time of 1 minute, G_{60}. The above moduli are all determined at a fixed level of shear strain of 1%. The Goose Lake clay tested had a plasticity index of 14 and compacted at three water contents as indicated in Fig. 8.9. The test specimens were consolidated to a confining stress of 670 kPa. Figure 8.9 shows that the shear modulus at a fairly large strain of 1% tends to increase significantly with increasing speed of loading to failure. The increase in modulus at the shortest loading time of 0.006 s is shown to be about 70% over the value obtained at the static loading of 1 minute.

The increase in the modulus appears to emerge from the viscous nature of deformation characteristics which prevails more or less in cohesive soils. The effects of viscosity can be explained qualitatively by referring to the Kelvin model

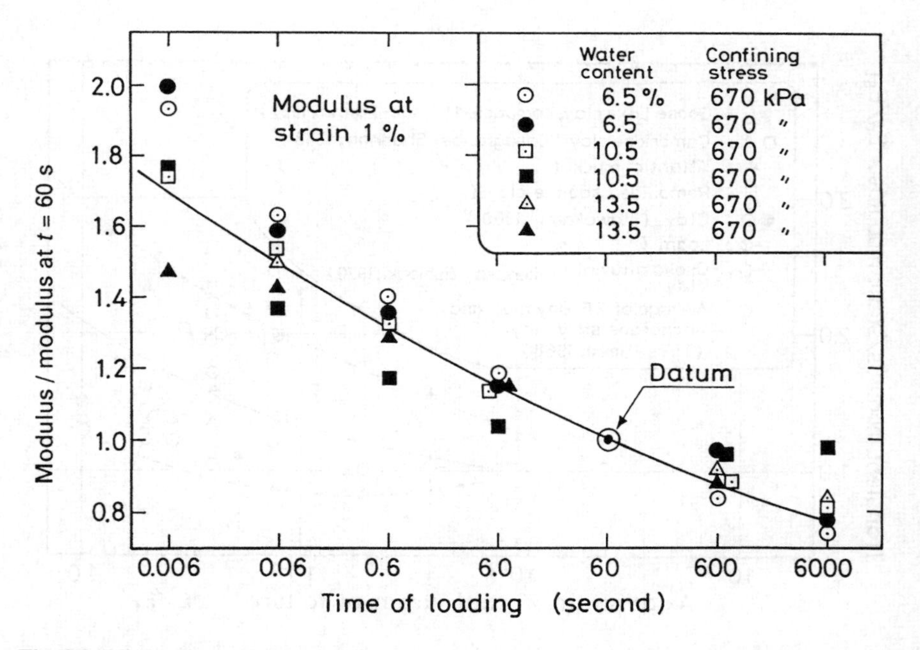

Fig. 8.9 Effects of time of loading on the shear modulus of cohesive soils at a strain level of 1% (Olson and Parola, 1967).

representation shown in eqn (3.18). Let a monotonic loading be represented by $\tau = \dot{\tau}_p t$ with a loading speed $\dot{\tau}_p$, and let the load application be started at $t = 0$ with $\tau = 0$. Integration of eqn (3.18) under the above initial condition yields a stress–strain–time relation as

$$\gamma = \frac{\tau}{G} - \frac{\dot{\tau}_p \bar{t}}{G'}(1 - e^{-t/\bar{t}}) \tag{8.1}$$

where $t = G'/G$ denotes the relaxation time, and G' is the dashpot constant explained in Fig. 3.4(a). The relation of eqn (8.1) is displayed schematically in Fig. 8.10. It can be seen that the intercept of asymptotes with the τ axis becomes

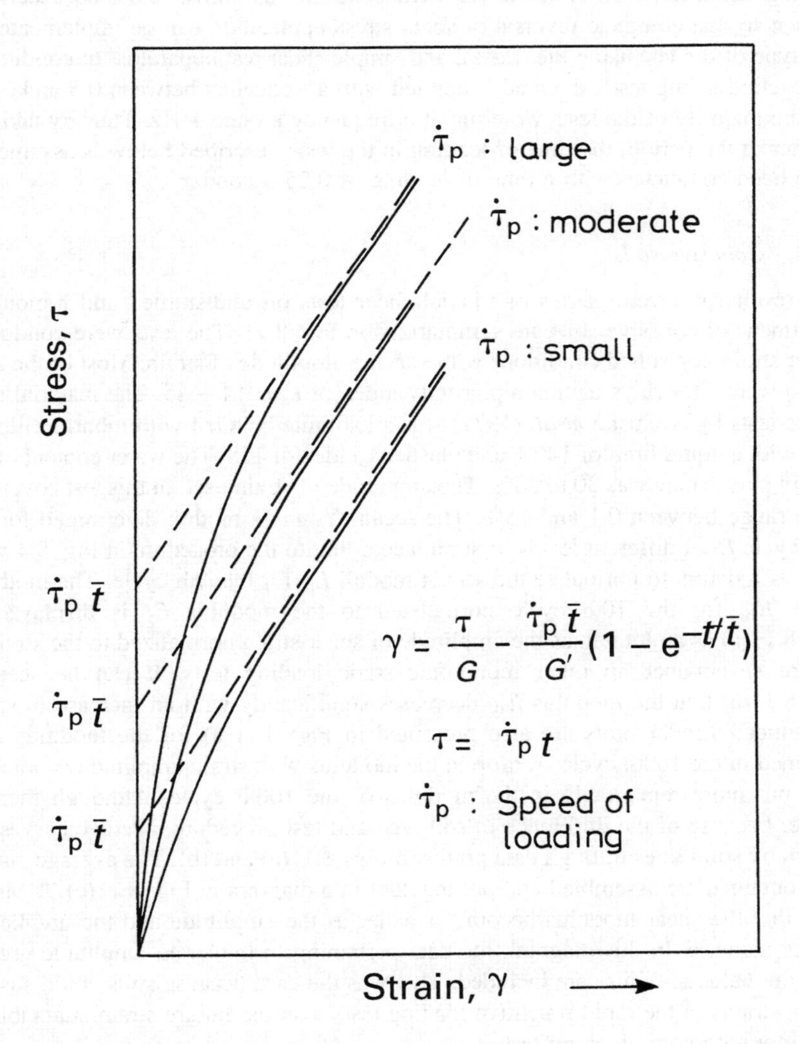

Fig. 8.10 Interpretation of loading speed on the modulus based on a viscoelastic model.

larger with increasing speed of load application and hence the stress–strain curve becomes steeper as the loading speed increases. Although the Kelvin model may not fit perfectly, it may be deemed to offer a physical interpretation for the effects of loading speed on the deformation characteristics of cohesive soils.

8.3 Deformation characteristics of soils under cyclic loading

In view of the symmetric nature of load application on soil elements in level ground during earthquakes, extensive studies have been made through laboratory tests to investigate the stiffness degradation characteristics of cohesive soils under cyclic loading conditions. To simulate the field condition as above, the load scheme is chosen so that complete reversal of shear stress application can be implemented in any type of the test using the triaxial and simple shear test apparatus. In conducting the cyclic loading tests, the load is applied with a frequency between 0.5 and 5 Hz, but the majority of the tests were run at a frequency around 1 Hz. Thus by taking a quarter of the period, the time of loading in the tests described below is assumed to have been conducted with a time of loading of 0.25 second.

8.3.1 *Cohesive soils*

The results of several series of triaxial shear tests on undisturbed and remoulded specimens of cohesive soils are summarized in Fig. 8.11. The tests were conducted under strain-controlled conditions with varying amplitude of strain. Most of the soils tested were silty clays having a plasticity index of $I_p = 14 - 45$. The material used in the tests by Kokusho *et al.* (1971) was a kaolinite blended with montmorillonite clay with a liquid limit of 140% and plasticity index of 118. The water content of the highly plastic clay was 30 to 80%. The amplitude of strain used in this test covered a wide range between 0.1 and 15%. The secant Young's moduli determined for the first cycle E_1 at different levels of strain according to the procedure in Fig. 8.4 were used as a datum to normalize the secant moduli E_N for the Nth cycle. The modulus value E_{10} for the 10th cycle normalized to the modulus E_1 is displayed in Fig. 8.11(a) as a plot versus the amplitude of shear strain normalized to the strain at failure ε_f obtained in other monotonic static loading tests. It can be seen in Fig. 8.11(a) that the modulus E_{10} decreases significantly with an increase in strain amplitude. Similar plots are also presented in Fig. 8.11(b) for the modulus E_{100} obtained in the 100th cycle. A drop in the modulus with strain amplitude is noted as well but more remarkably in the modulus of the 100th cycle. Although there is scatter because of the difference in soil type and test procedure, average curves are drawn by solid lines through data points in Figs 8.11(a) and (b). The average curves thus obtained are assembled and put together in a diagram in Fig. 8.11(c). It can be seen that the shear modulus becomes smaller as the amplitude and the number of cycles increase. In this diagram, the data pertaining to the strain amplitude greater than the value at failure are included. This was the case because soils could sustain larger strains in the rapid repetitive loading tests than the failure strain attainable in the slow monotonic loading tests.

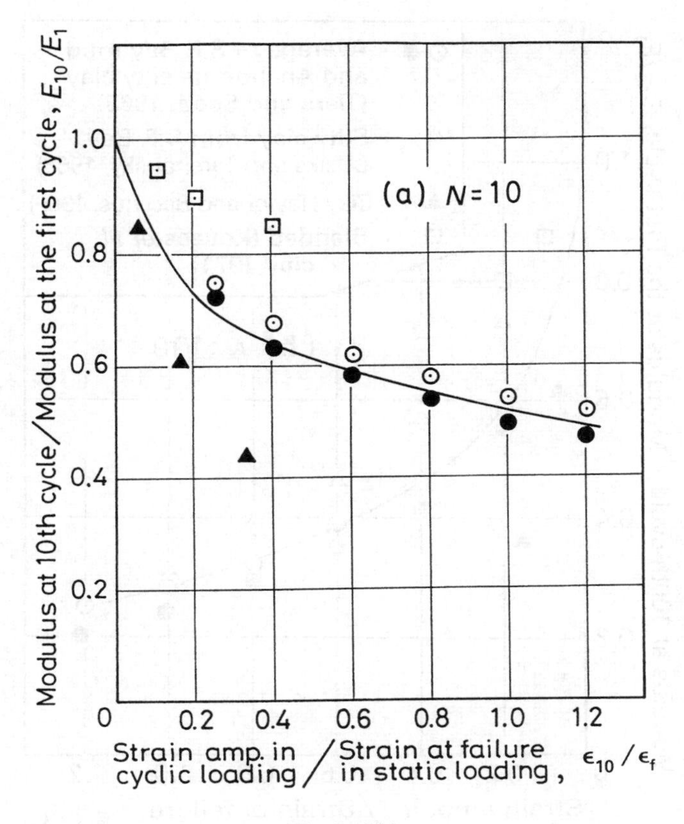

Fig. 8.11(a)

As mentioned in the foregoing, the period of cyclic load application employed in the above tests was about 1.0 second on average, giving an equivalent time of loading as being approximately 0.25 second. Thus the first-cycle shear modulus E_1 adopted as a datum in Fig. 8.11(c) is deemed to represent the modulus value which would be obtained with a time of loading of 0.25 second. Therefore this modulus is to be interpreted as equivalent to the dynamic modulus E_d shown previously in Fig. 8.8. If the reciprocal of the ratio E_d/E_{st} is taken in the ordinate of Fig. 8.8 while leaving the abscissa unchanged, this modified plot can be interpreted to provide a relation between E_{st}/E_1 and the normalized strain amplitude, which is the same kind of representation method as that of Fig. 8.11(c). The average curve indicated in Fig. 8.8 was modified in this way and transferred in the plot of Fig. 8.11(c) as indicated by a dashed line. Note that this line is indicative of strain dependency of the static modulus E_{st}. Since the data in Fig. 8.8 are from different sources, exact comparison may not be possible. However it is of interest to notice that the strain dependency of the static modulus can be interpreted in the general framework of deformation characteristics under cyclic loading conditions. Based on

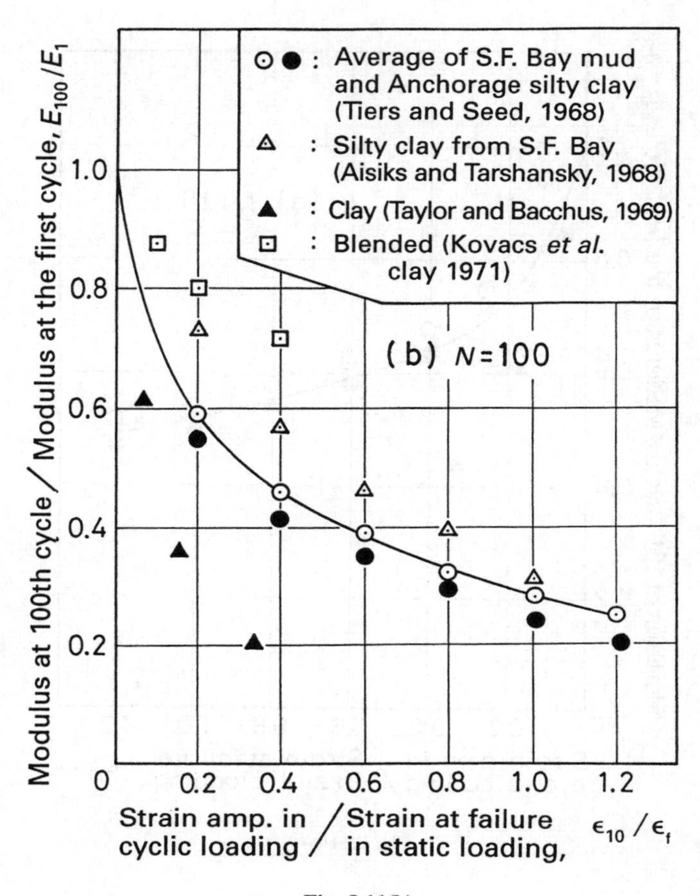

Fig. 8.11(b)

the data arranged in Fig. 8.11(c), the characteristic change in modulus can be interpreted as follows:

1. In the first cycle of load application, the shear modulus is equal, over the whole range of strain, to the modulus value obtained in the transient loading with a time of loading of about 0.25 second. This modulus E_1 is apparently greater than the static modulus E_{st} at varying degrees depending upon the strain level, taking a value which is 2.2 times as large as the static modulus at the level of failure strain.

2. As the number of cycles increases, the stiffness of cohesive soils is degraded, resulting in progressive reduction in the modulus. Apparently, the degradation takes place more notably with increasing amplitude of cyclic shear strain.

3. At the stage where the cyclic load has been applied about 20 times, the modulus becomes coincident, over the whole strain range, with the value obtained in the static loading where the time of loading is about 1.0 minute. This implies the fact

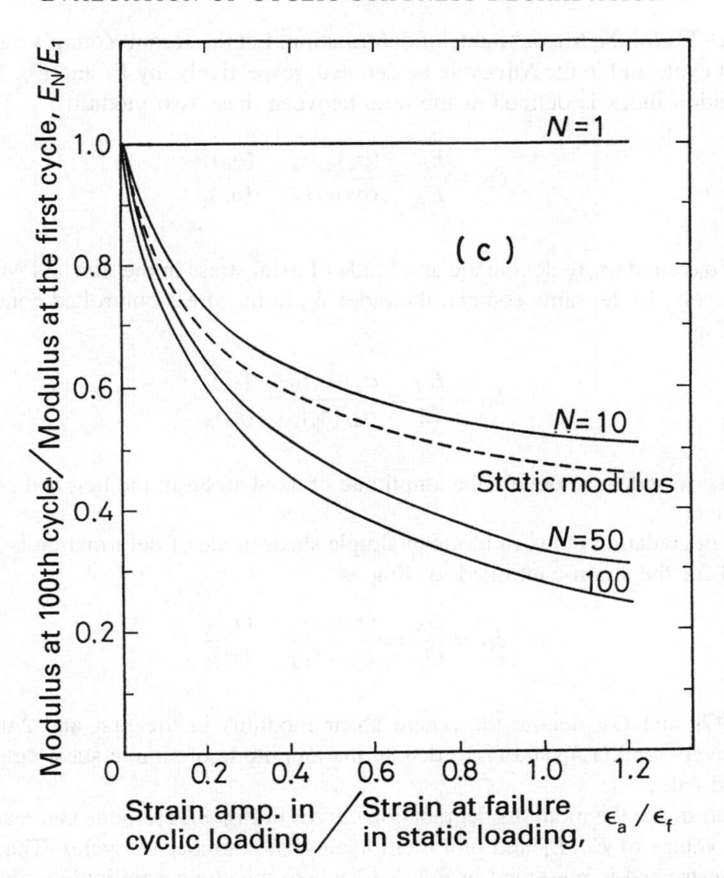

Fig. 8.11 (c) Reduction of shear modulus with increasing strain at the 10th and 100th cycle of load application.

that the gain in the modulus produced by the rapidity of loading has been gradually lost due to the degradation effect of load repetition, and complete offsetting takes place after the 20 cycles of loading, thereby bringing the soils to a state of deformation which is analogous to the static loading condition.

4. With continued application of cyclic loads, the modulus is further reduced because of the still persistent degradation effect. However, it is highly likely that the modulus degradation becomes negligibly small and discontinues after about 100 cycles of load application.

8.4 Evaluation of cyclic stiffness degradation

The effect of cyclic degradation as above under strain-controlled conditions can be expressed quantitatively in terms of the degradation index δ_D proposed by Idriss *et al.* (1978) for the triaxial mode of deformation. Let the secant Young's modulus in the first cycle and in the *N*th cycle be denoted, respectively, by E_1 and E_N. Then, the degradation index is defined as the ratio between these two moduli:

$$\delta_D = \frac{E_N}{E_1} = \frac{(\sigma_d)_N/\varepsilon_a}{(\sigma_d)_1/\varepsilon_a} = \frac{(\sigma_d)_N}{(\sigma_d)_1} \tag{8.2}$$

where $(\sigma_d)_1$ and $(\sigma_d)_N$ denote the amplitude of axial stress in the first and *N*th cycles, respectively. In the same context, the index δ_D in the stress-controlled conditions is defined as

$$\delta_D = \frac{E_N}{E_1} = \frac{\sigma_a/(\varepsilon_a)_N}{\sigma_a/(\varepsilon_a)_1} = \frac{(\varepsilon_a)_1}{(\varepsilon_a)_N} \tag{8.3}$$

where $(\varepsilon_a)_1$ and $(\varepsilon_a)_N$ denote the amplitude of axial strain in the first and *N*th cycle, respectively.

The degradation index in terms of simple shear mode of deformation is similarly defined for the strain-controlled loading as

$$\delta_D = \frac{G_N}{G_1} = \frac{(\tau_a)_N/\gamma_a}{(\tau_a)_1/\gamma_a} = \frac{(\tau_a)_N}{(\tau_a)_1} \tag{8.4}$$

where G_1 and G_N denote the secant shear modulus in the first and *N*th cycles, respectively, and $(\tau_a)_1$ and $(\tau_a)_N$ denote the amplitude of simple shear stress in the first and *N*th cycle.

Returning to the modulus degradation curves in Fig. 8.11(c), one can readily read off the values of E_N/E_1 and plot them against the number of cycles. The diagram thus constructed is presented in Fig. 8.12 where the strain amplitude ratio γ_a/γ_f is taken as a parameter. The summarized data in Fig. 8.12 indicate that the degradation index δ_D tends to decrease linearly with the number of cycles if plotted in log–log scale. If the data points for each strain amplitude are represented by a straight line, it can be written as

$$\log E_1 - \log E_N = d \log N. \tag{8.5}$$

Then with reference to the definition of δ_D in eqn (8.2), the above relation is rewritten as

$$\delta_D = \frac{E_N}{E_1} = N^{-d} \tag{8.6}$$

where d indicates the slopes of the straight line and is called the *degradation parameter* by Idriss *et al.* (1978). The degradation parameter can be defined alternatively in terms of the shear modulus,

$$\delta_D = \frac{G_N}{G_1} = N^{-d}. \tag{8.7}$$

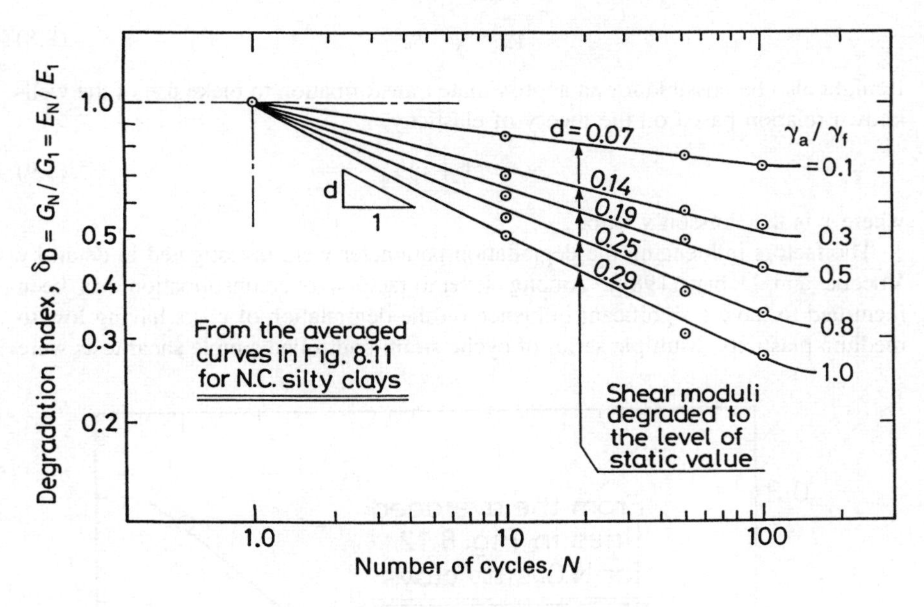

Fig. 8.12 Modulus degradation with cycles.

The value of d can be determined readily by taking slopes of the straight lines in the log–log plot, as indicated in Fig. 8.12. Since the slope becomes steeper with increasing strain amplitude as seen in Fig. 8.12, it is obvious that the degradation parameter tends to increase with an increase in strain amplitude. Then it becomes possible to represent the value of δ_D as a function of shear strain amplitude. The plot in this context is presented in Fig. 8.13 based on the data arranged in Fig. 8.12. In determining the value of strain amplitude γ_a in the abscissa, a failure strain of $\gamma_f = 3\%$ was chosen which appears representative of the data from various silty clays. Figure 8.13 shows a significant increase in the degradation parameter with increasing amplitude of shear strain, but as shown later, the strain dependency of the degradation parameter becomes much less pronounced, if the soil is over-consolidated. It should be remembered that, although not explicitly indicated in Fig. 8.14, there exists a lower limit in strain below which no degradation occurs, because of no effect of dilatancy to be manifested below this strain limit.

The value of the degradation parameter itself is determined uniquely, as defined by eqns (8.6) and (8.7), irrespective of whether it is based on triaxial mode or simple shear mode of deformation. However, the shear strains γ_a and ε_a take different values for these two modes of deformation. Therefore when the degradation parameters from different deformation modes are represented in one diagram, such as that in Fig. 8.13 for comparison's sake, one should be cautious about the transformation rule between γ_a and ε_a. This aspect was discussed by Vucetic and Dobry (1988). By considering a mode-independent parameter used in the theory of plasticity, they derived a relation as

$$\gamma_a = \sqrt{3}\,\varepsilon_a. \tag{8.8}$$

It might also be possible for an approximate transformation to make use of the well-known relation based on the theory of elasticity,

$$\gamma_a = (1 + \nu)\,\varepsilon_a \tag{8.9}$$

where ν is the Poisson's ratio.

The factors influencing the degradation parameter were investigated in detail by Vucetic and Dobry (1988). Among several factors, overconsolidation has been identified to have a significant influence on the degradation of clays having low to medium plasticity. Multiple series of cyclic strain-controlled simple shear tests were

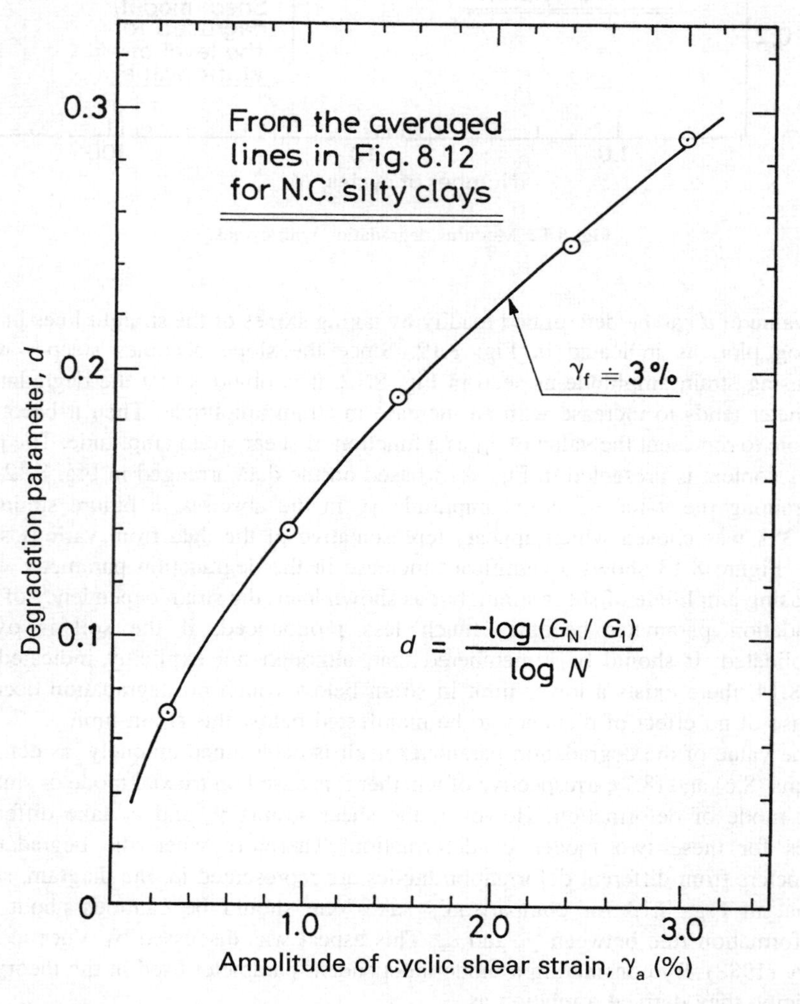

Fig. 8.13 An example of degradation parameters as a function of strain amplitude.

conducted by Vucetic and Dobry (1988) on clay specimens from a marine deposit off the coast of Venezuela (VNP clay). In this test program, undisturbed clay samples were consolidated to a vertical effective stress which was 1.5 to 2.0 times greater than the corresponding vertical effective overburden pressure in situ. Then the specimens were unloaded to a smaller vertical stress to achieve the desired overconsolidation ratio, OCR. This process was in compliance with the SHANSEP procedure proposed by Ladd and Foott (1974). The degradation index δ_D defined by eqn (8.6) was obtained as a function of the number of cycles N for various strain amplitudes γ_a employed in each test. Then the slope of δ_D versus N on the log–log plot was read off to determine the degradation parameter d defined by eqn (8.6). The outcome of the tests is presented in Fig. 8.14 in terms of the degradation parameter plotted versus the strain amplitude γ_a for each of the overconsolidated state with OCR values of 1.0, 2.0, and 4.0. It may be seen that, even with a small increase in the degree of overconsolidation within the range of OCR = 4.0, the stiffness degradation is suppressed significantly by the increase in OCR.

Another important factor has been identified to be the plasticity of clay. A majority of cyclic triaxial and simple shear test data on various clays were collected from several geotechnical laboratories and values of d were established as a function of γ_a by Tan and Vucetic (1989). The outcome of such data compilation for normally consolidated marine clays is presented in Fig. 8.15 in which the number beside the curves indicates the plasticity index of the clays tested. The data from San

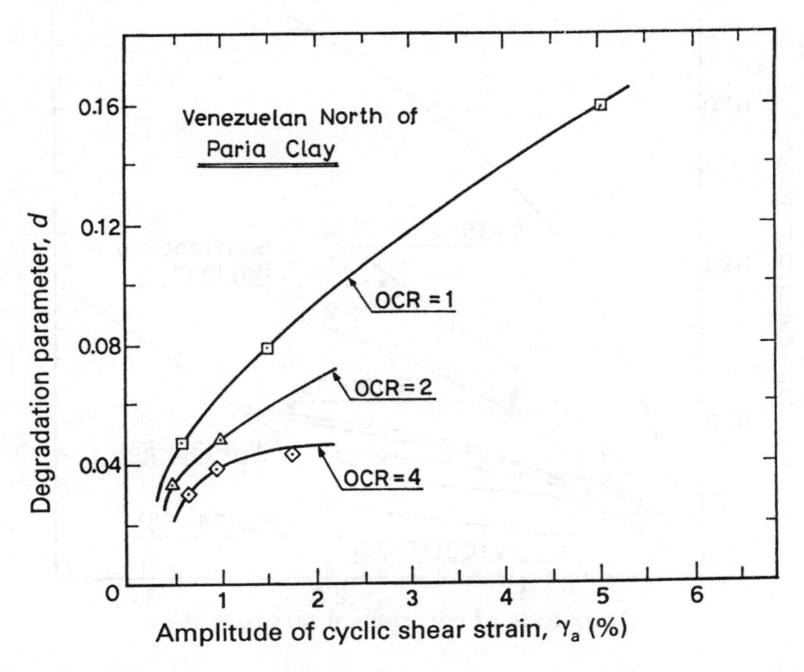

Fig. 8.14 Effects of overconsolidation ratio on the strain-dependent degradation parameter (Vucetic and Dobry, 1988).

Francisco Bay mud obtained from the cyclic triaxial test was converted through the use of eqn (8.8) and presented again in Fig. 8.15. The data on clays from Gulf of Alaska (GAL) were also secured from the cyclic triaxial test. The values of the degradation parameter on several marine clays from Venezuela were obtained from cyclic simple shear tests.

The compilation of test data in Fig. 8.15 indicates that the degradation parameter for clays with low plasticity index is larger than that for clays having higher plasticity index. This means that as clays are more plastic, they become less susceptible to the degradation in the course of cyclic load application.

In an attempt to examine the overall effects of OCR and plasticity index, multiple series of cyclic strain-controlled simple shear tests were conducted by Tan and Vucetic (1989) on six different clays with different plasticity indices and OCR

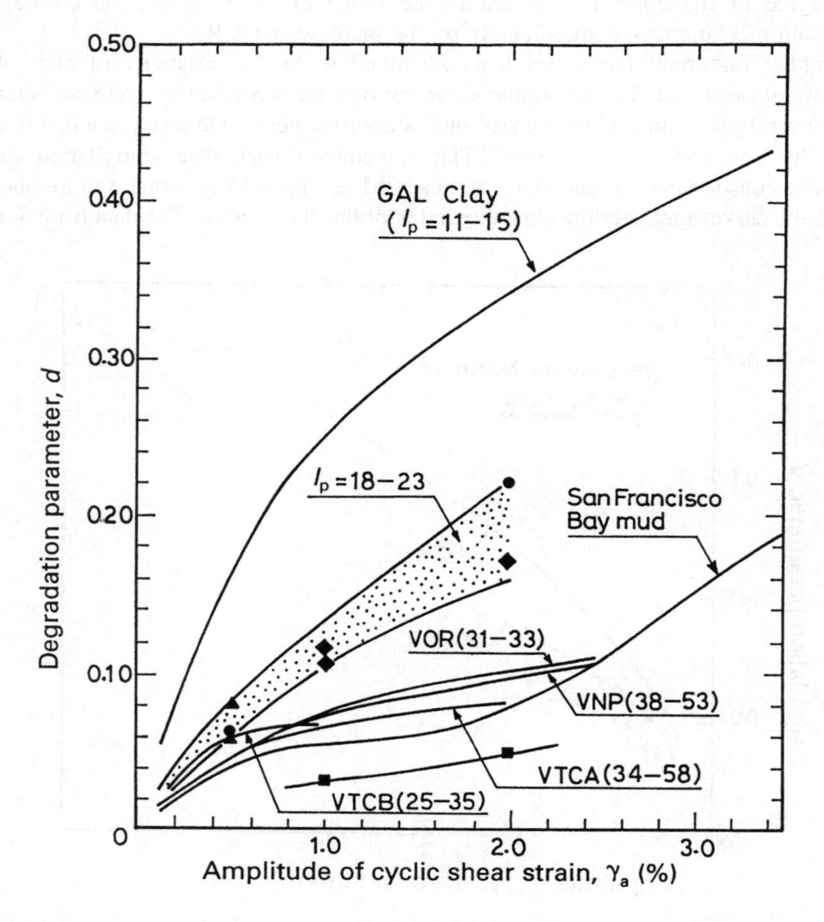

Fig. 8.15 Effects of plasticity index on the strain-dependent degradation parameter of normally consolidated clays (Tan and Vucetic, 1989).

values. The degradation parameter d obtained from these tests is plotted in Fig. 8.16 versus the amplitude of cyclic shear strain γ_a. The numbers next to the points or lines are the corresponding values of plasticity index I_p. If note is taken of the data points for low plasticity clays with $I_p = 11–15$, it is evident that the degradation parameter is notably reduced by increasing the OCR from 1.0 to 2.0. Similar reduction in the degradation parameter is also observed for medium plasticity clays with $I_p = 18–23$ for the change in OCR value ranging from 1 to 4. It can thus be mentioned in summary that the cyclic degradation of stiffness is most pronounced in normally consolidated clays with low plasticity index, and in the clays with increased OCR having higher plasticity index, stiffness degradation becomes less and less significant.

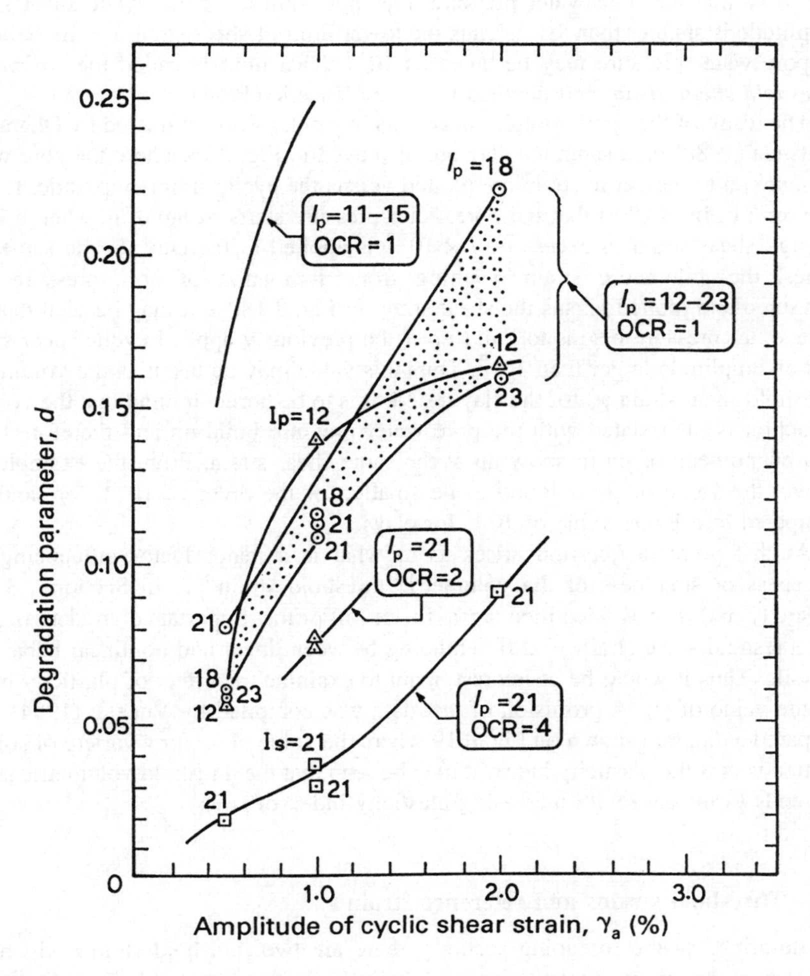

Fig. 8.16 Effects of plasticity index on the strain-dependent degradation parameter of overconsolidated clays (Tan and Vucetic, 1989).

8.5 Threshold strains for cyclic degradation

In the foregoing section, the stiffness degradation has been discussed by tacitly assuming the amplitude of cyclic strain to be of moderate to large magnitude. It is obvious however that there is a lower limit below which cyclic degradation will not occur. Inasmuch as the degradation accrues as a consequence of dilatancy as mentioned above, the lower limit of the cyclic strain amplitude ought to be equal to the threshold between where the dilatancy can and cannot occur in soils. The laboratory tests in this context were performed on saturated sands by Dobry *et al.* (1980). Figure 8.17 shows the outcome summarized by Dobry (1989) where the pore pressures developed during the 10 times application of cyclic loads are plotted versus the amplitude of cyclic shear strain employed in the tests. It is noted in Fig. 8.17 that the pore water pressure does not build up if the cyclic shear strain amplitude is smaller than 10^{-4}. Thus the lower limit of shear strain for the build-up of pore water pressure may be taken as 10^{-4}. This limit is called the *volumetric threshold shear strain* and denoted by γ_v by Vucetic (1994).

The result of the cyclic simple shear tests in similar vein conducted by Ohara and Matsuda (1988) on a saturated clay are displayed in Fig. 8.18 where the pore water pressure and volumetric strain are plotted versus the cyclic strain amplitude. It may be seen in Fig. 8.18(a) that the pore water pressure starts to build up when a large enough shear strain in excess of 7×10^{-3} is applied to the clay. In the same test series, the volumetric strain resulting from dissipation of pore pressure was measured and plotted versus the shear strain in Fig. 8.18(b). It may be seen that the pore water pressure begins to build up if the previously applied cyclic shear strain had an amplitude larger than 10^{-3}. Thus this value may be taken as the volumetric threshold shear strain γ_v for the clay tested. It is to be borne in mind that the volume reduction is interrelated with the pore water pressure build-up and therefore these two phenomena begin to show up at the same shear strain. From the examples as above, the value of γ_v is found to be smaller, of the order of 10^{-4}, for sands as compared to a larger value of 10^{-3} for clays.

At this point, a question arises as to what is a major factor influencing the largeness or smallness of the volumetric threshold strain γ_v. In Section 7.3, the plasticity index was identified as a factor of prime importance in determining the threshold shear strain γ_ℓ differentiating between linear and nonlinear behaviour of soils. Thus it would be of interest again to examine the effect of plasticity index on the value of γ_v. A profusion of test data was compiled by Vucetic (1994) who prepared a diagram shown on Fig. 8.19 where the value of γ_v for a variety of soils is plotted versus the plasticity index. It may be seen that the threshold volumetric strain γ_v tends to increase with increasing plasticity index of soils.

8.6 Threshold strains and reference strain

As described in the foregoing sections, there are two threshold strains which are used as a datum to characterize soil behaviour; the threshold shear strain γ_ℓ differentiating between linearity and nonlinearity and the volumetric threshold shear

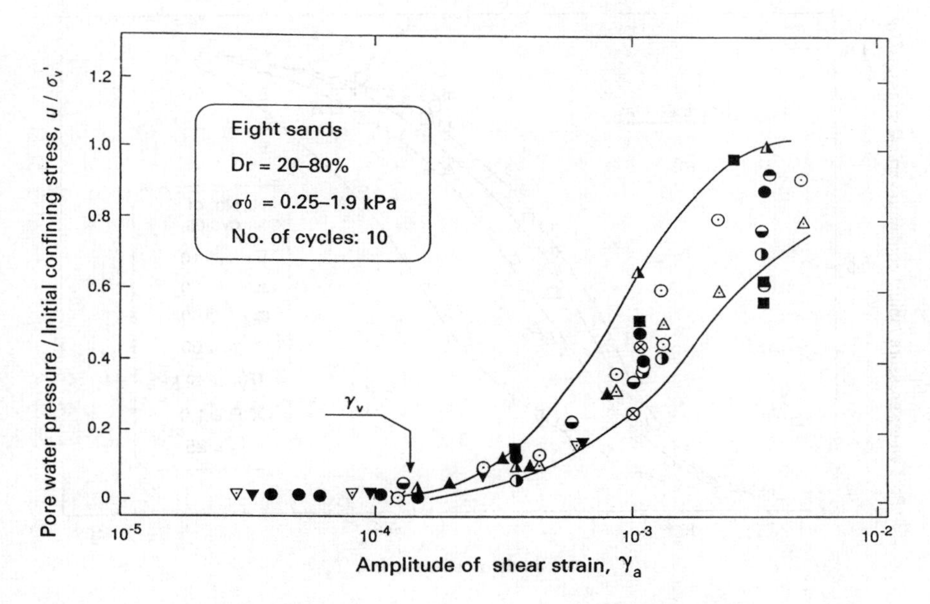

Fig. 8.17 Pore water pressure versus shear strain (Dobry, 1989).

strain γ_v separating condition on whether degradation can or cannot occur. It was also found that these two datum strains have a tendency to increase as the plasticity index of the soil increases. The curves correlating the datum strain and I_p are quoted from Figs 7.22 and 8.19 and shown together in Fig. 8.20. It is of interest to note that these two curves have a shape akin to each other, indicating the similarity of underlying mechanism characterizing the behaviour of soils at these two critical stages of shear strain development. In fact, these characteristic curves are parallel and separated by about 1.5 log cycles (Vucetic 1994).

The influence of plasticity index on the stiffness reduction characteristics was studied by Kokusho *et al.* (1982) who arrange several series of test data as shown in Fig. 8.21 where the amplitude of cyclic shear strain required to reduce the stiffness to varying fractions of its initial value is plotted versus the plasticity index. The characteristic curves shown in Fig. 8.20 were then quoted and laid off in Fig. 8.21 by shifting the abscissa so that the quoted curve can best fit the data points presented by Kokusho *et al.* (1982).

It may be seen that the curve for γ_ℓ and γ_v is well suited as a backbone curve for representing the general trend of variation of test data. Thus, it may be mentioned that the characteristic curve for γ_ℓ and γ_v can be utilized to specify various degrees of modulus reduction as a function of plasticity index. The family of curves established in Fig. 8.21 may also be deemed as those indicating an equal degree of cyclic stiffness reduction or degradation. Thus, it is worth while examining these curves in the light of the threshold curves for γ_ℓ and γ_v. In Fig. 8.22, the family of curves from Fig. 8.21 is displayed together with the threshold curves. If the family of

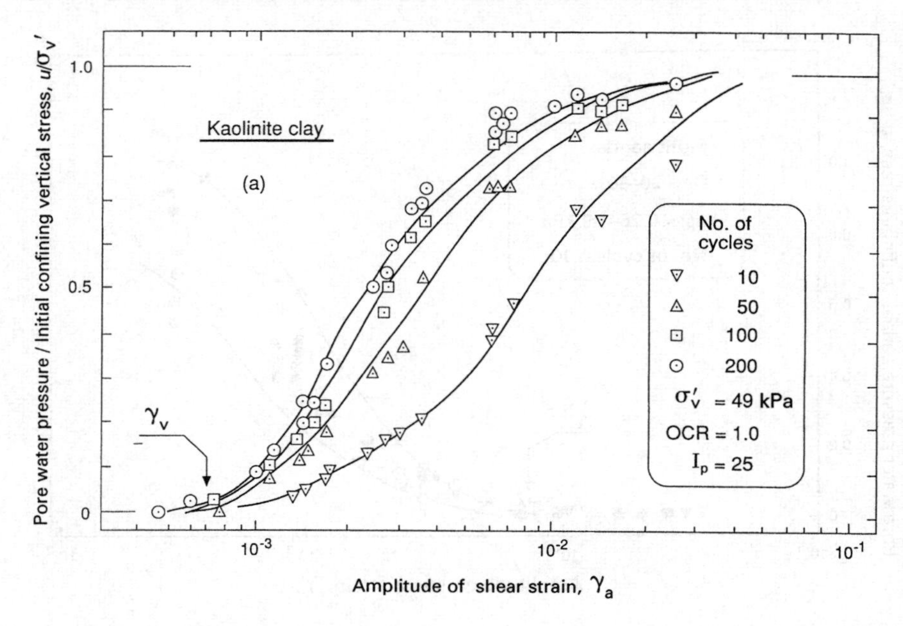

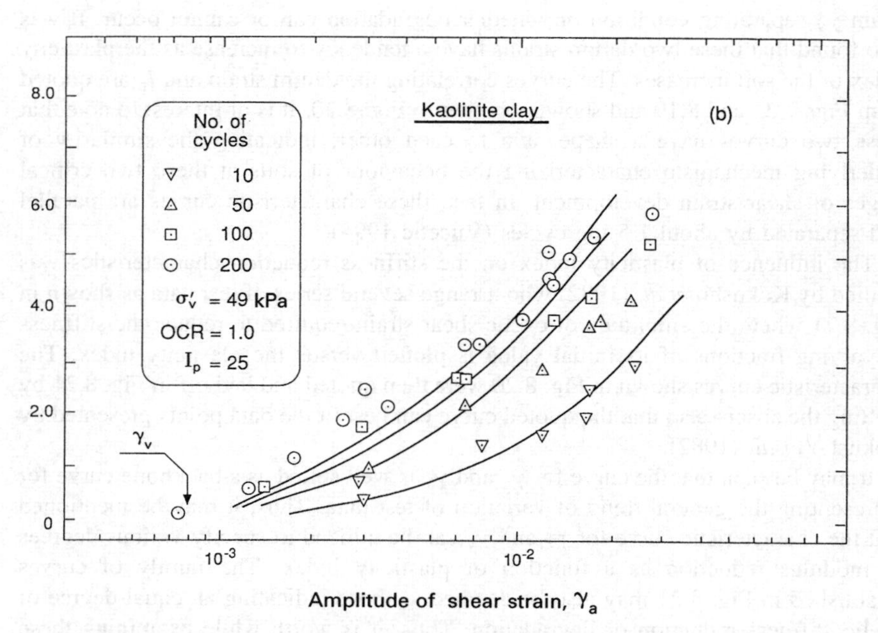

Fig. 8.18 Pore water pressure during undrained shear, and volumetric strain during the following reconsolidation plotted versus the amplitude of shear strain (Ohara and Matsuda, 1988).

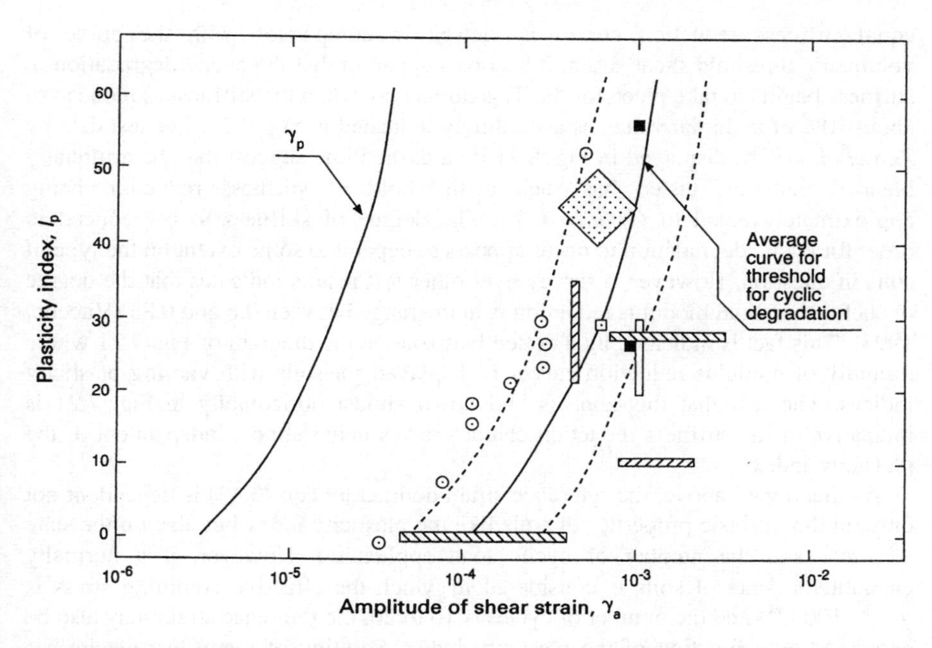

Fig. 8.19 Effects of plasticity index on the threshold strain inducing cyclic degradation (Vucetic, 1994).

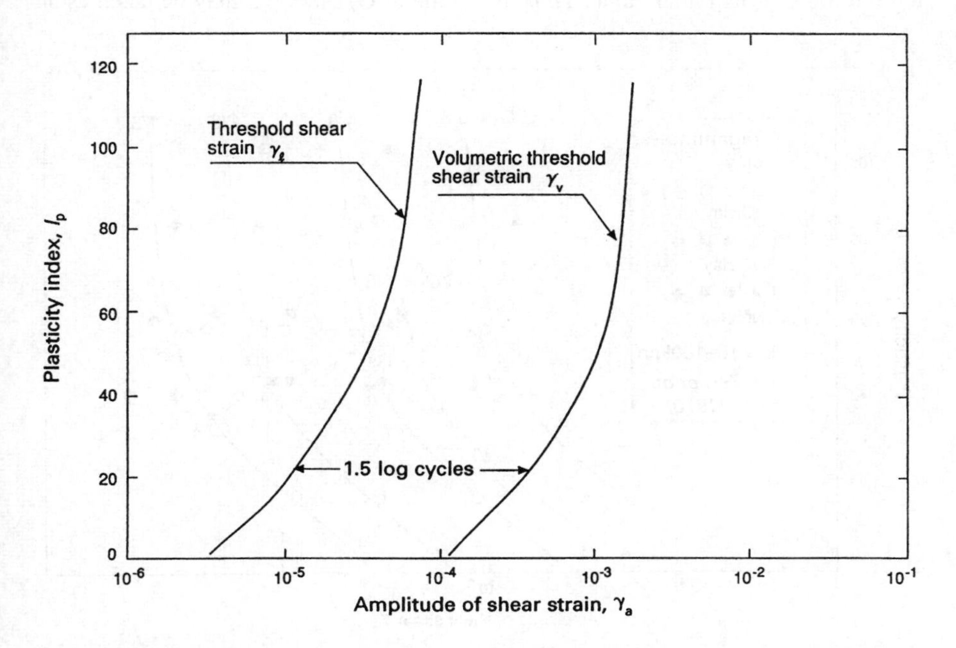

Fig. 8.20 Effects of plasticity index on the two threshold strains associated with nonlinearity and cyclic degradation.

equal stiffness reduction curves is viewed in comparison with the curve of volumetric threshold shear strain, it becomes apparent that the cyclic degradation in stiffness begins to take place for the Teganuma clay when the stiffness is reduced to about 70% of its initial value, as accordingly indicated in Fig. 8.22. The test data by Zen *et al.* (1978) displayed in Fig. 8.21 by a dashed line suggest that the artificially blended materials tested has such a threshold of stiffness reduction being approximately equal to $G/G_0 = 0.75$. The degree of stiffness to be reduced in order for cyclic degradation to occur appears to depend to some extent on the type of soils in question. However, a survey over other test results indicates that the degree of such a threshold modulus reduction is in the range between 0.6 and 0.85 (Vucetic, 1994). This fact is indicated by a dotted belt zone in the diagram of Fig. 7.21 where a family of modulus reduction curves is displayed for soils with varying plasticity indices. The fact that this zone is laid down almost horizontally in Fig. 7.21 is indicative of the stiffness reduction characteristics being almost independent of the plasticity index.

As mentioned above, the reference strain defined by eqn (3.43) is dependent not only on the intrinsic properties of soils like the plasticity index but also on the state of stress and the number of cyclic load application. However, if a normally encountered state of soils is considered in which the effective confining stress is $\sigma_0{}' \approx 100$ kPa and the number of cycles is 10 to 20, the reference strain may also be expressed as a function of the plasticity index. Scrutiny of many test results has indicated that the reference strain is manifested when the shear modulus is reduced to 40 to 60% of its initial value. Thus, the value of $G/G_0 = 0.5$ may be taken as an

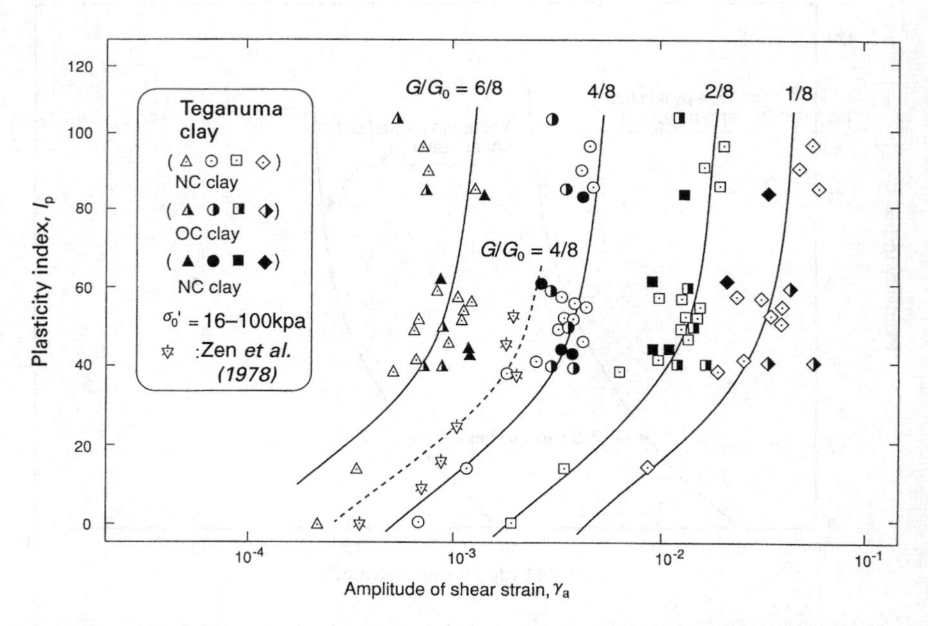

Fig. 8.21 Effects of plasticity index on the strain-dependent stiffness degradation (Kokusho *et al.* 1982).

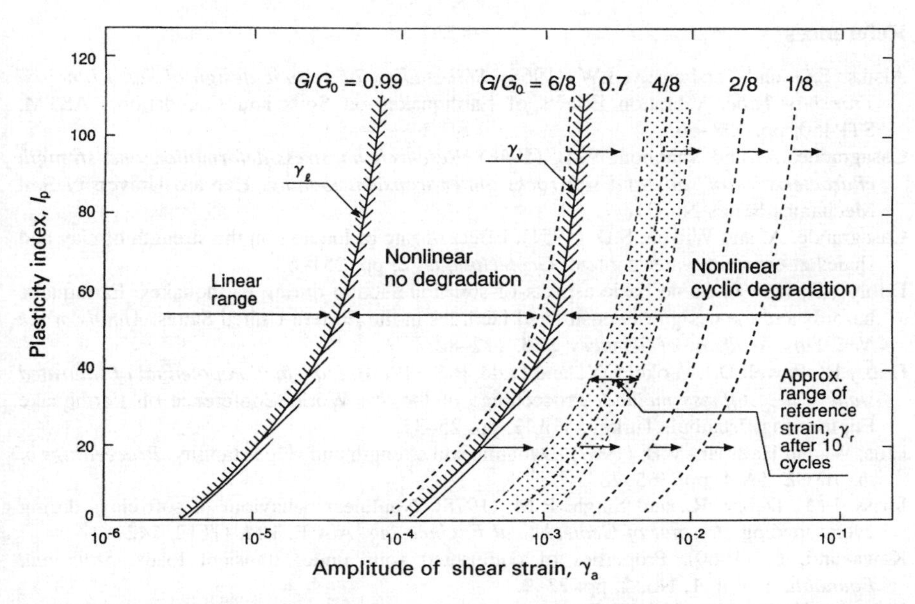

Fig. 8.22 Summary chart showing the influence of plasticity index on the strain-dependent stiffness degradation.

average required for the modulus to be decreased to induce the reference strain. The range of the reference strain as above together with its average is also shown in Fig. 8.22 as a function of the plasticity index. It is to be noticed that the range of reference strain in the normally encountered conditions as above lies, in the diagram of Fig. 8.22, in the domain to the right of the γ_v curve where cyclic degradation always takes place. In other words, where soils are subjected to cyclic loads with a strain amplitude equal to the reference strain, the soils are already deforming cyclically, accompanied by gradual changes in their properties such as accumulation of volumetric strains or build-up of pore water pressures.

References

Aisiks, E.G. and Tarshansky, I.W. (1968). *Soil studies for seismic design of San Francisco Transbay Tube*. Vibration Effects of Earthquakes on Soils and Foundations, ASTM, STP450, pp. 138–66.

Casagrande, A. and Shannon, W.L. (1948). *Research on stress-deformation and strength characteristics of soils and soft rocks under transient loading*. Harvard University Soil Mechanics Series No. 31.

Casagrande, A. and Wilson, S.D. (1951). Effect of rate of loading on the strength of clay and shales at constant water content. *Geotechnique*, **2**, pp. 251–63.

Dobry , R. (1989). Some basic aspects of soil liquefaction during earthquakes. Earthquake hazards and the design of constructed facilities in the Eastern United States. *Annals of the New York Academy of Sciences*, **558**, 172–82.

Dobry R., Powell D.J., Yokel F.Y., and Ladd, R.S. (1980). *Liquefaction potential of saturated sand – the stiffness method*. Proceedings of the 7th World Conference on Earthquake Engineering, Istanbul, Turkey, Vol. 3, pp. 25–32.

Ellis, W. and Hartman, V.B. (1967). Dynamic soil strength and slope stability. *Proceedings of the ASCE*, SM4, pp. 355–73.

Idriss, I.M., Dobry, R., and Singh, R.M. (1978). Nonlinear behaviour of soft clays during cyclic loading. *Journal of Geotechnical Engineering*, ASCE, **104**, GT12, 1427–47.

Kawakami, F. (1960). Properties of compacted soils under transient loads. *Soils and Foundation*, Vol. 1, No. 2, pp. 23–9.

Kokusho, T., Yoshida, Y., and Esashi, Y. (1982). Dynamic properties of soft clays for wide strain range. *Soils and Foundations*, **22**, 1–18.

Kovacs, W.D., Seed, H.B., and Chan, C.K. (1971). Dynamic moduli and damping ratios for a soft clay. *Journal of Soil Mechanics and Foundations*, ASCE, **97**, SM1, 59–75.

Ladd, C.C. and Foott, R. (1974). New design procedure for stability of soft clays. *Journal of Geotechnical Engineering*, ASCE, **100**, No. GT7, 763–86.

Ohara, S. and Matsuda, H. (1988). Study on the settlement of saturated clay layer induced by cyclic shear. *Soils and Foundations*, **28**, 103–113.

Ohsaki, Y. (1964). *Dynamic properties of soils and their application. Japanese Society of Soil Mechanics and Foundation Engineering*, 29–56.

Ohsaki, Y., Koizumi, Y., and Kishida, H. (1957). Dynamic properties of soils. *Transactions of the Architectural Institute of Japan*, **54**, 357–9.

Olson, R.E. and Parola, J.F. (1967). *Dynamic shearing properties of compacted clay*. Proceedings of the International Symposium on Wave Propagation and Dynamic Properties of Earth Materials, University of New Mexico, 173–81.

Richardson, A.M. and Whitman, R.V. (1963). Effect of strain rate upon undrained shear resistance of a saturated remoulded fat clay. *Geotechnique*, 310–24.

Shimming, B.B., Hass, H.J., and Sax, H.C. (1966). Study of dynamic and static failure envelops. *Proceedings of the ASCE*, SA2, 105–23.

Takenaka, J. and Nishigaki, Y. (1970). *Dynamic behaviour of diluvial soil deposits in Ohsaka*. Proceedings of the 5th Annual Convention of Japanese Society of Soil Mechanics and Foundation Engineering, pp. 89–91 (in Japanese).

Tan, K. and Vucetic, M. (1989). *Behaviour of medium and low plasticity clays under simple shear conditions*. Proceedings of the 4th International Conference on Soil Dynamics and Earthquake Engineering, Mexico City, pp. 131–41.

Taylor, P.W. and Bacchus, D.R. (1969). *Dynamic cyclic strain tests on a clay*. Proceedings of the 7th International Conference on Soil Mechanics and Foundation Engineering, Vol. 1, pp. 401–9.

Tiers, G.R. and Seed, H.B. (1968). Strength and stress–strain characteristics of clays subjected to seismic loading conditions. *Vibration Effects of Earthquakes on Soils and Foundations*, ASTM, STP450, 3–56.

Vucetic, M. and Dobry, R. (1988). Degradation of marine clays under cyclic loading. *Journal of Geotechnical Engineering*, ASCE, **114**, GT2, 133–49.

Vucetic, M. (1994). Cyclic threshold shear strains in soils. *Journal of Geotechnical Engineering*, ASCE, **120**, GT12, 2208–27.

Whitman, R.V. (1957). *The behaviour of soils under transient loading*. Proceedings of the 4th International Conference on Soil Mechanics and Foundation Engineering, Vol. 1, pp. 207–10.

Zen, K., Umehara, Y., and Hamada, K. (1978). *Laboratory tests and in situ seismic survey on vibratory shear modulus of clayey soils with various plasticities*. Proceedings of the 5th Japanese Earthquake Engineering Symposium, pp. 721–8.

9

STRENGTHS OF COHESIVE SOILS UNDER TRANSIENT AND CYCLIC LOADING CONDITIONS

9.1 Load patterns in dynamic loading tests

Several types of dynamic loading tests have been attempted thus far to determine the dynamic strength of soils. These may be classified into four types, as illustrated in Fig. 9.1, according to whether the loading is rapid or slow and also whether the loading is monotonic or cyclic. Monotonic loading tests can be conducted with varying speeds of loading. The conventional static loading tests employ a rate of loading to failure on the order of a few minutes. The monotonic loading tests conducted in less than a few seconds to failure are classified as rapid loading tests. The rapid loading test or transient test has been performed to evaluate the strength of soils exhibited under blast loading such as detonation of explosives or falling of bombs. The second type of dynamic loading consists of a cyclic load application that is executed following the static monotonic loading as illustrated in Fig. 9.1(b). This type of test has often been conducted to evaluate the strength of soils during earthquakes. The initial phase of the static monotonic shear stress application is envisaged as representing a sustained static pre-earthquake state of stress which exists in a soil element beneath sloping surfaces. After applying the sustained shear stress, soils samples are subjected to a sequence of cyclic stress until failure occurs. The phase of loading is considered as simulating the cyclic shear stress application during earthquakes. The third type of loading tests is performed, as illustrated in Fig. 9.1(c), primarily to investigate the effects on strength and stiffness deterioration of soils due to seismic shaking. Soil samples are softened and weakened at the end of a certain number of cycles, so that the static strength and deformation properties are changed from the initial values. These soil properties are necessary in making post-stability analysis of dams and embankments. The fourth type of loading test illustrated in Fig. 9.1(d) is sometimes used, although not common, to study the static strength of soil while it is being subjected to vibration. Static resistance of soils in the ground in close proximity to piles or sheet piles may be reduced to some extent due to the vibration caused by pile driving. The strength properties in such cases can be obtained in the laboratory tests in which a monotonically increasing shear stress is applied to the soil sample placed on a shaking table.

9.2 Definition of dynamic strength of soil

Among the four types of dynamic loading tests illustrated in Fig. 9.1, the second type of test has been commonly used to determine the strength of soils under seismic

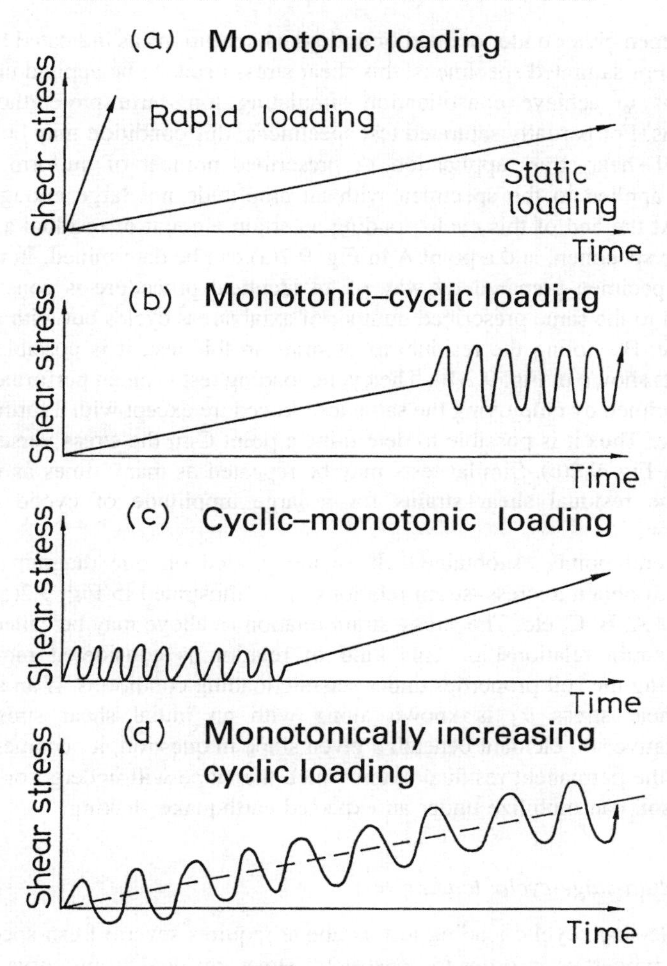

Fig. 9.1 Types of dynamic loading tests.

loading conditions. As such, detailed consideration will be given in the following pages to the definition of dynamic strength for tests using a cyclic triaxial test apparatus. The loading scheme may be divided into two classes depending upon whether the test employs a single specimen or multiple specimens to determine the strength characteristics of the soil.

9.2.1 *Single-stage cyclic loading test*

The loading scheme employed in this test is schematically illustrated in Fig. 9.2 in terms of shear stress versus shear strain plots. A specimen is first consolidated under an appropriate isotropic pressure σ'_0. An axial stress σ_s is then applied statically to the specimen either under drained or undrained conditions as indicated by point P in Fig. 9.2. For saturated specimens, this shear stress needs to be applied under drained conditions to achieve consolidation simulating long-term pre-earthquake stress conditions. For partially saturated test specimens, this condition may be replaced by undrained shear stress application. A prescribed number of uniform cyclic axial stress is applied to the specimen with an amplitude not large enough to induce failure. At the end of this cyclic loading a certain amount of residual axial strain is left in the specimen, and a point A in Fig. 9.2(a) can be determined. In the next test, a fresh specimen prepared by way of an identical procedure is consolidated and subjected to the same prescribed number of axial stress cycles but with an increased amplitude. By noting the residual axial strain in this test, it is possible to locate a point B as shown in Fig. 9.2(b). The cyclic loading test is again performed on a third fresh specimen by employing the same test procedure except with a further increased amplitude. Thus it is possible to determine a point C in the stress versus strain plot shown in Fig. 9.2(c). Similar tests may be repeated as many times as necessary to obtain the residual shear strains for a large amplitude of cyclic shear stress application.

If several points as obtained above are plotted on one diagram, it becomes possible to obtain a stress–strain relationship as illustrated in Fig. 9.2(d) by linking the points A, B, C, etc. The stress–strain relation as above may be called the stress–residual strain relationship. This kind of relation is considered most useful in representing the soil properties under seismic loading conditions. If an amplitude of cyclic shear stress σ_d is known along with an initial shear stress σ_s for a representative soil element beneath a given slope in question, it becomes possible to estimate the permanent residual strain which the slope will undergo or the strength that the soil can mobilize under an expected earthquake shaking.

9.2.2 *Multi-stage cyclic loading test*

The single-stage cyclic loading test as above requires several fresh specimens with identical properties in order to construct a stress–residual strain curve. In the case where the number of available test specimens is limited, the multi-stage cyclic loading test may be use. The loading scheme in this type of test is illustrated in Fig. 9.3. A specimen is consolidated first and subjected to the initial shear stress σ_s in the same manner as in the single-stage cyclic loading test. A sequence of prescribed shear stress cycles is then applied to the test specimen with a relatively small amplitude. In the course of this cyclic loading the specimen deforms to a shear strain as designated by point A in Fig. 9.3. The amplitude of the cyclic load is then increased and the load is cycled by the same number. The specimen is deformed to a strain as indicated by point B in Fig. 9.3. Similarly, some additional sequences of uniform cyclic axial stresses with the same number of cycles but with increased amplitude are further superimposed on the specimen under undrained conditions. When the points reached at the end of each sequences are connected in the plot of the stress–strain diagram, it is possible to obtain a curve such as the dashed curve shown on Fig. 9.3. This curve may be considered as representing the stress–residual

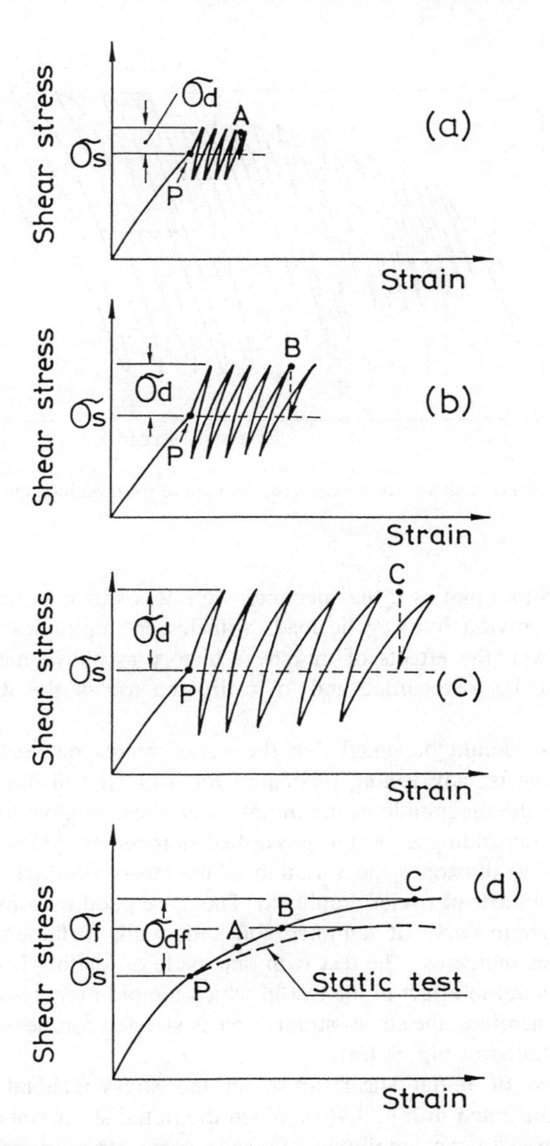

Fig. 9.2 Construction of a shear stress versus residual strain curve.

strain relationship similar to that obtained in the single-stage cyclic loading test.

In the above loading scheme, cyclic loads with amplitudes increasing step-by-step are applied to a single specimen in sequence. Therefore it is likely that the response of the specimen to a certain sequence of loading cycles is affected by the other preceding sequences with lesser amplitudes. Since the preceding sequences act to increase residual strains in the test specimen, their effects would appear in the

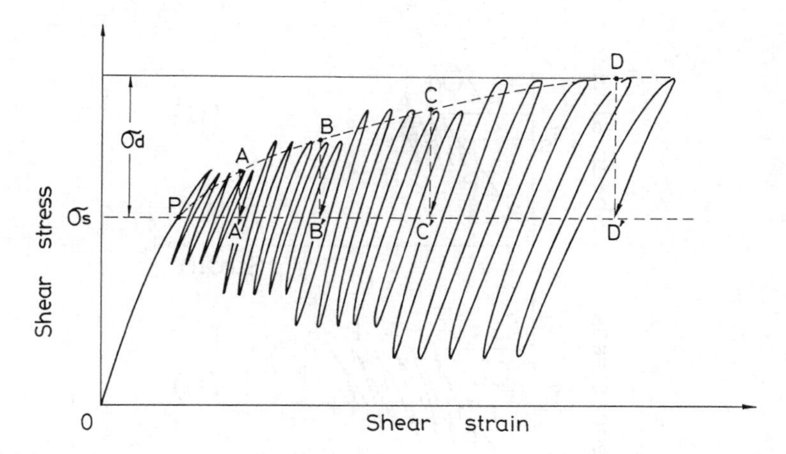

Fig. 9.3 Construction of a shear stress versus residual curve from multi-stage loading test results.

stress–residual strain plot as if the specimen were less stiff than that which had not been subjected previously to cyclic loads with lesser amplitudes. For all practical purposes, however, the effects of the stress history as above may be considered small enough to be disregarded and to justify the use of the multi-stage cyclic loading test.

In passing, it should be noted that the stress versus residual strain curve as constructed above by any testing procedure for a given soil has different shapes depending upon the magnitude of the initial shear stress relative to the cyclic shear stress, and also depending upon the prescribed number of cycles employed in the tests. Figure 9.4(a) illustrates the variation of the stress–residual strain curve with changes in the number of cycles employed. There are good reasons to infer that the stress–residual strain curve of saturated soft soils tends to flatten as the specified number of cycles increases. The test with one cycle of loading is considered to be approximately equal in effect to the test in which a rapid monotonic load is applied to the sample. Therefore, the stress–strain curve is steepest for the test with one cycle loading as illustrated in Fig. 9.4(a).

The influence of initial shear stress on the stress–residual strain curve is schematically illustrated in Fig. 9.4(b). When the initial shear stress σ_s is relatively large as compared to the amplitude of cyclic shear stress σ_d, the influences of loading speed and load repetition are feeble and the shape of the stress–residual strain curve becomes similar to the stress–strain relation obtained in the static test.

9.2.3 *An example of the stress–residual strain relationship*

An example of the stress–residual strain relation is demonstrated in Fig. 9.5. The undisturbed test specimens used in the laboratory cyclic triaxial tests were obtained in block from a place near the scarp of a landslide that occurred in Shiroishi, south of

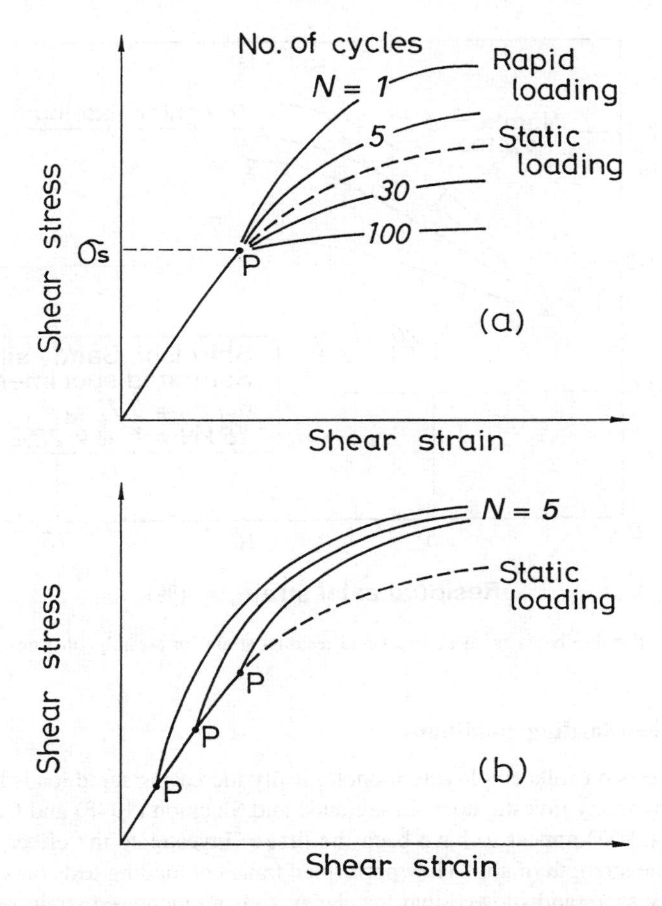

Fig. 9.4 Effects of the number of load cycles and the initial shear stress on the shear stress versus residual strain curve.

Sendai, Japan, at the time of the Miyagiken-oki earthquake of June 12, 1978. The single-stage cyclic loading was employed. The test specimens were saturated and consolidated under a confining pressure of $\sigma'_0 = 50$ kPa. A static axial stress of $\sigma_s = 144$ kPa was imposed on each test specimen under drained conditions, and then cyclic loads with varying amplitudes were applied to the specimens. The results of the tests are summarized in Fig. 9.5. in which the cyclic component plus the static component of the axial stress normalized to the undrained static axial stress at failure is plotted versus the residual axial strain. It should be noted here that the total axial stress $\sigma_s + \sigma_d$ required to cause a certain magnitude of residual strain is greater for the loading with a smaller number of cycles than for the loading involving a larger number of cycles. This is due to the stiffness degradation caused by the load repetition. In this particular silty soil, the dynamic strength is seen increasing to a value which is 1.5 times the corresponding static strength.

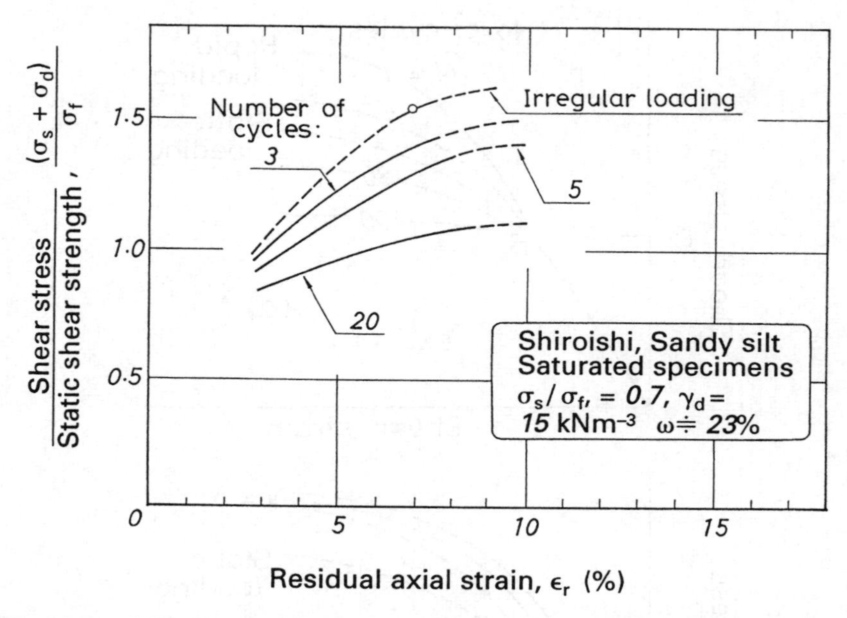

Fig. 9.5 Relationship between shear stress and residual strain for a sandy silt from Shiroishi.

9.3 Transient loading conditions

Laboratory tests on soils employing monotonically increasing rapid loads have been carried out by many investigators. Casagrande and Shannon (1948) and Casagrande and Wilson (1951) appear to have been the first to investigate the effect of rate of loading on the strength of soil. They performed transient loading tests on samples of clays and dry sand, and showed that for clayey soils an increased strain rate from 1 to 8000% per minute caused the strength to increase by about 50% and the modulus of deformation to increase by almost 100%. Whitman (1957) subsequently performed transient loading tests on cohesive soils and found similar effect of loading rate both on the strength and stiffness of soils. Similar strain rate effects for other cohesive soils were noted by several other investigators such as Ohsaki *et al.* (1957), Kawakami (1960), Schimming *et al.* (1966) and Olson and Parola (1967). The results of rapid and static loading tests on undisturbed cohesive soils in Japan are presented in Fig. 9.6 (Ohsaki *et al.*, 1957). Slow loading tests were performed with a time of about 100 s to failure whereas a time to failure as short as 0.1 s was employed in the rapid loading test. The rapidity of loading may be considered as a cause of the increase in strength by about 15% as compared to the strength in the static test. Ohsaki (1964) summarized a comprehensive amount of data on the strength of cohesive soils obtained in the rapid loading tests as shown in Fig. 9.7, where the ratio of the strengths obtained in the rapid loading test and the slow loading test is plotted versus the time of loading to failure. In Fig. 9.7, several other data by Schimming *et al.* (1966) and Olson and Parola (1967) are added to the

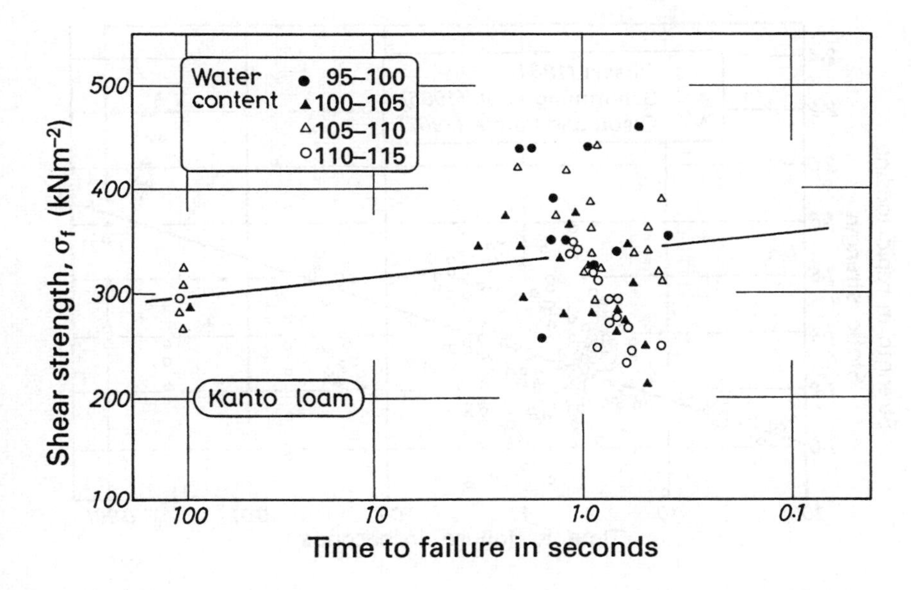

Fig. 9.6 Effects of time of loading on the strength of soils (Ohsaki *et al.*, 1957).

Ohsaki's original compilation. Although there is scattering of data points probably due to other factors influencing the strength, a general trend towards increasing strength in rapid loading may be visualized, as opposed to slow loading. An average straight line drawn in the figure appears to roughly indicate that the strength of cohesive soils obtained at a loading time of 0.25 s is about 40% greater than the strength in the static test with a time of loading of 100 s.

9.4 Combined static and cyclic loading

It was shown in the foregoing section that the strength of soil under combined static and cyclic loading conditions can be obtained from the stress–residual strain curve as shown in Figs 9.2 and 9.3. Many laboratory test results have been reported on the cyclic strengths of cohesive soils. Although they are not reported consistently in the form of the stress–residual strain relationship, it is generally possible to visualize the test schemes within the same framework as set forth in the foregoing section and to summarize the results accordingly. The outcome of the triaxial cyclic loading tests by Seed (1960), Seed and Chan (1966), and Ellis and Hartman (1967) is presented in Fig. 9.9. The tests were performed on compacted partially saturated cohesive soils. In this type of plot the cyclic strength defined as the static plus cyclic axial stress causing failure in test specimens divided by the corresponding static strength is plotted versus the initial static axial stress divided by the static strength. The cyclic strength normalized to the static as above will be referred to as the *cyclic strength ratio* for brevity.

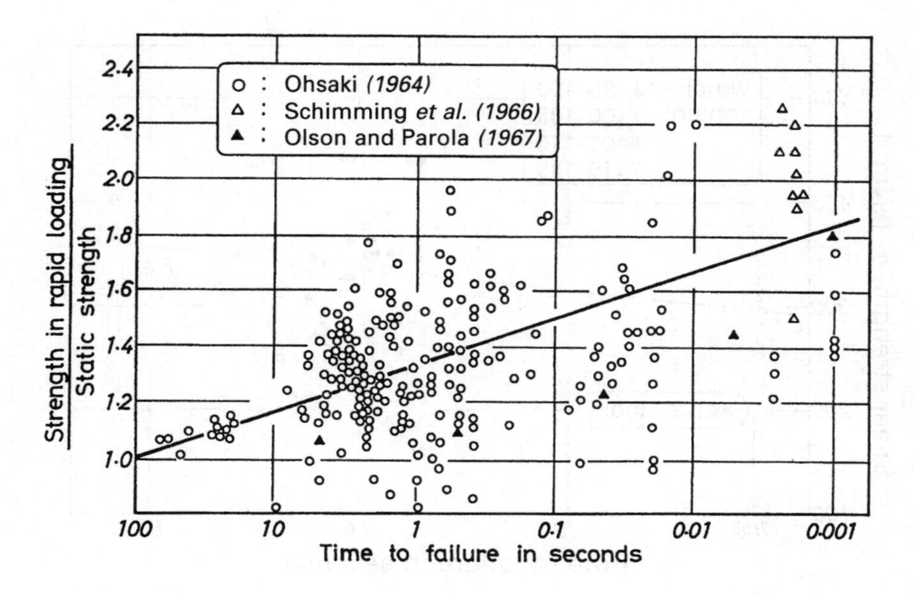

Fig. 9.7 Effects of time of loading on the strength of cohesive soils.

It should be kept in mind that when the initial shear stress does not exist, the cyclic loading completely reverses its directions between the triaxial compression and the triaxial extension. However, as the initial shear stress increases, the degree to which the load is reversed during each cycle becomes partial and incomplete as illustrated in Fig. 9.8. When the initial shear stress is increased to a value corresponding to point B in Fig. 9.9, the static component becomes equal to the cyclic component of shear stress, and no reversal of the deviator shear stress occurs during each cycle of the loading. For further increase in the initial shear stress, the load pattern becomes a one-way loading, and involves no reversal of the shear stress as illustrated in Fig. 9.8(a). Even when the initial shear stress is less than the value corresponding to point B in Fig. 9.9, it is possible to conduct the one-way loading test by adjusting the apparatus in such a way that no state of triaxial extension whatsoever can occur within a test specimen as shown in Fig. 9.8(c). The test results shown in Fig. 9.9 are those obtained for just one cycle of loading executed with 1 Hz frequency. Therefore the time of loading calculated as one quarter of the period of cycle is 0.25 s. It may be noted in Fig. 9.9 that in the absence of the initial sustained stress ($\sigma_s/\sigma_f = 0$), the cyclic strength ratio was approximately 1.4 for the loading involving only one cycle. It should be noted that without the initial shear stress, one cycle of loading is a complete reversal of shear stress between the triaxial compression and extension. Since the failure was produced always on the triaxial compression side, the cyclic strength in this type of loading was determined as the maximum axial stress causing failure on the triaxial compression side. This implies that the one-cycle load causing failure executed with 1 Hz frequency was practically

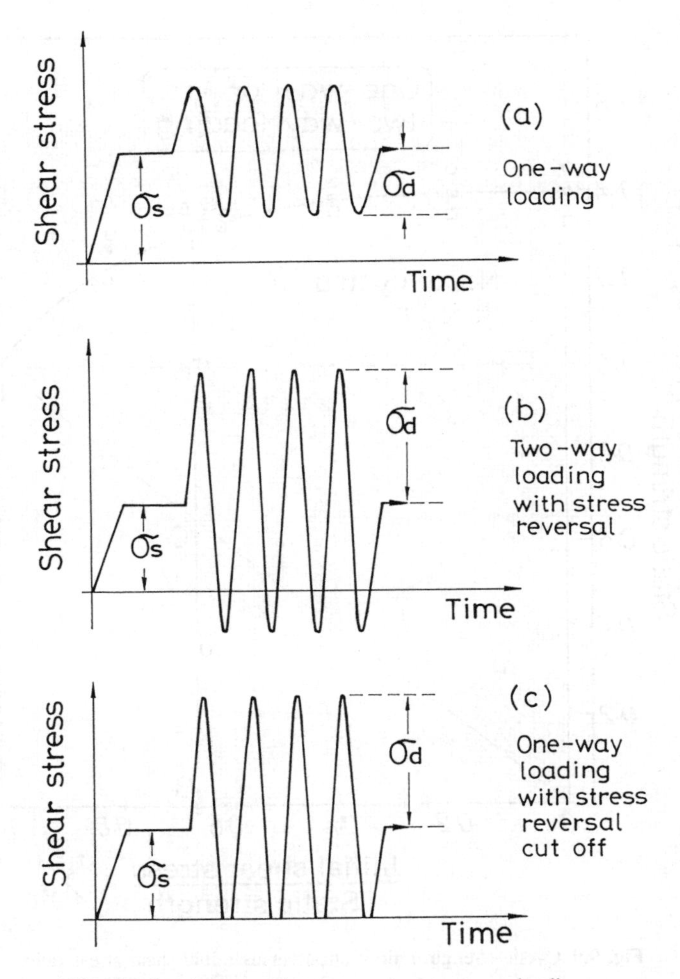

Fig. 9.8 Load patterns in one-way and two-way loadings.

equivalent to a monotonic transient load applied with a loading time of 0.25 s. With this fact in mind it is of interest to note that the 40% increase in cyclic strength for one-cycle of loading as demonstrated in Fig. 9.9 is coincident with the equal percentage of increase in strength in the transient loading test executed with a corresponding time of loading of 0.25 s, as demonstrated in Fig. 9.7.

Turning attention back to the plot in Fig. 9.9, one can see that the cyclic strength ratio decreases gradually to a value of 1.0 as the initial sustained shear stress increases to the static strength value of the soil in question. This could be the case because the increase in initial shear stress causes the decrease in the relative magnitude of cyclic component, and therefore the state of stress becomes more and more analogous to that of the static test. The results of one-way loading tests in the

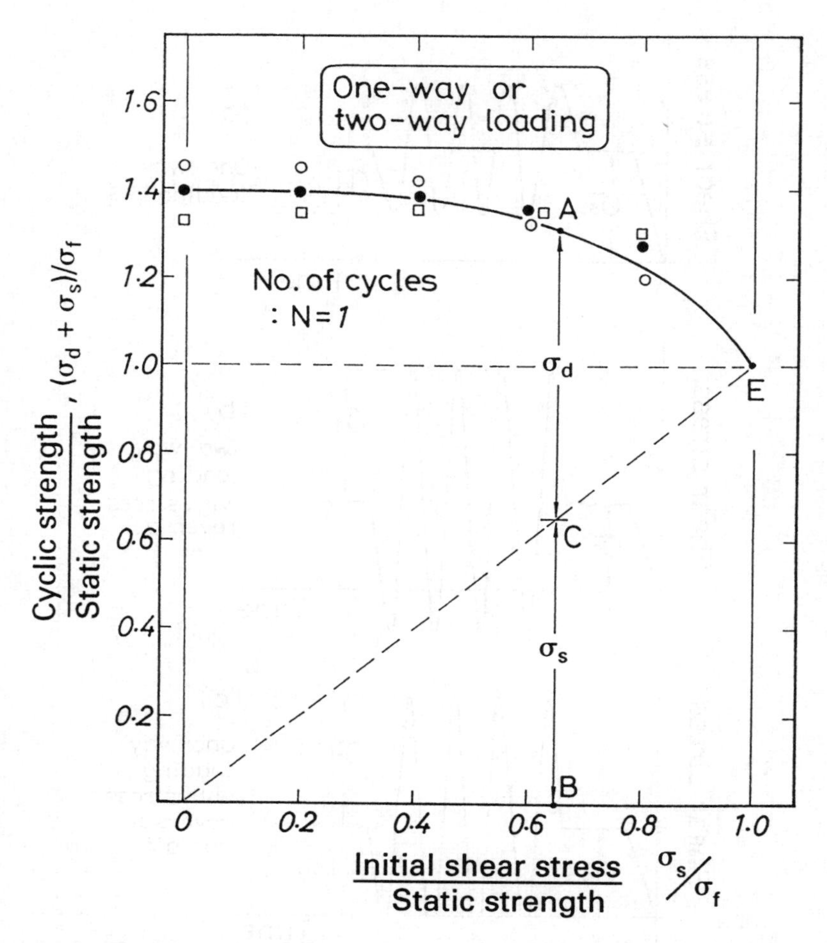

Fig. 9.9 Cyclic strength ratio plotted versus initial shear stress ratio.

same series as shown in Fig. 9.9 employing 50 load cycles are summarized in Fig. 9.10. All the test results by the same investigators including those shown in Figs 9.9 and 9.10 are summarized in Fig. 9.11 with the number of cycles taken as a parameter. It may be seen in Fig. 9.11 that in the cyclic one-way loading, the cyclic strength gradually decreases as the specified number of load cycles increases and it eventually becomes equal to the static strength when the number of cycles becomes close to about 100. The fact that the cyclic strength is largest at one cycle of loading comes from the corresponding soil characteristics tending to show a higher strength in the rapid loading test. The higher cyclic strength due to the effect of loading rate tends to fade out as the specified number of cycles increases because of the effect of load repetition bringing about strength deterioration within the test specimen. It is to be emphasized that because the cyclic loading is envisioned as the repetition of rapid

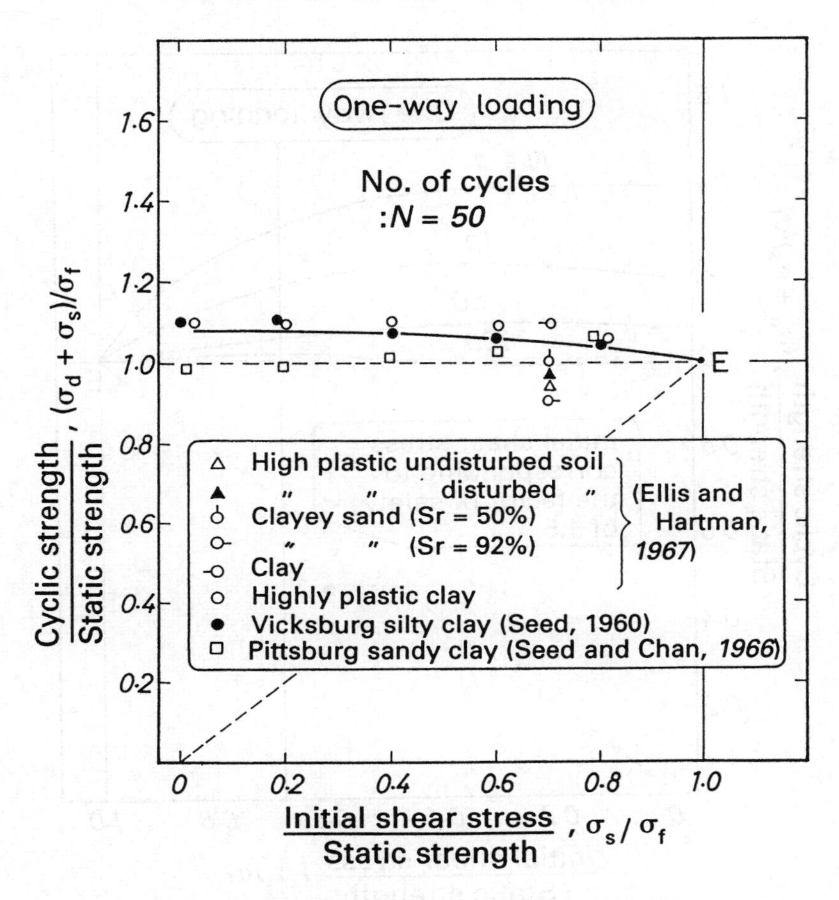

Fig. 9.10 Cyclic strength ratio plotted versus initial shear stress ratio.

loads, the tendency to increase the strength due to the rapidity of load application is manifested concurrently with the tendency to decrease the strength due to the cyclic nature of the load application. The smaller the number of cycles specified, the more predominantly manifested is the effect of increasing strength due to the rapidity of load application. The offsetting detrimental effect due to the cyclic nature of the load application becomes predominant as the specified number of load cycles is increased, bringing the specimen's strength to a level of static strength or even smaller.

When the stress reversal is allowed to occur on every cycle producing a totally or partially two-way loading condition in the test specimen as illustrated in Fig. 9.8, the cyclic strength shows entirely different features. A comparison of the results between the one-way and the two-way tests are demonstrated in Fig. 9.12 for the case of 10 load cycles. A summary of the two-way cyclic loading tests is presented in Fig. 9.13 with dashed lines together with the summary curve of the one-way

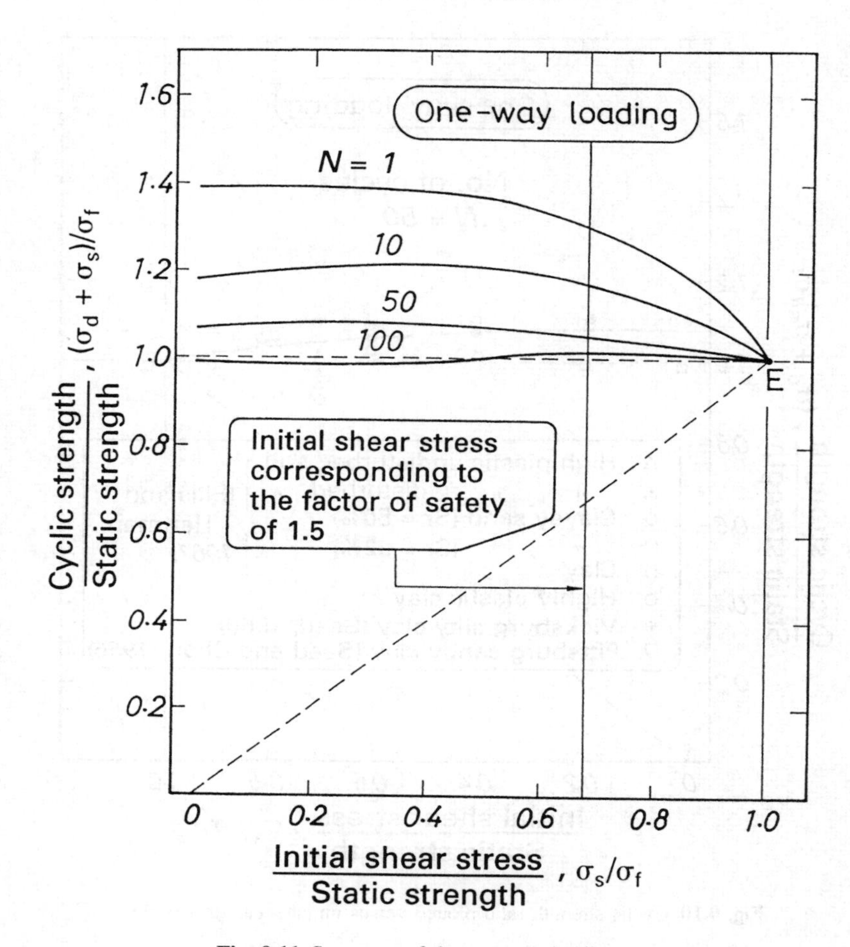

Fig. 9.11 Summary of the one-way loading tests.

loading tests. It is apparent from Fig. 9.13 that the decrease in cyclic strength occurs more appreciably in the two-way loading than in the one-way loading conditions. Thus, it may be argued that the detrimental effect due to cyclic loading is more severe for the cyclic loading in two directions than for the cyclic loading in only one direction.

9.5 Irregular loading conditions

9.5.1 *Test performance and results*

The conventional triaxial test apparatus was incorporated into an electrohydraulic servo loading system by which any form of axial load history could be applied to test specimens. The irregular time histories stored in the tape recorder were retrieved

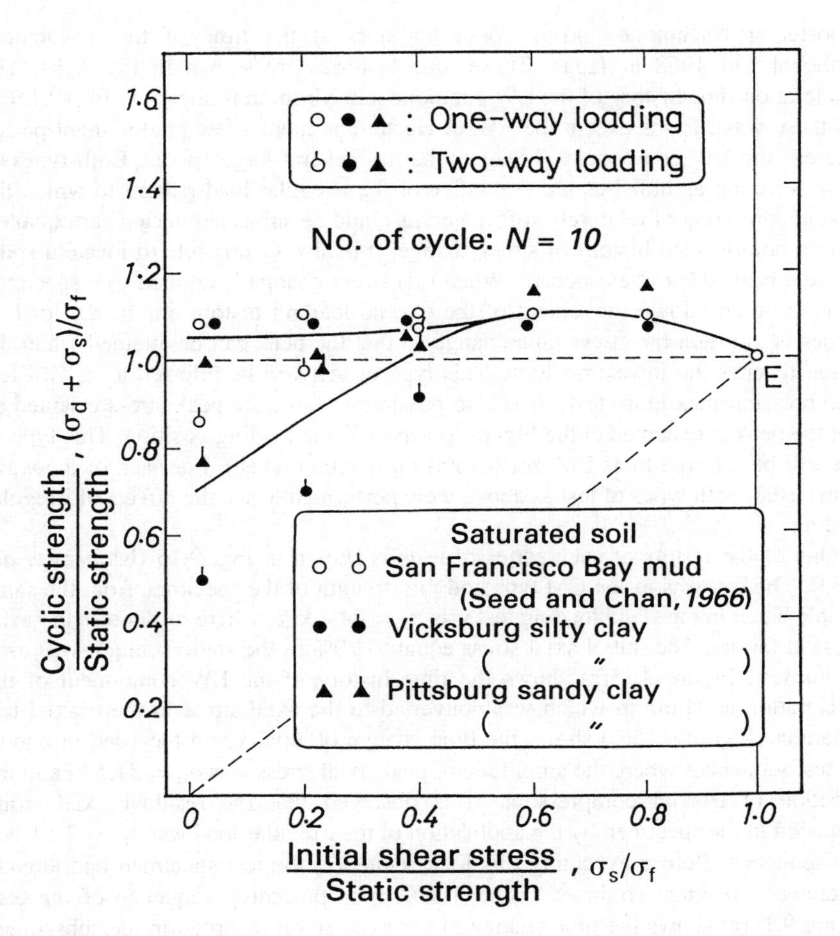

Fig. 9.12 Comparison of the test results between the one-way loading and the two-way loading tests.

and transmitted to the actuator to produce controlled motions in the triaxial loading piston.

Undisturbed samples of volcanic clay were procured from an intact surface exposed on a mountain slope which had suffered a large-scale landslide at the time of the January 14, 1978, Near–Izu earthquake with $M = 7.0$ (Ishihara and Nagao, 1983). The plasticity index of the clay was about 30. The water content varied between 110 and 140% and the saturation ratio ranged from 85 to 90%. The cohesion and angle of internal friction of this clay determined from the conventional triaxial tests were $C = 48$ kPa and $\phi = 17\,^{\circ}$, respectively.

The irregular load pattern used in the tests are four different time histories of the horizontal accelerations that were obtained on the surface of medium dense sand deposits at Hachinohe and Muroran harbours at the time of the Tokachioki earthquake of 1968 in Japan. These time histories are shown in Fig. 9.14. The acceleration time history of the EW component at Muroran is shown in Fig. 9.15(a). All these wave forms except the EW of Hachinohe have a few predominant peaks, whereas the EW component at Hachinohe has several large spikes. Both types of wave form are assumed as representative of the irregular load pattern to which the ground consisting of relatively stiff materials could be subjected during earthquakes.

In a complicated history of stress change, it is always possible to locate a spike where a peak shear stress occurs. When this stress change is applied to a specimen through up-and-down movement of the triaxial loading piston, one of the loading modes is to orient the stress time change so that the peak can be attained when the piston reaches the lowest position. This type of test will be referred to as CM test (compression maximum test). It is also possible to have the peak stress oriented so that the peak is executed at the highest position of the loading position. This type of test will be referred to as EM test (extension maximum test). For each of the wave forms used, both types of test as above were performed to see the effects of irregular loading.

One of the results of such series of tests is shown in Fig. 9.15 (Ishihara *et al.*, 1983). The test was of the CM type and the strength of the specimen from the same sample batch in the static loading test was $\sigma_f = 84.4$ kPa, where σ_f denotes the axial stress at failure. The initial axial stress equal to 70% of the static strength was used in this test. Figure 9.15(a) shows the time history of the EW component of the acceleration at Muroran which was converted to the axial stress in the triaxial test apparatus. Figure 9.15(b) shows the time change of axial strain recorded in one of the test sequences where the amplitude of peak axial stress was $\sigma_d = 87.5$ kPa in the direction of triaxial compression. It is observed that the residual axial strain produced in the specimen by the application of the irregular load was $\varepsilon_{re} = 2.12\%$ in this sequence. Before executing this load sequence, the test specimen had already sustained a residual strain of $\varepsilon'_{re} = 1.88\%$ in a preceding sequence of the test. Figure 9.15(c) shows the time change of the axial strain recorded in the subsequent test sequence in which the amplitude of the irregular load was raised to $\sigma_d = 110.7$ kPa. The specimen having sustained an axial strain of $\varepsilon'_{re} = 4.0\%$ in the preceding sequences experienced an additional residual strain of $\varepsilon_r = 5.85\%$ in the course of the current loading sequence. In the last sequence, the specimens underwent a residual strain as large as 10.9% as indicated in Fig. 9.15(d).

An example of similar test sequences employing an oppositely oriented wave form (EM test) is demonstrated in Fig. 9.16. The specimen used in this test showed approximately the same static strength of $\sigma_f = 84.4$ kPa. The time change in the axial strain shown in Figs 9.16(b), (c), and (d), is constructed in the same way as in the case of the results of the CM test shown in Fig. 9.16. Note that the amplitude of the peak σ_d indicated in Fig. 9.16 refers always to the maximum spike on the side of triaxial compression. In the type of triaxial test procedures described here, the initial shear stress σ_s is applied the triaxial compression side, and therefore the key quantities such as residual strains and failure of test specimens are always induced on the side of triaxial compression. Thus the difference in the CM and EM tests is

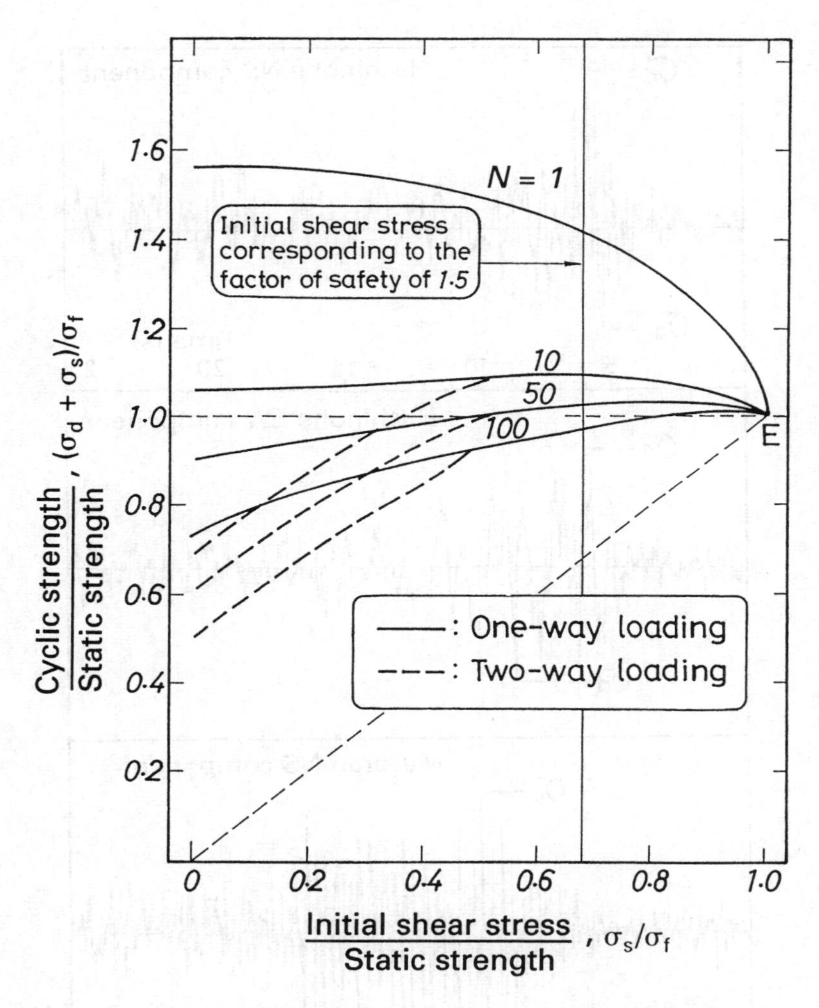

Fig. 9.13 Summary of the one-way and the two-way loading tests.

merely the difference in irregular time history of load application. In both cases, the initial stress condition $\sigma_s/\sigma_f = 0.7$ is realized in the triaxial compression side.

The time histories of the recorded axial strain shown in Figs 9.15 and 9.16 indicate that the major part of the residual strain is produced when the peak axial stress is applied to the specimen, and the irregular load after the advent of the peak does exert virtually no influence on the development of additional residual strain. This appears to imply that major displacement or failure produced in clay slopes during earthquakes takes place almost at the same time as the peak shear stress is applied to soil elements in the field.

In order to establish the stress–residual strain relationship as illustrated in Fig. 9.3, values of the total residual strain, $\varepsilon'_{re} + \varepsilon_{re}$, accumulated up to the current

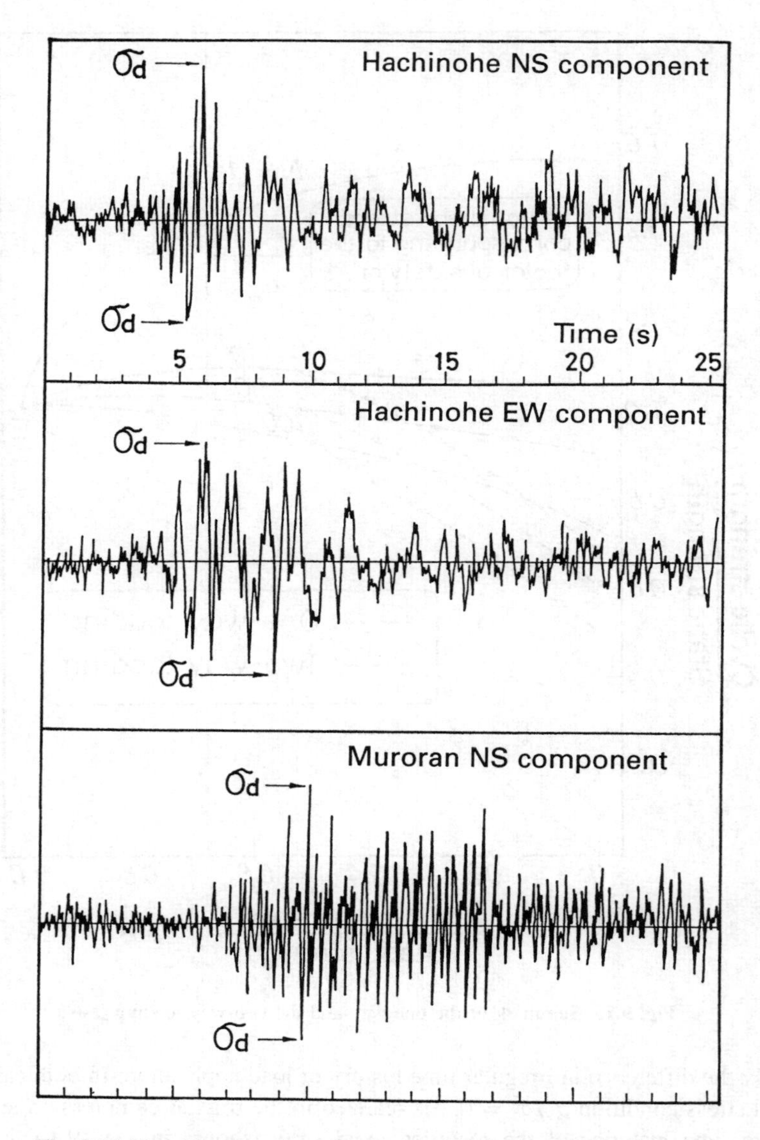

Fig. 9.14 Time histories of acceleration used in the irregular loading test.

sequence of irregular loading tests were read off from the test records such as those shown in Figs 9.15 and 9.16, and these values were plotted versus the peak amplitude of the current irregular load σ_d plus the initial axial stress σ_s. Note that the peak amplitude σ_d plotted refers to the peak value on the triaxial compression side. The results of such data compilation for the test data shown in Figs 9.15 and 9.16 are

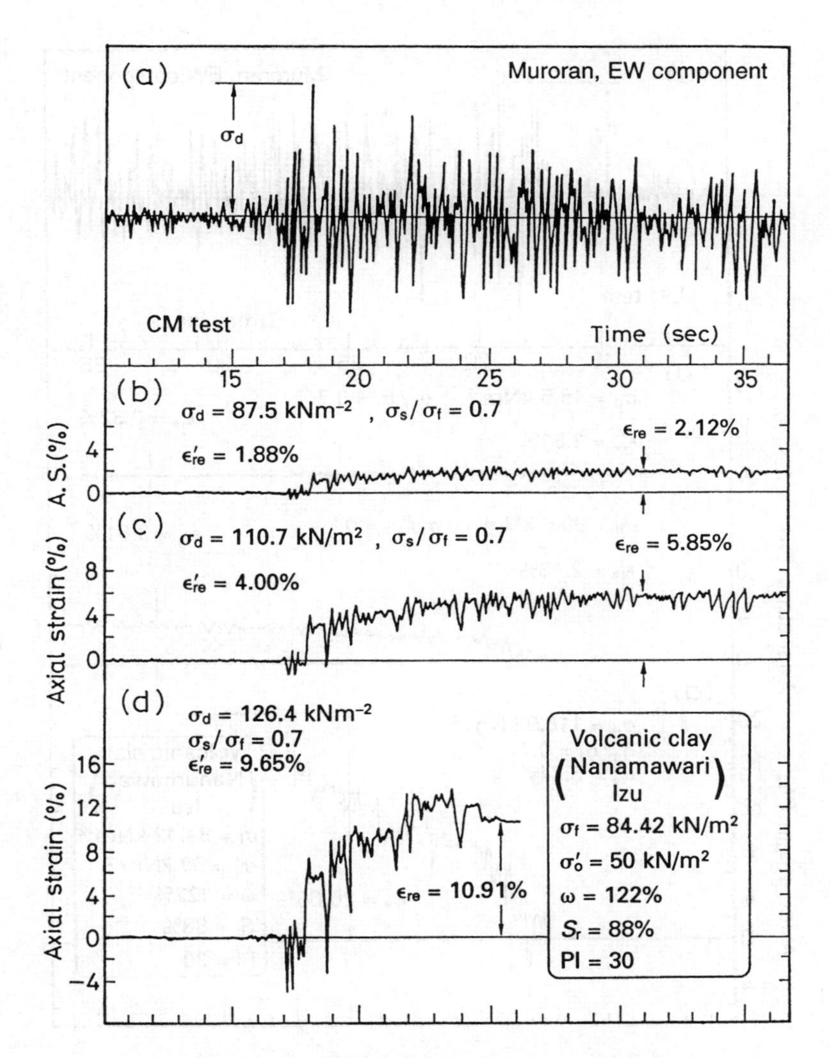

Fig. 9.15 Evolvement of residual strains in the irregular loading test.

presented in Fig. 9.17. In this plot, the combined static and dynamic axial stress, $\sigma_s + \sigma_d$ is shown normalized to the static strength σ_f in order to discern the effect of dynamic loading as against the static behavior. In Fig. 9.17 the data points indicated by arrows are those which were read off directly from the test results shown in Figs 9.15 and 9.16. It may be seen in Fig. 9.17 that there are some differences in the stress–residual strain relationship between the CM and EW tests, because of the difference in load time history, but viewed overall both test results give a consistent trend. It is of interest to note that the residual strain begins to increase abruptly when

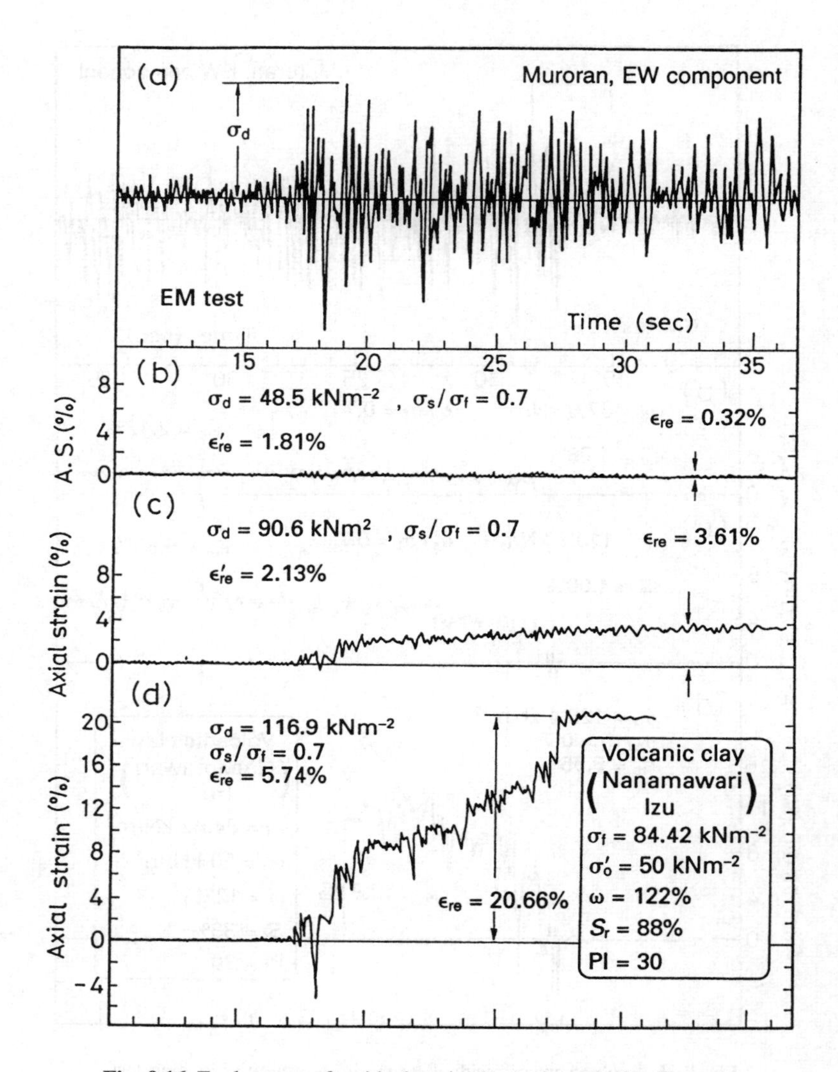

Fig. 9.16 Evolvement of residual strains in the irregular loading tests.

the axial stress $\sigma_d + \sigma_s$ reached a value about 80% of the static strength. Also indicated in Fig. 9.17 are the stress–strain curves for the static phase of loading until the initial shear stress σ_s is increased to 70% of the static strength. The static stress–strain curves which would have been obtained if the loading had been continued further up are shown in Fig. 9.17 by dashed lines. It may be seen in Fig. 9.17 that the stress–strain curve for static–dynamic loading is located far above the stress–strain curve for static loading alone. This fact indicates that if the soil specimen is subjected to a dynamic load after it has deformed statically to some extent, the

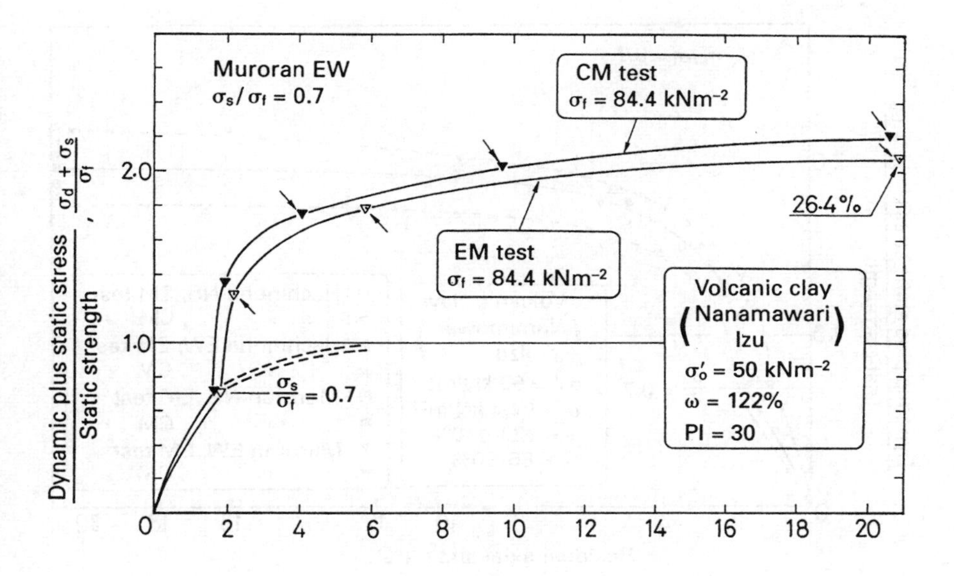

Fig. 9.17 Shear stress–residual strain relationship.

specimen tends to show a larger stiffness and higher strength than it is loaded to failure all the way in static conditions. For a volcanic clay soil as tested here, the increase in strength in the dynamic loading over that in the static loading condition amounts to almost 100% as indicated in Fig. 9.17. Such an increase in soil strength appears to emerge from highly rate-dependent nature of cohesive soils when subjected to rapid loads such as those used in the present test scheme.

9.5.2 *Effects of initial shear stress*

In order to examine the effects of initial shear stress on the stress–residual strain relation, several series of tests were conducted on the volcanic clay samples by employing the initial static axial stresses varying from $\sigma_s/\sigma_f = 0.2$ to 0.9. In these test series, all the test specimens were consolidated under a confining stress of $\sigma'_0 = 50$ kPa, and the four different time histories of axial load shown in Fig. 9.14 were used as the irregular load time history for each test series.

The result of a test series employing an initial shear stress 70% of the static strength are presented in Fig. 9.18. Although there exists some scatter which is due to varying time histories and orientation (CM or EM), all data points fall in a narrow zone enclosed by dashed lines in Fig. 9.18. A reasonable average curve is, therefore, drawn through the entire set of data. Figure 9.18 shows that the axial stress required to cause failure strain is about 1.95 times as much as the strength under the static loading conditions.

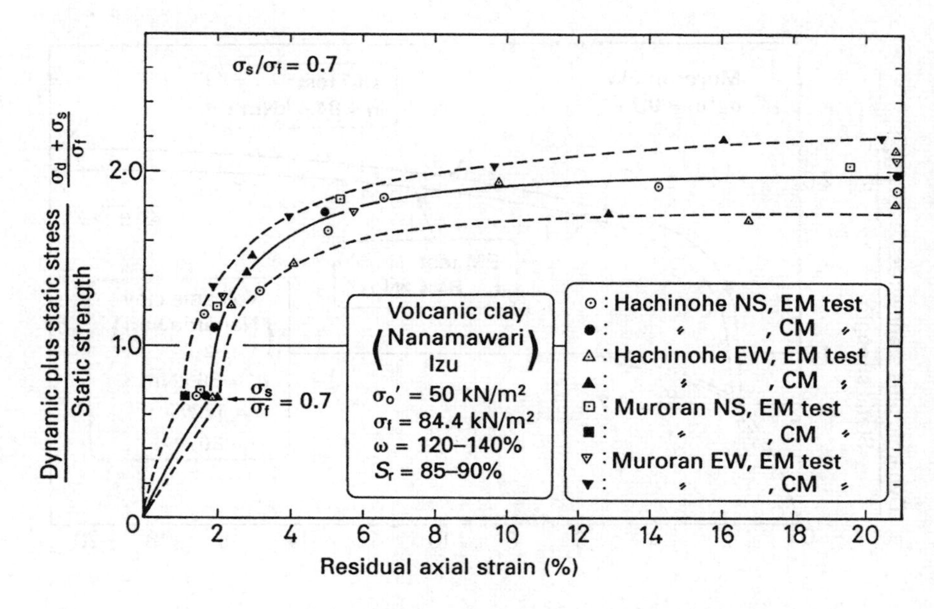

Fig. 9.18 Shear stress–residual strain relationship.

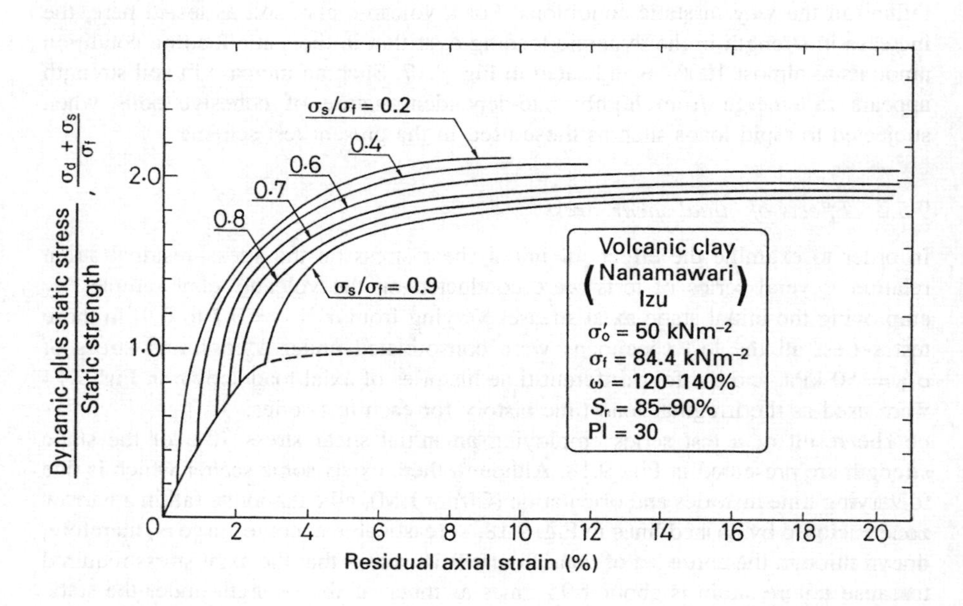

Fig. 9.19 Effects of initial shear stress on the shear stress–residual strain relationship.

The outcome of all other test series was similarly represented by average curves and brought together in Fig. 9.19. The set of summary curves in Fig. 9.19 shows that the stress–residual strain curve tends to flatten to some extent, as the initial shear stress is increased from 20% to 90% of the static strength value. It is somewhat surprising to see that even at a large level of the initial shear stress of $\sigma_s/\sigma_f + 0.9$, the stress–residual strain curve still stays far above the static stress–strain curve. It is likely that with increasing initial shear stress above 90% of the static strength, the stress–residual strain curve would drop sharply to coincide eventually with the static stress–strain curve. One of the important conclusions drawn from the plot of Fig. 9.19 is that the stress–residual strain relation is not appreciably affected by the initial shear stress, if it stays within the range of $\sigma_s/\sigma_f = 0.5$ and 0.8, which is generally the case with stress conditions encountered in in situ deposits of soils under slopes. Consequently the effect of initial sustained shear stress on the dynamic stress–strain relation may well be represented by the test results for which $\sigma_s/\sigma_f = 0.7$.

9.5.3 *Effects of confining stress*

For the purpose of examining the effects of confining stress, several series of tests were conducted on the volcanic clay samples by employing three confining pressures of $\sigma'_0 = 20$, 50, and 80 kPa. In each series of these tests, specimens were consolidated under a specified confining stress and then subjected to the initial shear stress equal to 70% of the static strength. The dynamic tests employing the four different time histories were then carried out.

The results of a test series employing a confining stress of 20 kPa are shown in Fig. 9.20(a). A reasonable average curve is drawn through the entire set of data points both for static and dynamic portions of the stress–residual strain curves. It may be seen that, as soon as the loading is switched into the dynamic phase, the stress–residual deformation curve becomes steeper and eventually converges to a horizontal line corresponding to a dynamic strength value which is about 2.15 times as much as the static strength. The stress–residual strain curve presented in Fig. 9.18 can also be regarded as a case of the present test series in which the confining stress is $\sigma'_0 = 50$ kPa. The same characteristic behaviour is observed in Fig. 9.18 as in the case of $\sigma'_0 = 20$ kPa, but the degree of increase of the dynamic strength over the static strength is seen to be about 1.95 times the static strength, a smaller value as against 2.15 times for the case of the confining stress of 20 kPa. The results of still other test series employing a confining stress of 80 kPa are presented in Fig. 9.20(b). The average curve in this figure indicates that the strength in the dynamic loading conditions is approximately 65% greater than the strength attained under the static conditions. The average stress–residual strain curves shown in Figs 9.20(a), 9.18 and 9.20(b) are assembled and plotted in Fig. 9.21 for comparison purposes. Also indicated by dashed lines in Fig. 9.21 are the stress–strain curves which would have been obtained if the loading had been continued further under static conditions. It can generally be seen that the effect of the dynamic phase of loading is conspicuous and acts towards increasing the stiffness as well as the ultimate strength of the soil.

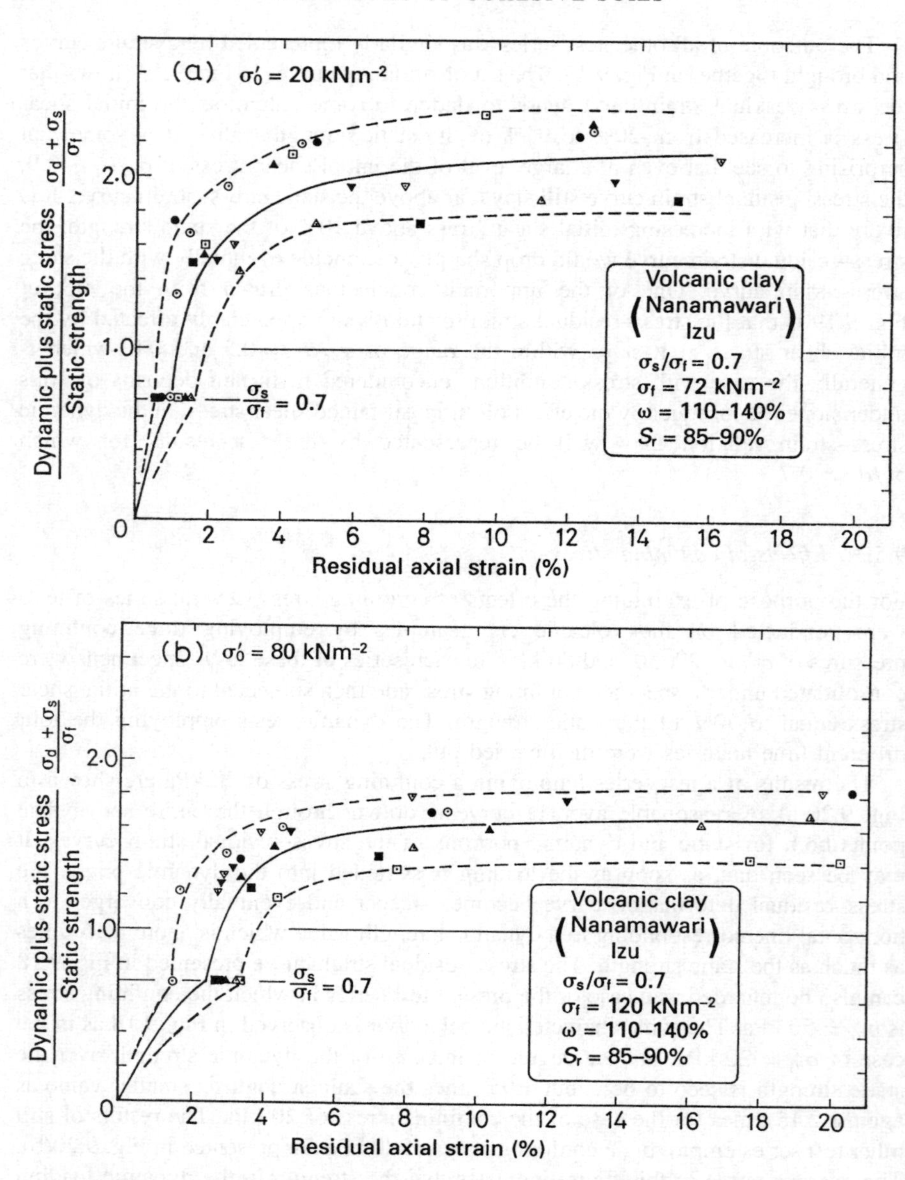

Fig. 9.20 Shear stress–residual strain relationship under different confining stress.

9.5.4 *Relationship between dynamic and static strength*

Several series of the tests described in the foregoing sections revealed that the magnitude of static initial shear stress does not significantly influence the subsequent behavior of the soil subjected to dynamic loads, if the initial shear stress lies within

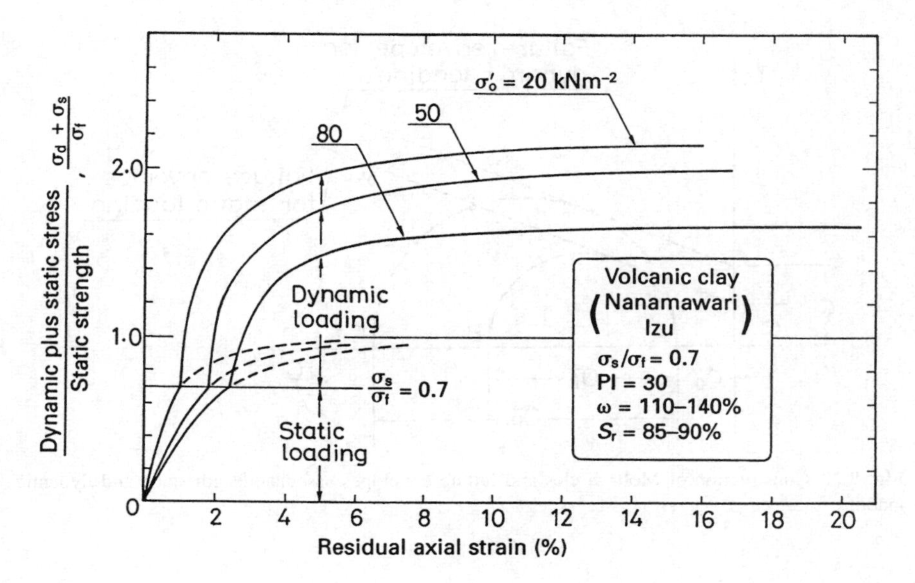

Fig. 9.21 Effects of confining stress on the shear stress–residual strain relationship.

the range of 40 to 90% of the strength value obtained in the static loading conditions. Therefore, the influence of variation in the initial shear stress will be disregarded in the following analysis.

It was discovered in the tests that the effect of confining stress is appreciable as seen in Fig. 9.21 and can not be neglected in evaluating the residual strains and strength in the dynamic loading conditions. This consequence may be taken for granted because the strength of partially saturated cohesive soils in static conditions has been known generally to vary with the magnitude of the confining stress. The effect of the confining stress on the static strength of soils has been evaluated in terms of the apparent angle of internal friction defined as an angle of slope in the Mohr circle representation of failure state in the stress space. Therefore it is of interest to establish a Mohr–Coulomb type failure criterion for the dynamic loading condition as well and compare it with the conventional failure criterion for the static loading condition. The method adopted herein for establishing the dynamic failure criterion is illustrated in Fig. 9.22. The value of the confining stress σ'_0 is first laid off at point A in abscissa and then the static and dynamic strength values are laid off towards the right as AB and AC, respectively. The circle drawn through points A and B is the Mohr circle associated with failure in the static loading. Likewise the Mohr circle for failure in the dynamic loading can be constructed by drawing a circle through points A and C, as shown in Fig. 9.22.

Using the above procedures, two sets of Mohr circles specifying failure in static and dynamic loading conditions were established for the test data on the volcanic clay presented in Fig. 9.20. The Mohr circles thus constructed are shown in

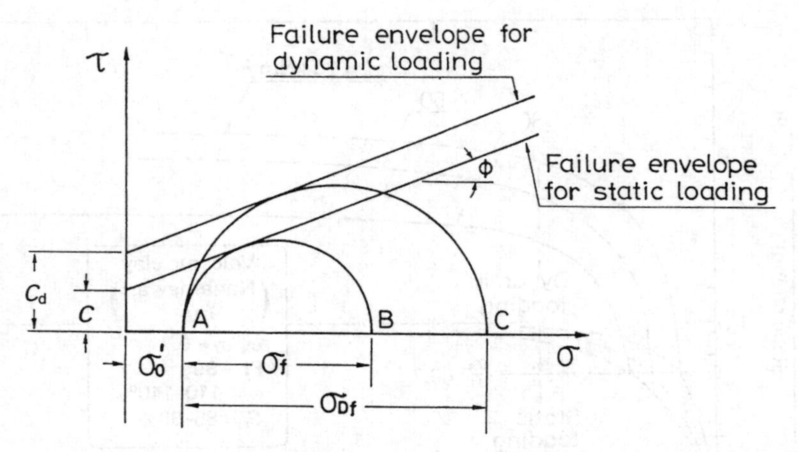

Fig. 9.22 Construction of Mohr circles and failure envelopes associated with static and dynamic loading.

Fig. 9.23. It is then possible to draw a straight envelop line for each set of the Mohr circles shown in Fig. 9.23. For the volcanic clay from the Izu area, the cohesion component for static loading C was 20 kPa and the cohesion for dynamic loading C_d was 48 kPa. It is of particular interest to note that the angle of internal friction was practically the same both for the static and dynamic loading conditions. The angle of internal friction ϕ for the clay tested was 17°. The fact that the effect of dynamic loading on the failure strength is manifested only through the cohesion component may be taken as reasonable if one is reminded of the fact that the increase in strength due to rapid loading such as seismic irregular loading emerges mainly from the viscous nature of soil materials. This aspect of soil properties was discussed in some detail is Section 8.2.

On the basis of the observation that the angle of internal friction is identical both in the static and dynamic loading condition, it becomes possible to deduce some correlation between the strength parameters pertaining to these two loading conditions. Suppose the angle of internal friction ϕ and cohesion C in the static loading are known for a given soil, then the axial stress required to cause failure σ_f under a confining stress σ'_0 is given by,

$$\sigma_f = \frac{2 \sin \phi}{1 - \sin \phi} \sigma'_0 + \frac{2C \cos \phi}{1 - \sin \phi}. \tag{9.1}$$

This is the well-known equation expressing the straight-line failure envelop in Fig. 9.22. When a dynamic test is performed under the same confining stress, the axial stress causing failure σ_{Df} if given by,

$$\sigma_{Df} = \frac{2 \sin \phi}{1 - \sin \phi} \sigma'_0 + \frac{2C_D \cos \phi}{1 - \sin \phi}. \tag{9.2}$$

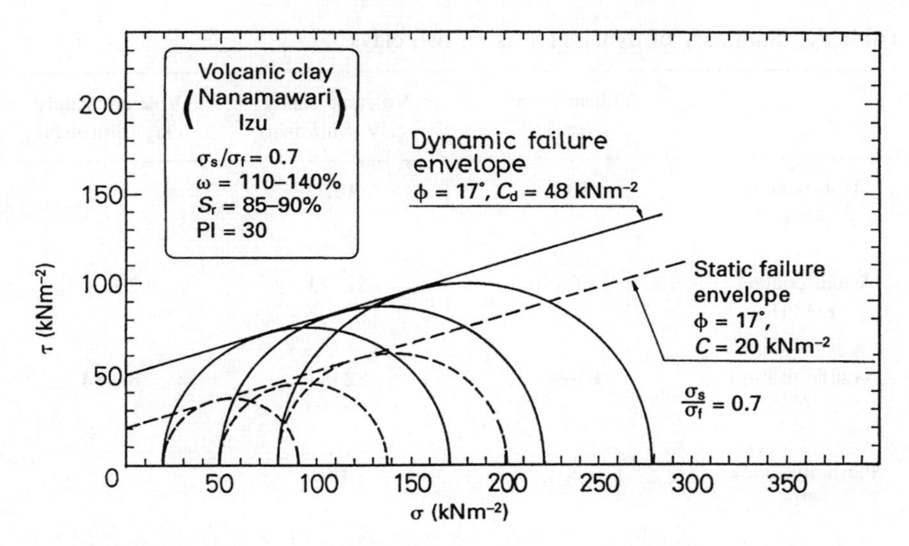

Fig. 9.23 Failure envelopes obtained from static and dynamic loading test results.

Combining the above two equations, one obtains,

$$\frac{C_{\rm D}}{C} - 1 = \left(1 + \frac{\sigma'_0}{C \cot \phi}\right)\left(\frac{\sigma_{\rm Df}}{\sigma_{\rm f}} - 1\right). \tag{9.3}$$

Thus, knowing the static strength parameter C and ϕ, one can estimate the dynamic cohesion value $C_{\rm D}$ from eqn (9.3), if a single dynamic test is run to determine the value of the dynamic strength $\sigma_{\rm Df}$ under an appropriate confining stress σ'_0. Once the value of the dynamic cohesion is thus known, it becomes in turn possible to estimate the dynamic strength $\sigma_{\rm Df}$ for any other value of the confining stress through the use of eqn (9.2).

The dynamic test program as described above has been implemented for another cohesive soil obtained from a site of man-made fills at Shiroishi which had suffered a large-scale landslide at the time of the June 12 1978, Miyagiken-oki earthquake ($M = 7.4$) in Japan (Ishihara and Kasuda, 1984). The soil, of volcanic origin, consists of a mixture of 13% gravel, 47% of sand, 12% silt, and 28% clay content. The specimens for the triaxial tests were prepared by compacting the material to two densities of about 18.7 and 19.0 kNm^{-3}. Test conditions and test results are summarized in Table 9.1, based on the method of data interpretation presented in Fig. 9.22. Although no definitive conclusion can be drawn from the tests on only two clays, it may be noted that the clay exhibits 1.6 to 2.4 times increase in cohesion under dynamic loading over that under static loading.

Table 9.1 Summary of dynamic tests on two clays

	Volcanic clay (Izu)	Volcanic sandy clay (Shiroishi)	Volcanic sandy clay (Shiroishi)
Unit weight γ_t (kNm^{-3})	13.3	18.7	19.0
Water content ω (%)	110–140	22–23	20–21
Saturation ratio S_r (%)	85–90	82–84	82–84
Plasticity index PI	30	18	18
Static cohesion C (kNm^{-2})	20	28	32
Angle of int. friction ϕ (degrees)	17	14	16
Dynamic cohesion C_D (kNm^{-2})	48	52	51
C_D/C	2.40	1.86	1.59

References

Casagrande, A. and Shannon, W.L. (1948). *Research on stress-deformation and strength characteristics of soils and rocks under transient loading*. Harvard University Soil Mechanics Series No. 31.

Casagrande, A. and Wilson, S.D. (1951). Effect of rate of loading on the strength of clay and shales at constant water content. *Geotechnique*, **2**, 251–63.

Ellis, W. and Hartman, V.B. (1967). Dynamic soil strength and slope stability. *Journal of Soil Mechanics and Foundations*, ASCE, SM 4, 355–73.

Ishihara, K. and Nagao, A. (1983). Analysis of landslides during the 1978 Izu–Ohshima–Kinkai earthquake. *Soils and Foundations*, **23**, 141–59.

Ishihara, K. and Kasuda, K. (1984). *Dynamic strength of a cohesive soil*. Proceedings of the 6th Conference on Soil Mechanics and Foundation Engineering, Budapest, pp. 91–8.

Ishihara, K., Nagao, A., and Mano, R. (1983). *Residual strain and strength of clay under seismic loading*. Proceedings of the 4th Canadian Conference on Earthquake Engineering, 602–13.

Kawakami, F. (1960). Properties of compacted soils under transient loads. *Soils and Foundations*, **1**, 23–9.

Ohsaki, Y., Koizumi, Y., and Kishida, H. (1957). Dynamic properties of soils. *Transactions of the Architectural Institute of Japan*, **54**, 357–9.

Ohsaki, Y. (1964). Dynamic properties of soils and their application. *Japanese Society of Soil Mechanics and Foundation Engineering*, 29–56.

Olson, R.E. and Parola, J.F. (1967). *Dynamic shearing properties of compacted clay*. Proceedings of the International Symposium on Wave Propagation and Dynamic Properties of Earth Materials, University of New Mexico, pp. 173–81.

Schimming, B.B., Haas, H.J., and Sax, H.C. (1966). Study of dynamic and static failure envelopes. *Journal of Soil Mechanics and Foundations*, ASCE, SM 2, 105–23.

Seed, H.B. (1960). *Soil strength during earthquakes*. Proceedings of the 2nd World Conference on Earthquake Engineering, Vol. 1, pp. 183–94.

Seed, H.B. and Chan, C.K. (1966). Clay strength under earthquake loading conditions. *Journal of Soil Mechanics and Foundations*, ASCE, SM2, 53–78.

Whitman, R.V. (1957). *The behavior of soils under transient loading*. Proceedings of the 4th International Conference on Soil Mechanics and Foundation Engineering, Vol. 1, 207–10.

10

RESISTANCE OF SAND TO CYCLIC LOADING

10.1 Simulation of field stress conditions in laboratory tests

Prior to shaking by an earthquake, an element of saturated soil under level ground has undergone a long-term consolidation process under K_0 conditions. This soil element is subjected undrained to a sequence of shear stress cycles during an earthquake as illustrated in Fig. 10.1(a). Note that the cyclic shear stress application is executed in such a way that lateral deformation is prohibited because the flat ground surface is assumed to extend infinitely in the horizontal direction. In the case of sloping ground, a soil element is considered to have been anisotropically consolidated approximately under K_0 conditions with an additional shear stress acting on the horizontal plane. During an earthquake, a sequence of shear stress cycles is applied to the soil element in undrained conditions as illustrated in Fig. 10.1(b). It is noteworthy, however, that the soil element is allowed to deform freely in the horizontal direction during the application of cyclic shear stress, because lateral movement of the ground is allowed to take place as a whole. When attempting to investigate behaviour of soils in the laboratory test, in situ stress conditions can be most fittingly reproduced in the sample tested in a torsional apparatus. In what follows, the principle and procedures for laboratory testing will be illustrated by referring to the stress conditions that can be produced in the torsional apparatus. In a typical test scheme, samples of saturated sand are consolidated first under K_0 conditions and then subjected to a sequence of torsional stress cycles under undrained conditions.

When simulating the level ground condition, the cyclic torsional stress must be applied while inhibiting lateral deformation as illustrated in Fig. 10.1(c). This type of test will be referred to as the anisotropically consolidated oedometer-conditioned torsion (ACOT) test (Ishihara and Li, 1972). When the stress condition in sloping ground is to be reproduced in the laboratory sample, effects of the initial shear stress acting on the horizontal plane shown in Fig. 10.1(b) can be represented implicitly in terms of the component of deviator stress $\sigma'_v - \sigma'_h$, which is applied to the sample at the time of the K_0 condition. The cyclic torsional stress is then applied while allowing the lateral deformation to occur freely as shown in Fig. 10.1(d). This type of test will be called the anisotropically consolidated torsion (ACT) test (Ishihara and Li, 1972).

10.2 The mechanism of liquefaction

The mechanism of liquefaction can be best understood by observing features of pore pressures and shear strains that are developed in each of the torsional tests as explained above. A typical apparatus used for this test is shown in Fig. 4.5.

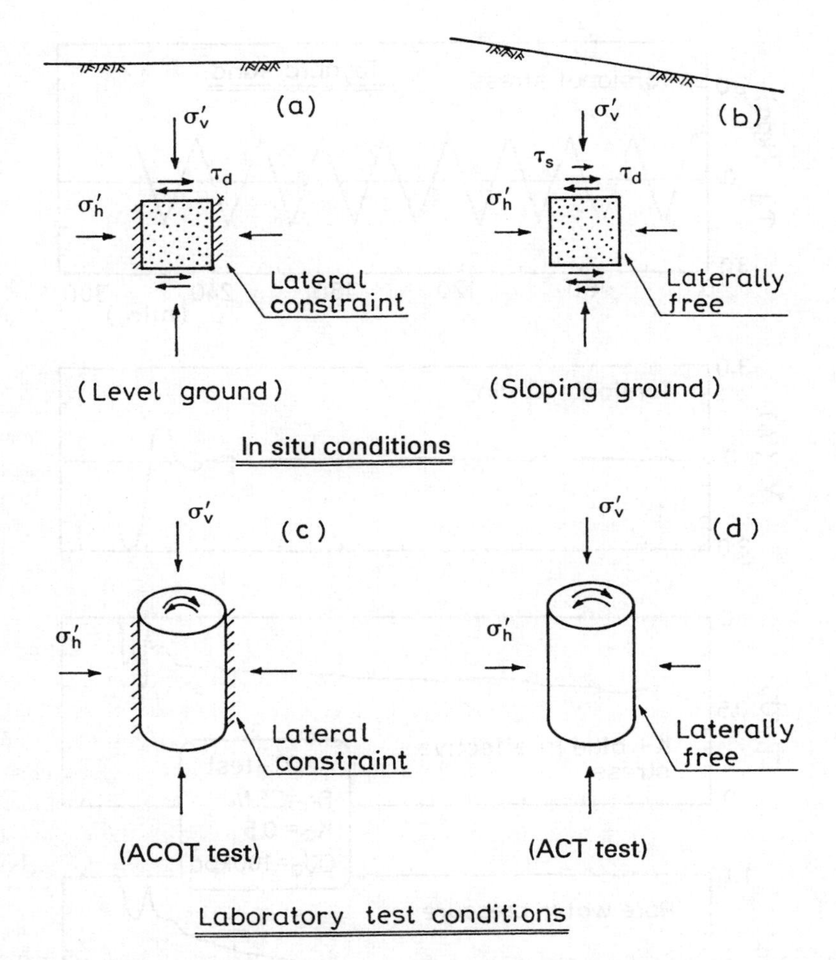

Fig. 10.1 Stress conditions in the field prior to and during seismic shaking and their simulation in the laboratory torsional test.

10.2.1 *The cyclic torsional test with lateral confinement (ACOT test)*

Since the cyclic torsional stress is applied undrained without permitting lateral deformation, both the axial and lateral strains are maintained at zero level all the way throughout the test. Thus the sample regains its original configuration whenever the torsionally deformed sample is brought back to its neutral position during cyclic torsional loading. Since in situ soil deposits are often under level ground conditions, it is apparent that the ACOT test is considered to best fittingly reproduce the field conditions prior to and during an earthquake.

If this test is performed under a special condition with $K_0 = 1.0$, the state of stress prior to cyclic loading becomes identical to that of the isotropically

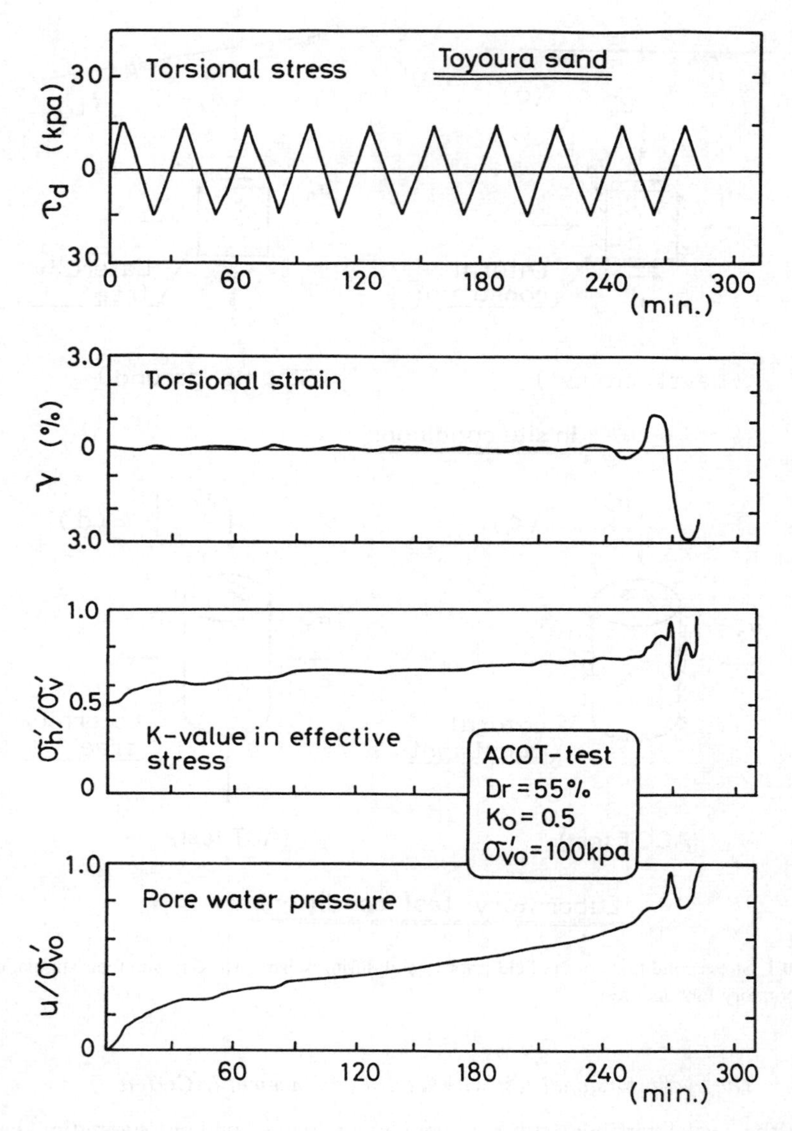

Fig. 10.2 Change in lateral stress and pore pressure build-up in the ACOT test with lateral confinement.

consolidated sample. If the cyclic torsional stress is applied undrained on such a sample, it will also remain undeformed in the triaxial mode without inducing any lateral deformation. Thus it is known that the isotropically consolidated sample undergoing cyclic torsional shear stress yields exactly the same stress conditions as

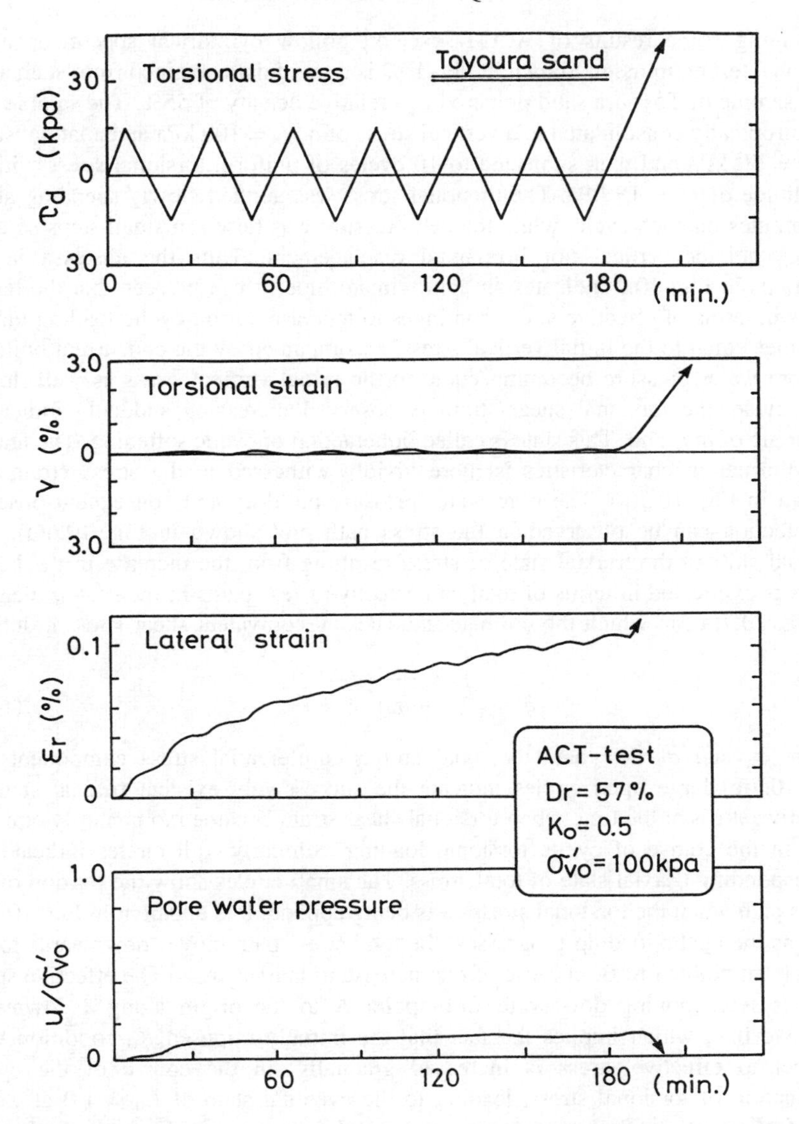

Fig. 10.3 Change in lateral strain and pore pressure in the ACT test without lateral confinement.

those in the triaxial test apparatus in which the cyclic axial stress is applied undrained on an isotropically consolidated sample. This is the basis on which the cyclic triaxial test was adopted first by Seed and Lee (1966) to explore the mechanism and factors influencing the onset of liquefaction.

The mechanism of sand response leading to liquefaction can be understood by examining some results of ACOT tests on hollow cylindrical specimens using torsional test equipment. Shown in Fig. 10.2 is a set of data obtained from such a test on a sample of Toyoura sand prepared at a relative density of 55%. The sample was anisotropically consolidated to a vertical stress of $\sigma'_{v0} = 100$ kPa and a lateral stress $\sigma'_{h0} = 50$ kPa and then subjected to 10 cycles of uniform torsional stress with an amplitude of $\tau_d = 15$ kPa. The torsional stress was applied slowly spending about 30 minutes on each cycle, while the cell pressure was raised in small steps so as to induce neither vertical nor horizontal displacement. Thus the abscissa in the diagrams in Fig. 10.2 indicates an approximate time. It can be seen that the lateral stress in terms of effective stress continues to increase during cyclic loading until it becomes equal to the initial vertical stress, accompanied by the concurrent build-up of pore water pressure becoming equal to the initial vertical stress as well. In the 10th cycle, the torsional shear strain is observed increasing suddenly indicating softening of the sand. This state is called liquefaction or cyclic softening. The feature of deformation characteristics is more vividly witnessed in the stress–strain plot shown in Fig. 10.5(a). The pore water pressure build-up and consequent onset of liquefaction can be observed in the stress path plot shown in Fig. 10.6(a). The gradual shift in the triaxial state of stress resulting from the increase in the lateral stress is explicated in terms of total and effective stress paths in the $p' - \overline{q}$ diagram in Fig. 10.7(a), in which the ordinate indicates an equivalent shear stress $\overline{q}$ defined by

$$\overline{q} = \sqrt{(\sigma - \sigma_\theta)^2 + 4\,\tau^2{}_d} \qquad (10.1)$$

where σ_a and σ_θ designate the axial and circumferential stress components. In Fig. 10.7(a) large open circles indicate the successively existent triaxial state of effective stress at the time when torsional shear strain became momentarily equal to zero in the course of cyclic torsional loading. Similarly, full circles indicate the corresponding triaxial state of total stress. The small circles show the portion of the stress path when the torsional stress τ_d is being applied. It is apparent in Fig. 10.7(a) that, as the cyclic loading progresses, the total stress path moves downwards to the right from point A to B, because of the increase in lateral stress. The effective stress path is seen moving downward from point A to the origin along a downward concave line, which implies the fact that the initially imposed K_0 condition with respect to effective stress is increased gradually, in the course of the cyclic application of torsional stress, leading to the eventual state of $K_0 = 1.0$ at which liquefaction sets in. The characteristic manner of increasing the K_0 value can be well understood if one is reminded of the unloading process in the conventional triaxial test where the K_0 value tends to increase if the lateral deformation is inhibited. Thus it can be mentioned that a sand deposit under the level ground condition can be brought to a state of liquefaction due to seismic shaking, accompanied by a change in lateral stress, where a complete loss of strength takes place with 100% pore water build-up. It is to be noted that the onset of liquefaction as above takes place without any change in the shape of the soil element whatsoever and accordingly it is to be considered as a completely different phenomenon from shear failure which is always accompanied by a significant distortion in the shape of the soil element.

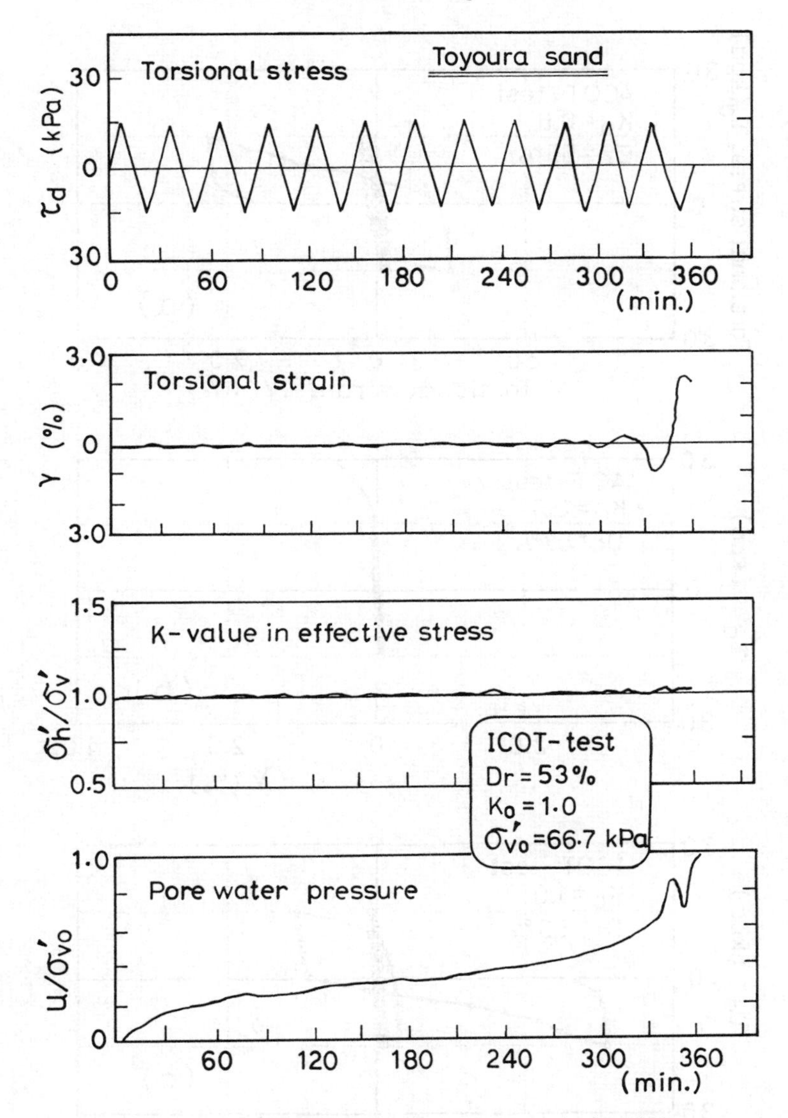

Fig. 10.4 Change in lateral stress and pore pressure in the ICOT test.

10.2.2 *The cyclic torsional test without lateral confinement (ACT test)*

The initially applied vertical and lateral stresses are maintained unchanged all the way through during undrained application of cyclic torsional shear stress. As no lateral constraint is imposed, the sample is deformed laterally as well as vertically. Thus the conditions in the ACT test are considered to be representative of those in the field where soil deposits are subjected to some initial shear stress such as those

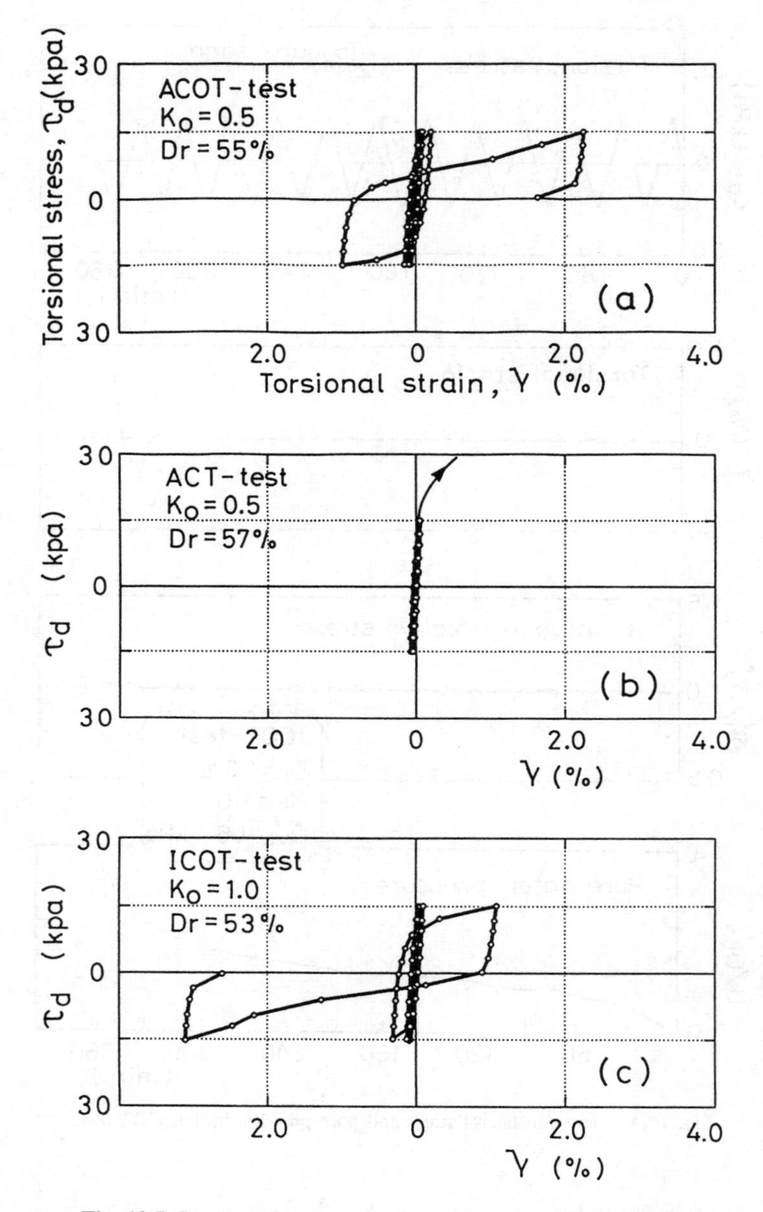

Fig. 10.5 Stress–strain curves for three types of cyclic torsional test.

under sloping grounds or under embankments or dykes as explained in Fig. 10.1(b). It is to be noted that the soil element under such conditions is always free to move in the horizontal direction.

The sand response in such a condition can be elucidated by examining some

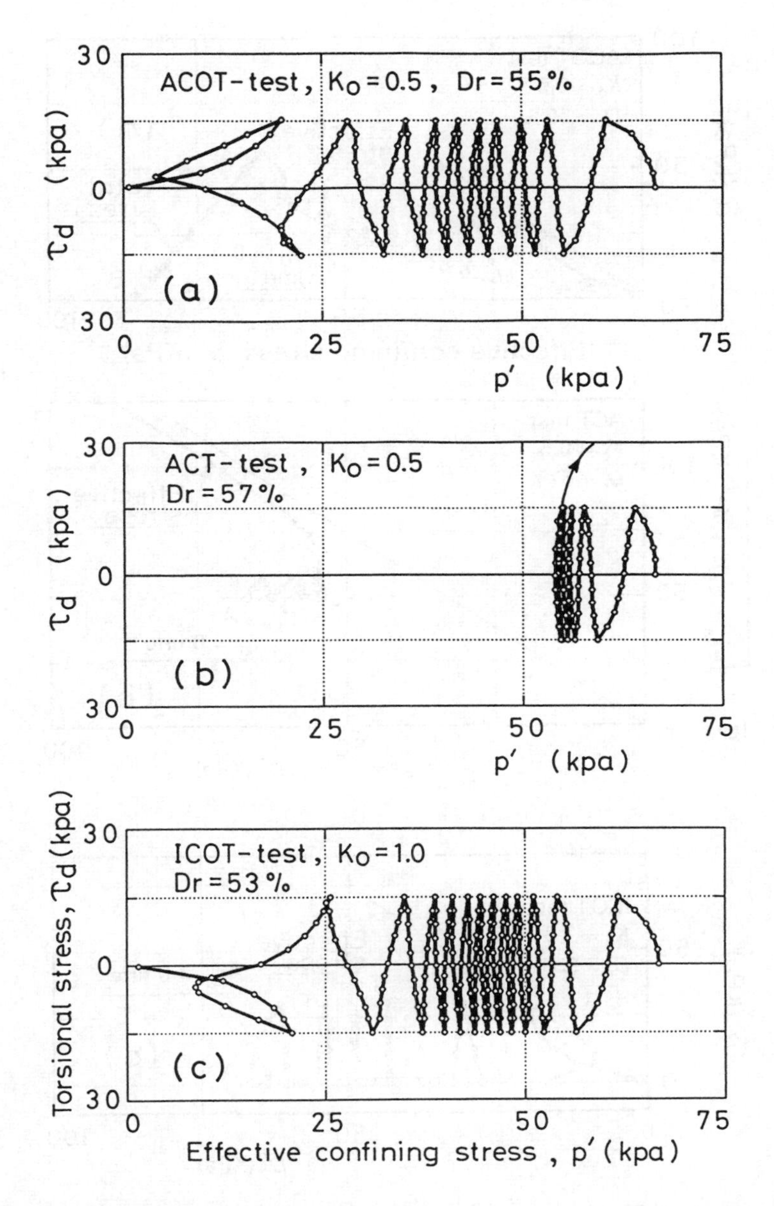

Fig. 10.6 Effective stress paths in torsional mode in three types of cyclic torsional test.

results of ACT test on a hollow cylindical specimen using the torsion test apparatus. A suite of data obtained from a test on a sample of Toyoura sand with a relative density of 57% is shown in Fig. 10.3. The sample was consolidated anisotropically to a vertical stress of $\sigma'_{v0} = 100$ kPa and a lateral stress of $\sigma'_{h0} = 50$ kPa, and then

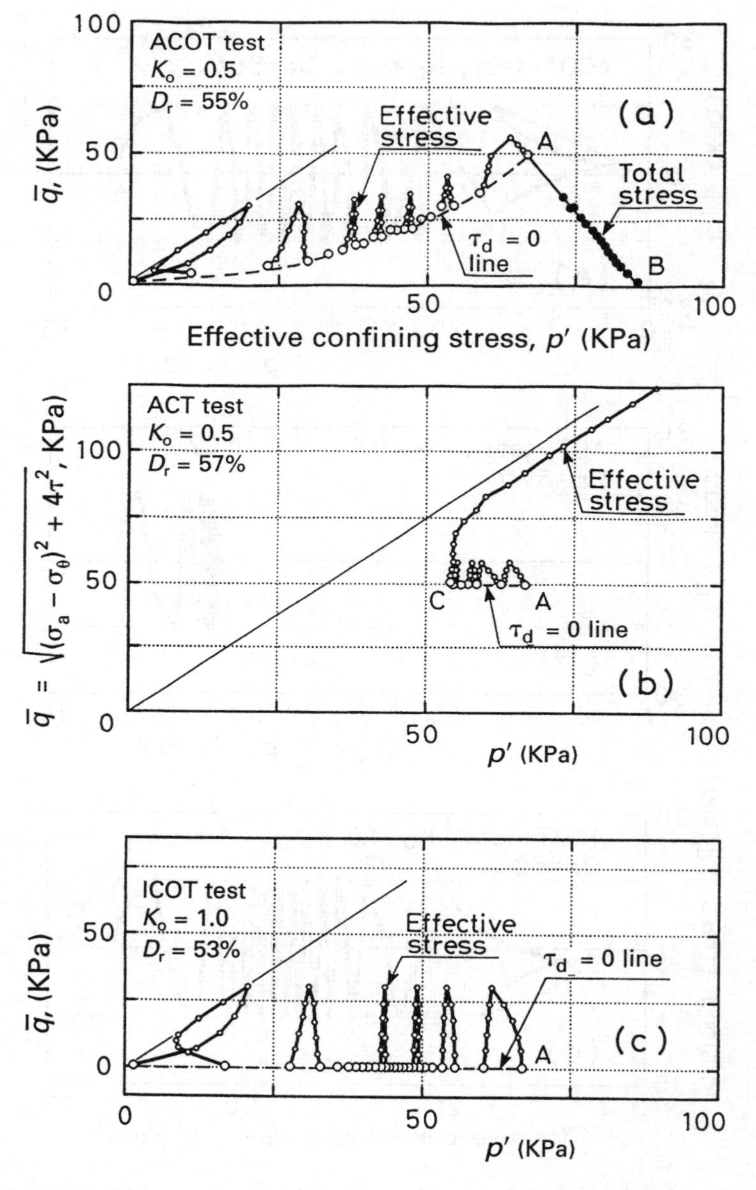

Fig. 10.7 Effective stress paths in triaxial mode in three types of cyclic torsional test.

subjected to 5 cycles of uniform torsional stress having an amplitude of $\tau_d = 15$ kPa. The test conditions were the same as those of the ACOT test described above, except for the condition associated with the lateral deformation. It can be seen in Fig. 10.3 that the lateral strain ε_r continues to develop during cyclic loading, producing a significant change in the shape of the sample. In unison to this, the pore water pressure is seen building up to some extent, but ceasing to develop partway at a certain stage nowhere close to the initial vertical stress. It is also to be noted that the amplitude of cyclic torsional strain does not grow suddenly at any stage of loading, indicating that the cyclic softening phenomenon has not taken place in contrast to the case of the ACOT test. The stress–strain relation and stress path shown in Figs 10.5(b) and 10.6(b) confirm this characteristic behaviour. This observation, taken together with the partial build-up of pore water pressure, leads to the conclusion that a complete softening or liquefaction can not occur in the sample in the ACT test. The gradual shift in the state of stress induced by the cyclic torsional loading can be observed in the stress path plot in p'–q space demonstrated in Fig. 10.7(b). It is apparent that the total stress path remains at point A but the effective stress path moves horizontally to the left as the cyclic torsional loading proceeds until it reaches a point C which lies nearer the failure line. Thus it can be mentioned that the pore water pressure build-up continues until the effective confining stress becomes small enough to match the failure condition and, while remaining at this state of stress, the sample continues to deform in the triaxial mode leading to a failure involving a significant amount of accumulated distortion in the sample. In the test shown in Fig. 10.7(b), the torsional stress was increased monotonically after the cyclic phase of loading. It is seen that a dilatant behaviour prevails during this monotonic loading, because the sample remains in a deformed state in the triaxial mode. It should be noticed that, in contrast to the liquefaction type failure in the ACOT test, the rupture in the ACT test is a shear type failure with some effective confining stress still persistent in the soil.

Cyclic behaviour of sand has been investigated sometimes by applying undrained cyclic axial stress on samples anisotropically consolidated in the triaxial test apparatus. This type of test originally explored by Lee and Seed (1967) should be regarded as being equivalent or alternative to the ACT test. Although the mode of cyclic loading is different in these two types of tests, they have an essentially common feature in that the sample tends to gradually develop a triaxial mode of deformation in the course of cyclic load application, while keeping the initially applied deviator stress unchanged. It appears that there has been some confusion among geotechnical engineers as to which condition, ACOT or ACT, does correspond to the cyclic triaxial test on anisotropically consolidated samples. It has now become clear that the latter type of cyclic triaxial test is essentially the same as the cyclic torsional test with free lateral constraint, i.e. the ACT test.

10.2.3 *The cyclic torsional test without lateral confinement on isotropically consolidated samples (ICOT test)*

This type of test is regarded as a particular case of the ACOT test in which the K_0 value is equal to unity. A suit of data from an ICOT test on a sample of Toyoura sand with a relative density of 53% is presented in Fig. 10.4. The sample was isotropically consolidated to a confining stress of $\sigma'_{v0} = \sigma'_{h0} = 66.7$ kPa and then subjected to 12 cycles of torsional stress having an amplitude of $\tau_d = 15$ kPa.

The result of the ICOT test shown in Fig. 10.4 indicates the gradual build-up of pore water pressure with the progression of cyclic loading and eventual occurrence of liquefaction where the pore water pressure becomes equal to the initial confining stress accompanied by the development of a large amplitude of shear strain. This characteristic behaviour is also observed in the stress–strain relation in Fig. 10.5(c) and the stress path in Fig. 10.6(c).

The three types of cyclic torsional tests as mentioned above were all carried out on those samples prepared with approximately the same relative density of $D_r = 55\%$ and consolidated to the same mean effective stress, $\sigma'_0 = (1 + 2K_0)\sigma'_{v0}/3 = 66.7$ kPa. The samples were all subjected to a cyclic torsional stress with the same amplitude of $\tau_d = 15$ kPa. Looking over the results of the three types of the tests, the following points can be addressed:

1. The degree of lateral constraint during cyclic loading could exert a profound influence on the behavior of sand. In particular, depending upon whether or not the sample is allowed to deform in the triaxial mode, either shear type failure or liquefaction type softening could take place at the end of cyclic loading.

2. If the mean principal stress is held constant, the ACOT test on a K_0-consolidated sample with lateral confinement shows practically the same behavior as that of the sample in the ICOT test.

10.3 Definition of liquefaction or cyclic softening

The basic mechanism of onset of liquefaction is elucidated from observation of the behavior of a sand sample undergoing cyclic stress application in the laboratory triaxial test apparatus. In the first attempt by Seed and Lee (1966), samples of saturated sand were consolidated under a confining pressure and subjected to a sequence of constant-amplitude cyclic axial stress under undrained conditions, until they deformed to a certain amount of peak to peak axial strain. This loading procedure creates stress conditions on a plane of 45° through the sample which is the same as those produced on the horizontal plane in the ground during earthquakes. This correspondence between the laboratory sample and in situ soils is the basis on which the cyclic triaxial test is warranted as a useful procedure for producing meaningful data to assess the resistance of sands to liquefaction. The stress conditions at each stage of loading in the cyclic triaxial test are illustrated in Fig. 10.8. When the axial stress σ_d is applied undrained, the shear stress induced on the 45° plane is $\sigma_d/2$. The normal stress $\sigma_d/2$ is also induced on this plane but this is purely compressive component of $\sigma_d/2$ which is mostly transmitted to pore water without inducing any change in the existing effective confining stress σ'_0.

Therefore the normal stress acting on the 45° plane can be disregarded. A typical result of the cyclic triaxial test is shown in Fig. 10.9. It is observed that the pore water pressure builds up steadily as the cyclic axial stress is applied, and eventually approaches a value equal to the initially applied confining pressure, thereby producing an axial strain of about 5% in double amplitude. Such a state has been referred to as *initial liquefaction* or simply *liquefaction*. For loose sand, the initial

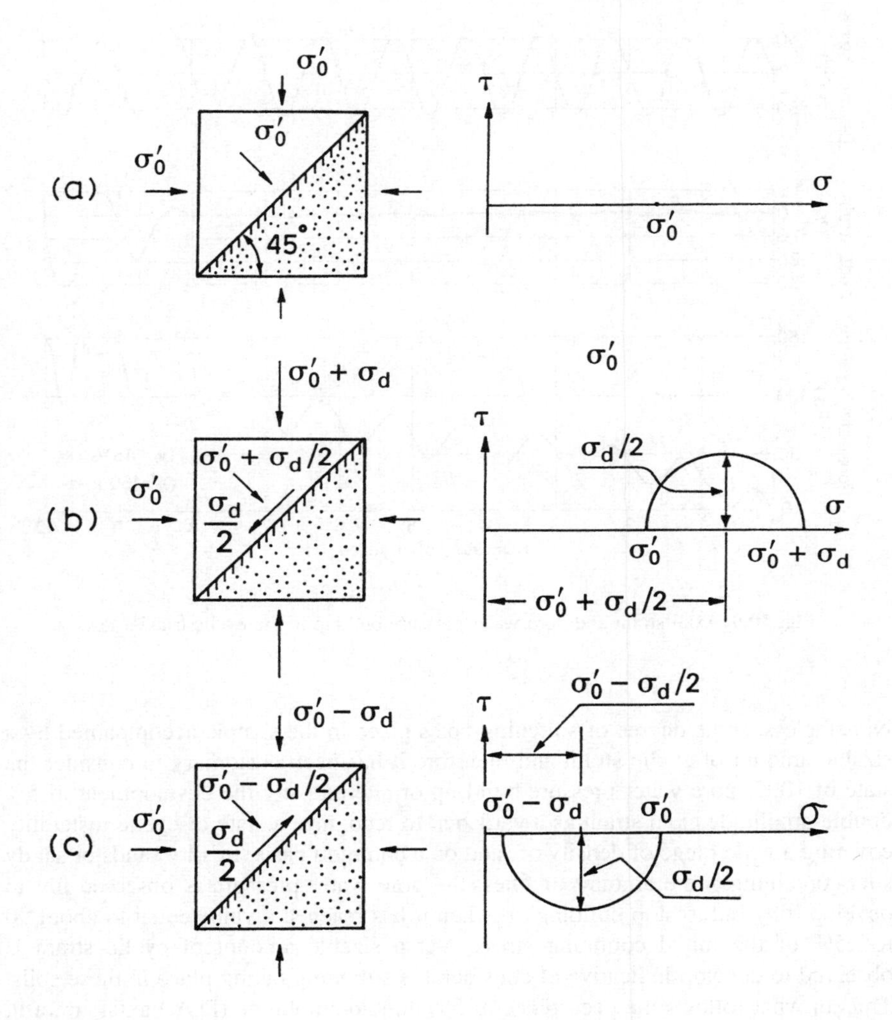

Fig. 10.8 Simulation of geostatic and cyclic stresses in the sample of cyclic triaxial load application.

liquefaction can certainly be taken as a state of softening, because infinitely large deformation is produced suddenly with complete loss of strength during or immediately following the 100% pore water pressure build-up. For medium dense to dense sand, a state of softening is also produced with the 100% pore water pressure build-up, accompanied by about 5% double-amplitude axial strain. However the deformation thereafter does not grow indefinitely large and complete loss of strength does not take place in the sample even after the onset of initial liquefaction.

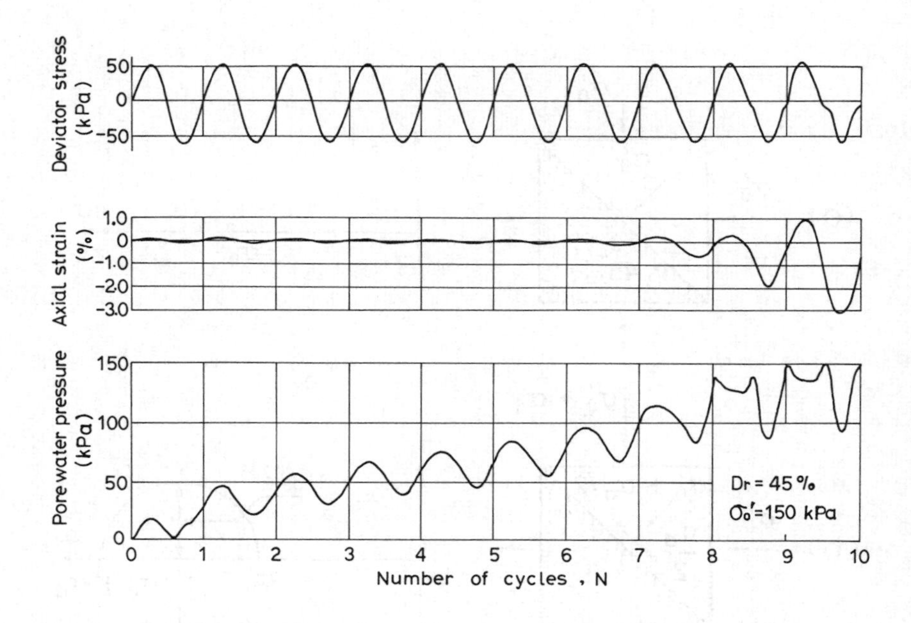

Fig. 10.9 Axial strain and pore water pressure buildup in the cyclic triaxial test.

Nonetheless, some degree of softening takes place in the sample accompanied by a sizable amount of cyclic strain and therefore it has been customary to consider the state of 100% pore water pressure build-up or alternatively the development of 5% double-amplitude axial strain as a yardstick to recognize a state of cyclic instability covering a wide range of density of sand on a common basis. In silty sands or sandy silts containing some amount of fines, the pore water pressure is observed not to develop fully but to stop building up when it has reached a value equal to about 90 to 95% of the initial confining stress. Yet a sizable amount of cyclic strain is observed to develop indicative of considerable softening taking place in these soils. Thus in what follows the occurrence of 5% double-amplitude (D.A.) axial strain in the cyclic triaxial test will be taken up as a criterion to coherently define the state of cyclic softening or liquefaction of the soils covering from clean sands to fines containing sands.

In order to specify the onset of liquefaction or development of 5% D.A. axial strain, the number of cycles needs to be specified in the constant-amplitude uniform cyclic loading. In principle the number of cycles may be set as arbitrary, if a pertinently chosen correction factor is incorporated to evaluate the irregular nature of seismic loading, but it has been customary to consider 10 or 20 cycles in view of the typical number of significant cycles being present in many of actual time histories of accelerations recorded during past earthquakes. Thus the onset condition of liquefaction or cyclic softening is specified in terms of the magnitude of cyclic stress ratio required to produce 5% double-amplitude axial strain in 20 cycles of uniform load application. This cyclic stress ratio is often referred to simply as cyclic strength.

10.4 Cyclic resistance of reconstituted clean sand

The potential for liquefaction of saturated sands under seismic loading conditions has been extensively investigated by many workers by means of cyclic triaxial tests, cyclic simple shear tests and cyclic torsional tests.

Reviews on the early stage of development in this aspect are described in a series of comprehensive state-of-the-art papers by Yoshimi *et al.* (1977), Seed (1979), and Finn (1981). The outcome of these studies has generally confirmed that the resistance to liquefaction of the samples of clean sand reconstituted in the laboratory is influenced primarily by the factors such as initial confining stress, intensity of shaking as represented by the cyclic shear stress, the number of cyclic stress application and void ratio or relative density. As a result of comprehensive laboratory studies, it has been recognized as reasonable and become customary to consider the combined effect of cyclic shear stress and initial confining stress in terms of the cyclic stress ratio which is defined as $\sigma_d/(2\sigma'_0)$ for the triaxial loading condition. Thus it has become a routine practice to take the cyclic stress ratio required to cause 5% D.A. axial strain under 20 load cycles as a factor quantifying the liquefaction resistance of sand under a given state of packing as represented by void ratio or relative density. This cyclic stress is represented by $\left[\sigma_{d\ell}/(2\sigma'_0)\right]_{20}$ and is referred to as the cyclic strength. It has also been observed that the resistance to liquefaction as quantified above tends to increase with an increase in relative density at which the sample is prepared for the laboratory tests. Typical examples of the tests confirming this effect of density were presented by Tatsuoka *et al.* (1986b) as shown in Fig. 10.10 where the cyclic stress ratio required to cause 5% double-amplitude (D.A.) axial strain in 20 cycles of loading is plotted versus the relative density of Toyoura sand. It may be seen that up to a relative density of 70%, the cyclic stress ratio tends to increase linearly with the relative density, but for the density in excess of 70%, the cyclic strength goes up sharply.

While the cyclic strength was shown to be uniquely related with the density, studies by Ladd (1974), Mulilis *et al.* (1977) and Tatsuoka *et al.* (1986a) have indicated that, even when the relative density is kept constant, samples prepared by different methods could exhibit different resistance to liquefaction. Shown in Fig. 10.11 are the results of cyclic triaxial tests which were performed by Mulilis *et al.* (1977) on saturated samples of Monterey No. 0 sand prepared by two different methods. In one method called air pluviation, oven-dry sand was continuously poured into the sample forming mould, and after saturation and consolidation, the sample was subjected undrained to cyclic axial stress until it softened to develop a significant amount of cyclic strain. In another method called moist tamping, the sand mixed with 8% moisture content was spread in the mould and compacted in layers with a tamping rod so as to attain a desired density. Among several other methods employed by Mulilis *et al.* (1977), the above two methods were shown to produce the samples having the lowest and greatest resistance to liquefaction. It can be seen

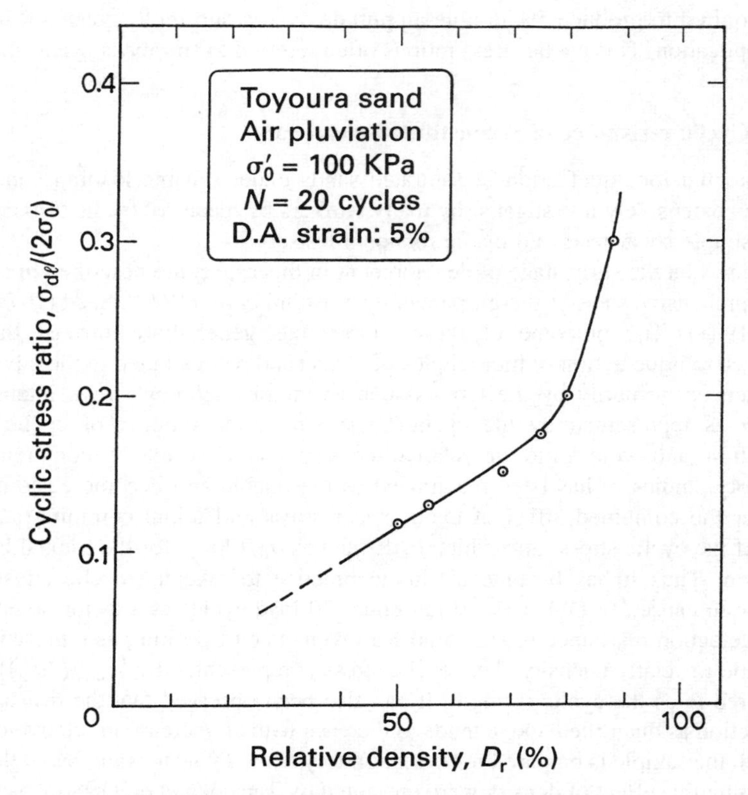

Fig. 10.10 Cyclic strength versus the relative density.

in Fig. 10.11 that there is a fairly large range in which the cyclic resistance of sand can vary depending upon the nature of fabric structure created by different methods of sample preparation. Thus, it has been recognized as important to specify a method of sample preparation if cyclic tests are to be run on reconstituted samples. In addition, it has been considered almost mandatory to conduct tests on undisturbed samples if the cyclic resistance of in situ sand deposits is to be evaluated with a reasonable level of confidence.

In view of the variability due to the sample preparation as above and also because of the diversity of test results due to other testing details, an effort was made by Silver *et al.* (1976) to implement a cooperative testing program in the US in which eight organizations were requested to conduct a series of cyclic triaxial tests under specified conditions regarding equipment, test performance, and data presentation. The Monterey No. 0 sand with $D_{50} = 0.36$ mm, $U_c = 1.5$, $e_{max} = 0.85$, and $e_{min} = 0.56$ was used to prepare test specimens of a relative density of 60% by means of the moist tamping as described above. The samples were consolidated to a confining stress of $\sigma'_0 = 100$ kPa and subjected undrained to cyclic loading. The summarized results of this cooperative test program are demonstrated in Fig. 10.12

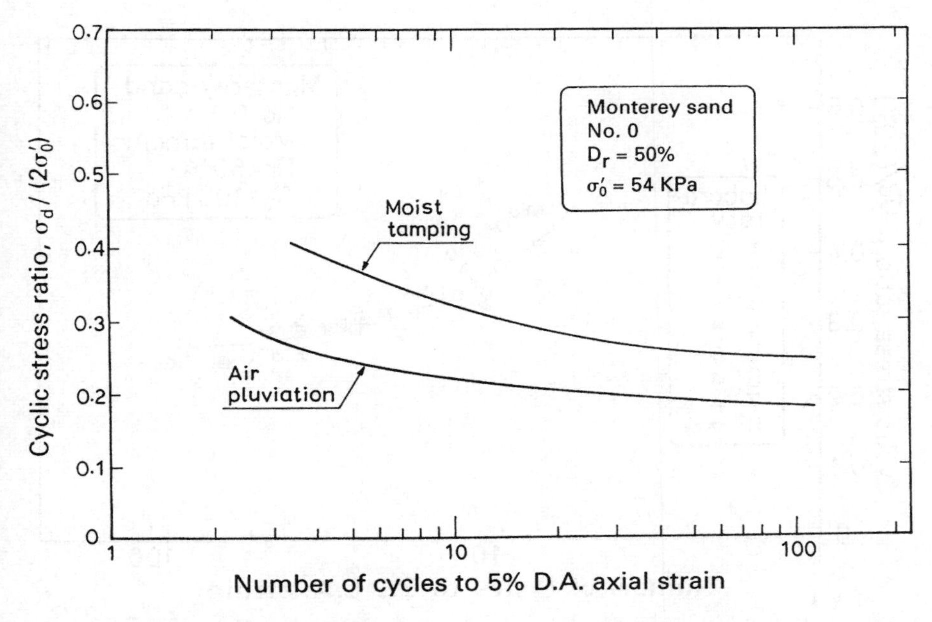

Fig. 10.11 Effects of sample preparation on cyclic strength of sand (after Mulilis *et al.* 1977).

in terms of the cyclic stress ratio plotted versus the number of cycles required to produce 5% double-amplitude axial strain. It may be seen that the test data lie in a rather narrow band indicating a reasonable level of consistency in the cyclic strength between various laboratories. From the average curve drawn in Fig. 10.12, the cyclic stress ratio causing 5% D.A. strain in 20 cycles of loading is read off as being 0.31. If this cyclic stress ratio is assumed to change in proportion to the relative density, the cyclic stress ratio corresponding to $D_r = 50\%$ may be evaluated as $0.31 \times 50/60 = 0.26$. This is the value of cyclic strength which can be compared against the value in Fig. 10.11 obtained by Mulilis *et al.* (1977). For the sample formed by moist tamping, the cyclic stress ratio causing 5% D.A. strain in 20 cycles of loading is read off from the curve of Fig. 10.11 as being 0.28. If allowance is taken for the difference in the initial confining stress, the cyclic strength obtained by the cooperative tests should be considered to be in good agreement with that obtained previously by Mulilis *et al.* (1977).

A similar attempt of coalescent tests was executed by the Research Committee of the Japanese Society of Soil Mechanics and Foundation Engineering (JSSMFE) with an aim to diffuse the use of some concerted testing procedures among many organizations involved in geotechnical testing. A detailed account of this undertaking is described by Toki *et al.* (1986) and Tatsuoka *et al.* (1986b). In this program, five laboratories participated and each was requested to perform a series of tests on samples of Toyoura sand prepared at relative densities of 50 and 80% by using the cyclic triaxial test equipment of their own. It was stipulated that

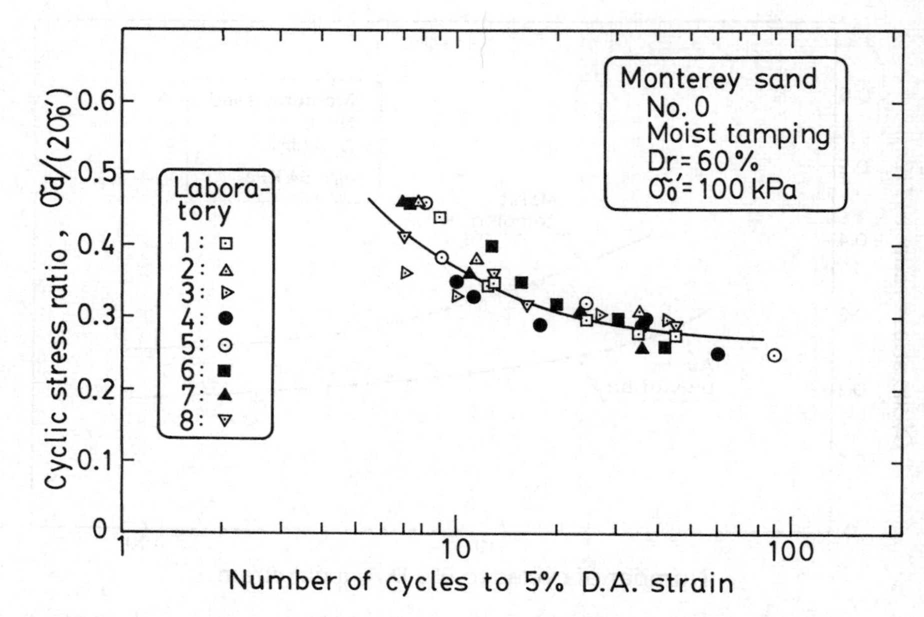

Fig. 10.12 Results of the co-operative tests in the US on the cyclic strength of sand (Silver *et al.* 1976).

the samples be prepared by the method of air pluviation where dried sand was poured in the sample forming mould from a funnel in a specified manner. All the samples were consolidated isotropically under a confining pressure of $\sigma'_0 = 98$ kPa and cyclic loads were applied undrained until they produced a state of cyclic softening with attainment of the D.A. axial strain of 5%.

Since each individual sample had been prepared with a density slightly deviating from the specified value, a correction was made to the test data for obtaining the cyclic shear strength corresponding to the relative density of 50% and 80%. The outcome of the cooperative tests thus codified is presented in Fig. 10.13, where the cyclic stress ratio is plotted versus the number of cycles required to produce 5% D.A. axial strain. In Fig. 10.13 the test data from five laboratories are indicated by different signs, but those on samples 5 cm in diameter are shown by open marks and those on samples 7–10 cm in diameter by full marks. It may be seen that the data from several sources fall within a relatively narrow band, indicating a reasonable degree of coincidence among the values of cyclic strength from various laboratories. There is a tendency, however, for the smaller-size specimens to show slightly greater resistance to cyclic softening as compared to the large samples. This difference is attributed to the effects of system compliance arising from membrane penetration, sample seating, or tubing in the test apparatus. The solid line in Fig. 10.13 gives an approximate boundary for the two groups of data from different sample sizes. This curve may also be taken as an average of the data from different laboratories. The

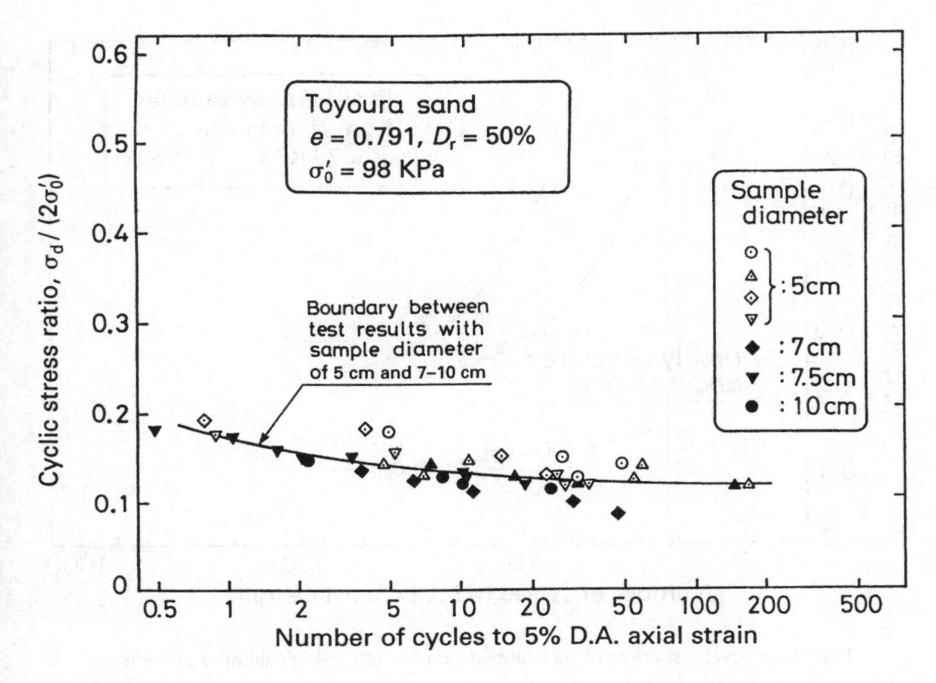

Fig. 10.13 Results of the co-operative tests in Japan on the cyclic strength of sand (Toki *et al.* 1986).

test data on samples with a relative density of 80% showed similar general trends, but there was considerable scatter in the range of large cyclic stress ratio corresponding to a small number of cycles. This scatter is believed to have resulted from the highly dilatant nature of the dense samples which are prone to fluctuation in their behaviour depending upon the details of the test procedures. The cyclic stress ratio causing 5% D.A. strain in 20 cycles of loading is read off from Fig. 10.13 as being 0.14 for the samples with a relative density of $D_r = 50\%$. Since all the samples were prepared by the air pulviation method, this value of cyclic strength is to be compared with the corresponding test data by Mulilis *et al.* (1977) shown in Fig. 10.11 which indicates a similarly defined cyclic strength of 0.21. Such a large difference can not be explained properly, but Toyoura sand with a much finer grain size appears to exhibit smaller resistance to liquefaction.

10.5 Cyclic resistance of in situ deposits of sands

In view of the diversity of cyclic strength of sand samples reconstituted by different methods of preparation, it has been recognized that deposits of sands in the field might exhibit varying resistance to seismic load application. Thus there has been an increasing demand for any effort to recover as perfectly undisturbed

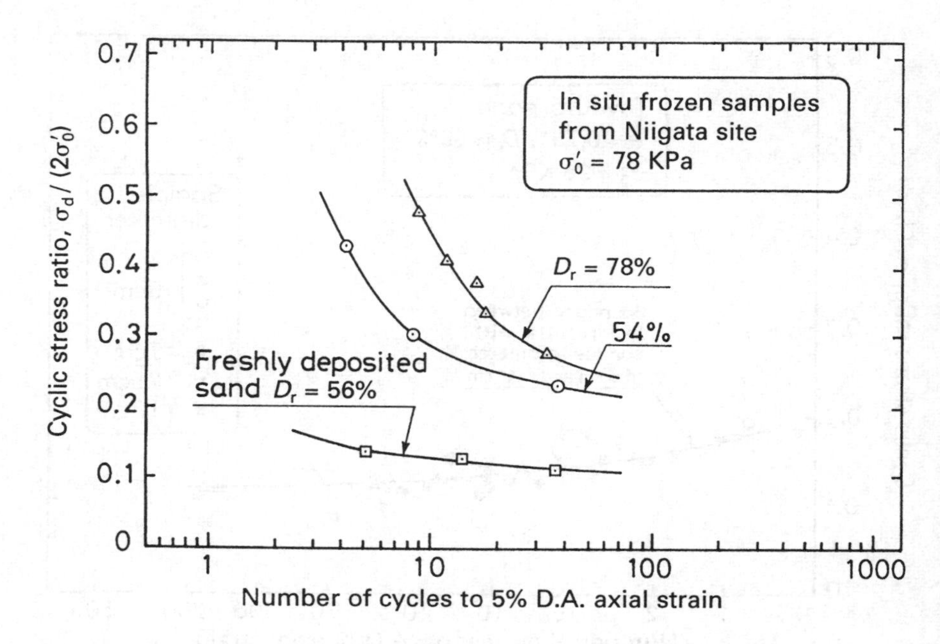

Fig. 10.14 Cyclic strength of undisturbed samples of sand (Yoshimi *et al.* 1989).

samples as possible from in situ sand deposits and to test them in the laboratory under conditions representative of those prevailing in the field. The in situ techniques used to recover undisturbed samples of sands from below the ground water table are divided into two: tube sampling and the ground freezing method. The tube sampling technique has been shown to be useful for recovering undisturbed samples from loose deposits of sands but adverse effects due to sample disturbance become pronounced with increasing density of sands (Ishihara, 1985).

The ground freezing method has been developed instead and used successfully to recover high-quality undisturbed samples even from dense deposits of sands. In a comprehensive investigation by Yoshimi *et al.* (1984; 1989), undisturbed samples were recovered by the freezing technique from in situ sand deposits in Niigata. At the same time, a man-made fill was provided in a large bin 4 m wide, 6 m long, and 5 m deep by letting the sand settle out underwater. Undisturbed samples were also secured from this freshly deposited fill by means of the freezing method. A series of cyclic triaxial tests were conducted on the undisturbed sample thus obtained (Yoshimi *et al.* 1989). The outcome of the test program is demonstrated in Fig. 10.14, where it can be seen that, for clean sands with relative densities of about 50% and 80%, the cyclic resistance of undisturbed samples from the in situ deposit is about twice as great as the cyclic resistance of the samples from the newly deposited sand fill. The cyclic strength of in situ deposits is thus considered to vary greatly depending upon aging and the inherent fabric structure of sands created under different depositional conditions. For this reason, the test results on reconstituted samples should be considered generally not to reflect true behaviour of in situ sands, and tests on high-quality undisturbed samples are always needed to precisely evaluate performances of in situ deposits of sands during earthquakes.

10.6 Cyclic resistance of silty sands

Liquefaction is a state of particle suspension resulting from release of contacts between particles of sands constituting a deposit. Therefore the type of soil most susceptible to liquefaction is the one in which the resistance to deformation is mobilized by friction between particles under the influence of confining pressures. When the soil is fine grained or contains some amount of fines, cohesion or adhesion tends to develop between fine particles thereby making it difficult for them to be separated from each other. Consequently, a greater resistance to liquefaction is generally exhibited by the sand containing some fines. However this tendency depends on the nature of the fines contained in the sand. If the fines consist of minerals with a dry surface texture free from adhesion, they will easily permit separation of individual particles and therefore the sand containing such fines will exhibit as great a potential to liquefaction as does the clean sand. A typical example of such fines is tailings materials produced as residue in the concentration process of ore in the mine industry. Since the tailings consist basically of ground-up rocks, it still preserves the hardness of parent rocks with a dry surface. Therefore the rock flour, in its water-saturated condition, does not possess significant cohesion and behaves as if it were clean sand. Thus tailings have been shown to exhibit as low resistance to liquefaction as the clean sand (Ishihara *et al.* 1980).

The degree of liquefiability of the sand containing more or less cohesive fines such as those found in the fluvial deposits has been investigated in the laboratory tests by Ishihara *et al.* (1978). The outcome of these studies equally disclosed that, with increasing content of fines, the cyclic resistance of sand tends to increase to a certain extent under its normally consolidated state, but to a greater extent if it is overconsolidated. This tendency may be taken for granted because adhesion between fine particles tends to prevent separation of individual particles when the sand is about to liquefy. Thus the sand containing such plastic fines generally exhibits a higher resistance to liquefaction. However, since the effects of fines are manifested variously depending upon the nature of the fines themselves, of prime importance among other things would be to seek a key parameter which is capable of more specifically quantifying the liquefaction characteristics of fines-containing sands.

It is commonly observed that the relative density fails to be an appropriate index parameter if the fines content is greater than about 50%. Compilation of several series of laboratory test data has shown that the most important index property influencing the cyclic strength is the plasticity index of the fines contained in the sand (Ishihara and Koseki, 1989). This is clearly demonstrated in the results summarized in Fig. 10.15, where the cyclic strength is plotted against the plasticity

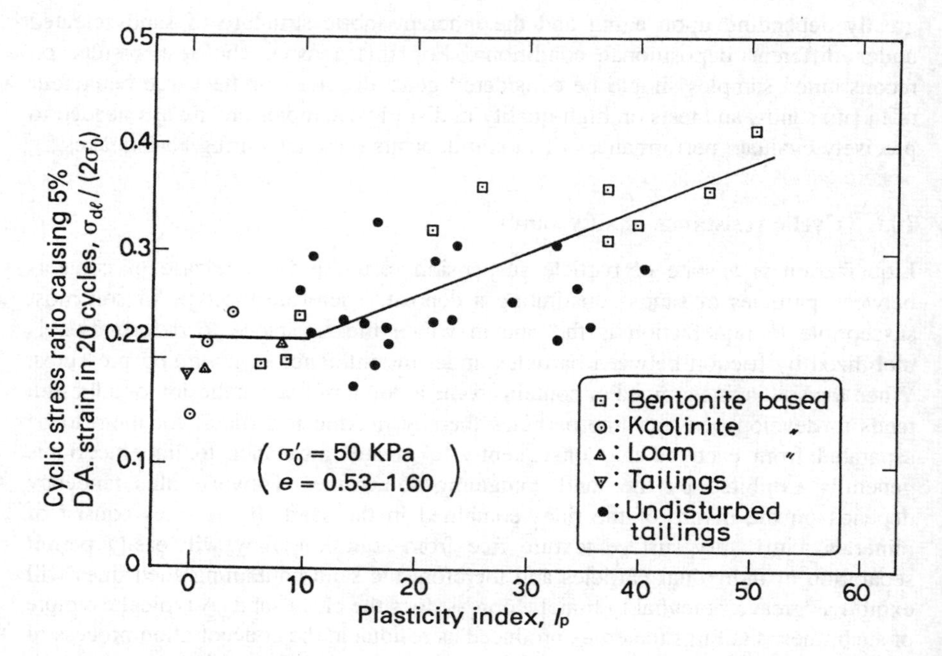

Fig. 10.15 Effects of plasticity index on the cyclic strength of fines-containing sand.

index I_p of the materials used in the tests. The cyclic strength does not change much for the low plasticity range, below $I_p = 10$, but increases thereafter with increasing plasticity index.

10.7 Cyclic resistance of gravelly soils

Cases have occasionally been reported where liquefaction-associated damage took place in gravelly soils. For instance, at the time of the Fukui earthquake of June 28, 1948 in Japan, signs of disastrous liquefaction were observed in an area of alluvial fan deposits near the epicenter where gravel was predominant. During the 1983 Borah Peak earthquake in the US, liquefaction was reported to have occurred in gravelly soil deposits at several sites, causing lateral spreading over the gently sloping hillsides. While the drainage conditions surrounding gravelly deposits may exert some influence on the dissipation of pore water pressure and hence on the liquefiability, it is of prime importance to clarify the resistance of gravelly sand itself to cyclic loading. One of the earlier endeavours in this context was made by Wong *et al.* (1975) who performed a series of cyclic triaxial tests on reconstituted specimens of gravelly soils with different gradation by means of a large-size triaxial test apparatus. The results of these tests indicated somewhat higher cyclic strength as compared to the strength of clean sands. However, whether the result of such tests reflects the cyclic strength of in situ deposits remained open to question.

More recently, an attempt was made by Kokusho and Tanaka (1994) to recover undisturbed block samples from a volcanic debris deposit containing gravels where liquefaction-associated damage took place in nearby private dwellings at the time of the Hokkaido–Nanseioki Earthquake of July 12, 1993 in Japan. The block samples recovered by the ground freezing technique were cut into pieces of cylindrical specimens 30 cm in diameter and 60 cm in length and tested by using the triaxial test apparatus. The grain size distribution curves of the specimens are shown in Fig. 10.16. The outcome of the cyclic triaxial tests is shown in Fig. 10.17 where it can be seen that the cyclic strength is surprisingly small, taking a value of 0.19 which is the same order of magnitude as the strength of loose to medium sand.

The above is the only one example of tests on undisturbed samples from loose deposits of gravelly soils and no other cases of detailed testing have been reported particularly for gravelly soils from loose deposits of alluvial origin.

On the other hand, many efforts have been expended recently in Japan to clarify the cyclic behaviour of moderately dense to dense deposits of gravelly soils. The incentive has been instigated by the need for developing techniques and methodologies to verify the safety of sites of nuclear facilities which ought to be designed considering a strong intensity of shaking. Figure 10.18 shows the grain size distribution curves of the gravelly soils tested in the laboratory to determine their cyclic strength. All the samples were recovered by the ground freezing technique and cut into specimens 30 cm in diameter and 60 cm in length. The samples from site B in Chiba were recovered from deposits at depths of 6–8 m. They are of Pleistocene origin and had a maximum particle size of 94 mm with a gravel content of 55%. The samples from site A were recovered from depths 5–15 m. This site is located near the Tone river and of alluvial origin. The samples had a maximum particle size of 105 mm with a gravel content of 66% (Goto *et al.* 1994). The samples from the Tokyo station were recovered from a Pleistocene deposit at depths 19–22 m. The range of the grading curve in Fig. 10.18 indicates that the maximum grain size is 90 mm and the gravel content is between 80 and 50%. (Hatanaka *et al.* 1988). The outcome of the cyclic triaxial tests on these samples is presented in Fig. 10.19 in terms of the cyclic stress ratio plotted versus the number of cycles required to induce 2% double-amplitude axial strain. Because of the stiffness of dense gravelly soils being generally high, the double amplitude of 5% was difficult to be induced in some of the test specimens. Therefore, the double amplitude of 2% is taken as a criterion to identify the state of cyclic softening. Figure 10.19 shows that the cyclic strength for 20 cycles is generally high, being of the order of 0.4 to 1.2, as compared to that of sandy soils. The results of other series of cyclic undrained triaxial test on stiff gravelly soils are shown in Fig. 10.20. The grading curve of the material from Site T is shown in Fig. 6.17. The gravelly soil from Site A has about the same grading characteristics as that of Site T. The test results indicate again high cyclic strength of the order of 0.3 to 0.5. Looking over the results of tests on undisturbed specimens as above, it may be mentioned that, for the gravelly deposits of Pleistocene era, the cyclic strength is significantly higher than that generally encountered in loose to medium loose sand deposits of alluvial origin. It may also be noted that there is a wide range of value of the cyclic strength because

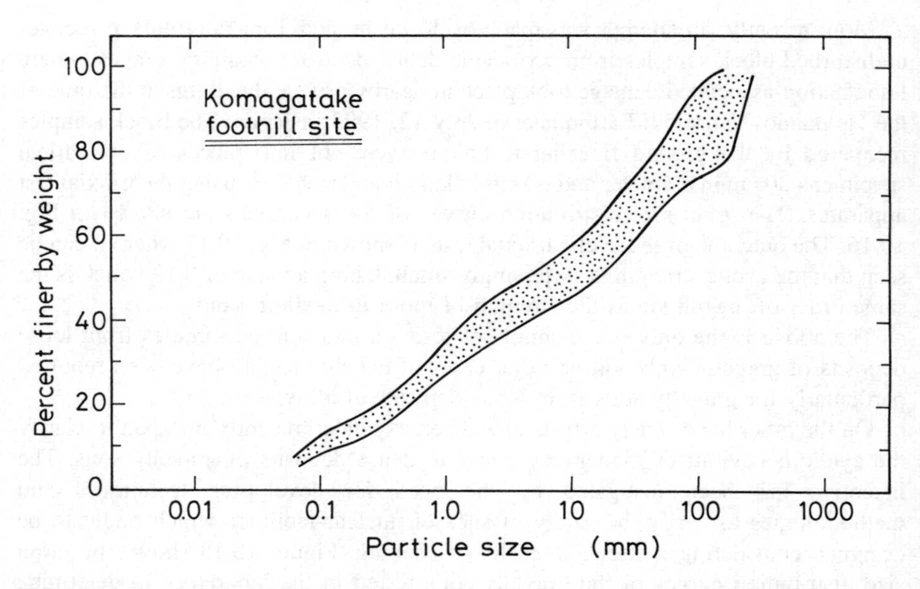

Fig. 10.16 Range of grading for gravelly soils from Komagatake foothill site (Kokusho and Tanaka, 1994).

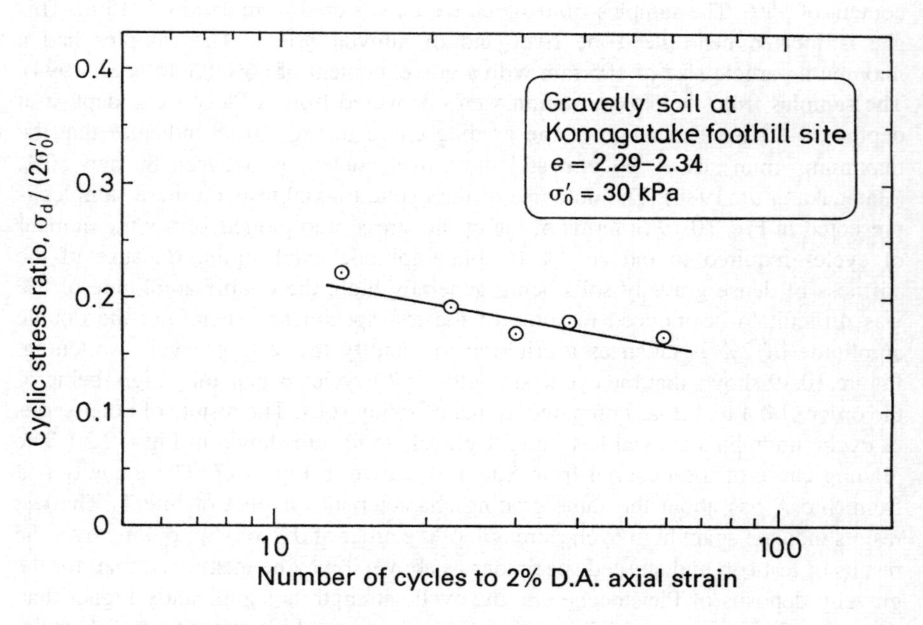

Fig. 10.17 Cyclic stress ratio versus the number of cycles for loose gravelly soils (Kokusho *et al.* 1995).

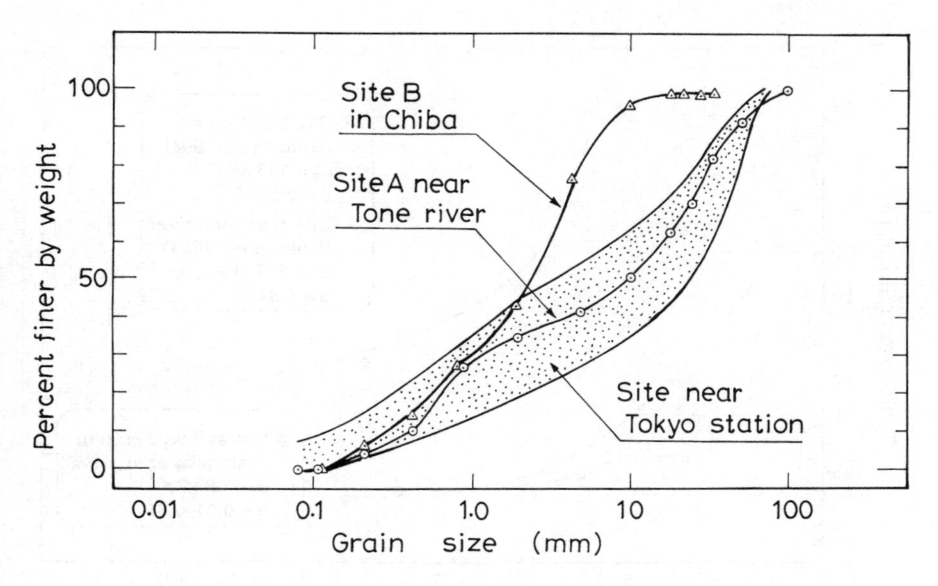

Fig. 10.18 Grain size distribution curves of gravelly soils tested.

of the diverse variation in the depositional environments affecting intact gravelly soils in the field.

10.8 Effects of K_0 conditions on liquefaction resistance of sand

To investigate the influence of K_0 conditions on liquefaction resistance, a series of torsional tests with lateral confinement (ACOT test) was carried out on hollow cylindrical samples of Fuji river sand ($D_{50} = 0.40$ mm, $e_{max} = 1.03$, $e_{min} = 0.48$) with a relative density of 55% using the torsional test apparatus. The observation in this series of tests confirmed the previously stated assumption that when saturated sand consolidated under K_0 conditions is subjected to cyclic shear stresses with the lateral deformation perfectly confined, the lateral stress will change to produce an isotropic state of stress upon liquefaction. In order to see the effect of K_0 consolidation on the liquefaction potential of sand, the cyclic stress ratio is now defined as the amplitude of cyclic torsional stress τ_d divided by the vertical effective confining stress $\sigma_v{}'$ at the time of consolidation and plotted in Fig. 10.21 versus the number of cycles required to cause 5% D.A. torsional strain or initial liquefaction. It is clearly seen that the larger the K_0 value at the time of consolidation, the greater the resistance to liquefaction. To examine the effects of K_0 conditions, the vertical confining stress $\sigma_v{}'$ was converted to the mean effective confining stress $\sigma_0{}'$ through the relation,

$$\sigma_0{}' = \frac{1 + 2K_0}{3}\sigma_v{}' = C_1\sigma_v{}'. \tag{10.2}$$

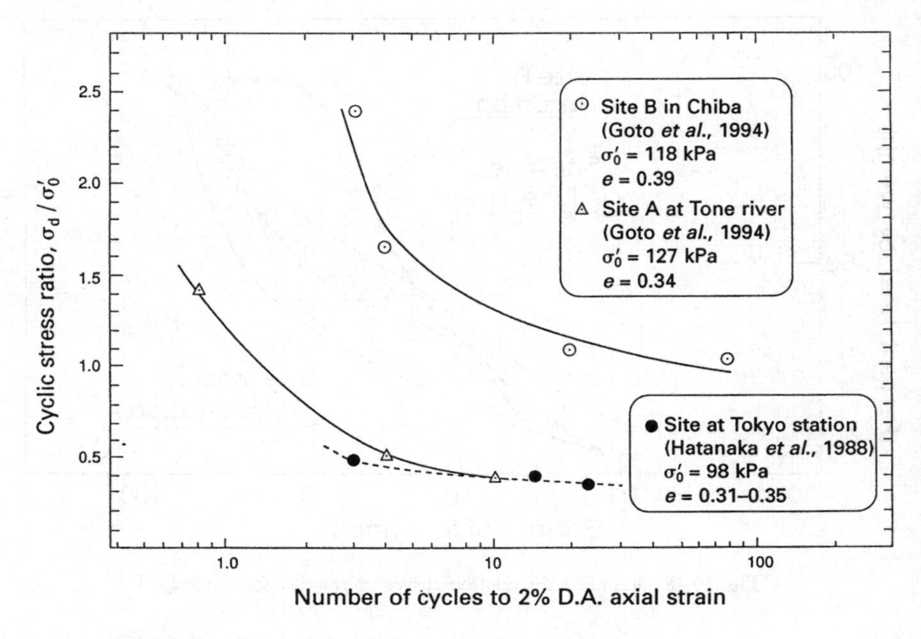

Fig. 10.19 Cyclic stress ratio versus number of cycles for gravelly soils.

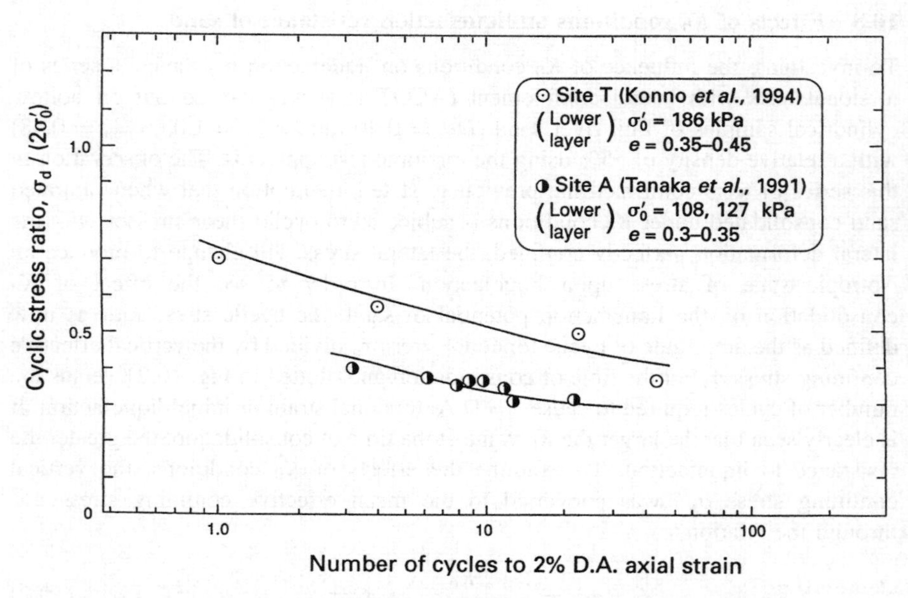

Fig. 10.20 Cyclic stress ratio versus number of cycles for gravelly soils.

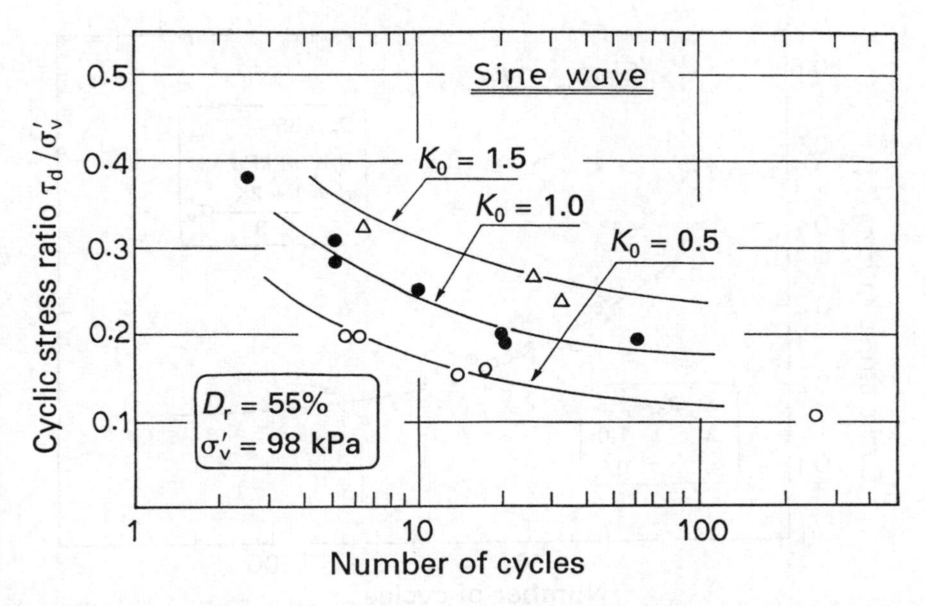

Fig. 10.21 Effects of K_0 consolidation on cyclic strength under the condition of lateral constraint.

If the cyclic stress ratio $\tau_d/\sigma_v{}'$ in the ordinate of Fig. 10.21 is changed to the cyclic stress ratio $\tau_d/\sigma_0{}'$, the test data can be rearranged as shown in Fig. 10.22, where it may be seen that all the data under different K_0 conditions are correlated uniquely with the number of cycles. Thus, it can be concluded that the effects of K_0 condition can be properly evaluated by taking the cyclic stress ratio with respect to the initial mean effective confining stress as given by eqn (10.2).

10.9 Cyclic resistance of sand under irregular seismic loading

Time histories in shear stress application due to the upward propagation of shear waves through the level ground are essentially irregular and multi-directional when viewed on the horizontal plane. In order to quantitatively represent the liquefaction characteristics of sand deposits in such a complicated loading environment, it has been a common practice to introduce some coefficients correcting for the cyclic strength under uniform cyclic loading. To evaluate the correction coefficient, multiple series of laboratory tests have been carried out using simple shear test equipment in which pairs of loads with any irregular time histories could be applied to the specimen in two mutually perpendicular directions (Ishihara and Nagase, 1988). The test apparatus used is shown in Fig. 4.14.

One of the results of tests on loose sand employing the NS- and EW-components of acceleration time histories recorded during the Niigata earthquake (1964) is demonstrated in Fig. 10.23. The two components of the acceleration time histories are shown in Fig. 10.23(a). A trajectory on the horizontal plane traced by

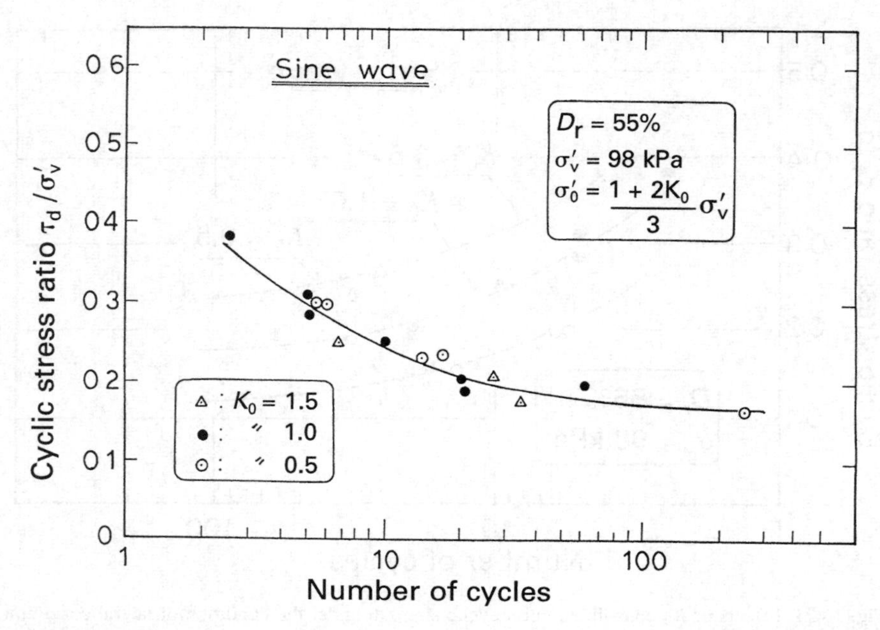

Fig. 10.22 Cyclic stress ratio versus the number of cycles for the ACOT tests with different K_0 conditions.

combination of these two components for a certain time span of main shaking is demonstrated in Fig. 10.24.

The two components were converted to the irregular patterns of simple shear stress alteration and applied to specimens of sand in two mutually perpendicular directions using the simple shear test device shown in Fig. 4.4. The tests were carried out first by setting the amplitude of irregular load at an appropriate level not large enough to induce liquefaction through the entire duration of load application. In the first sequence of the test, the amplitude of irregular shear stress was set equal to $\tau_{max}/\sigma'_0 = 0.112$ and changes in pore water pressure u and simple shear strains in two directions γ_{NS} and γ_{EW} were recorded. The recorded data are shown in Fig. 10.23(c), where it can be seen that the pore water pressure built up to about 60% of the initial confining pressure at the time when the maximum shear stress occurred, but it remained constant thereafter although the sample was still undergoing shear stress changes with lesser amplitude. The pore water pressure developed during this sequence of test will be termed the residual pore water pressure. It is to be noted that the change in shear strain in this test sequence is inconsequentially small. In the next sequence of the test, the same irregular load but with an increased amplitude of $\tau_{max}/\sigma'_0 = 0.154$ was applied to a new specimen prepared to identical density. The records obtained in this test sequence are shown in Fig. 10.23(b), where it can be seen that the pore water pressure rose to a value equal to the initial confining stress, indicating the onset of initial liquefaction in the test

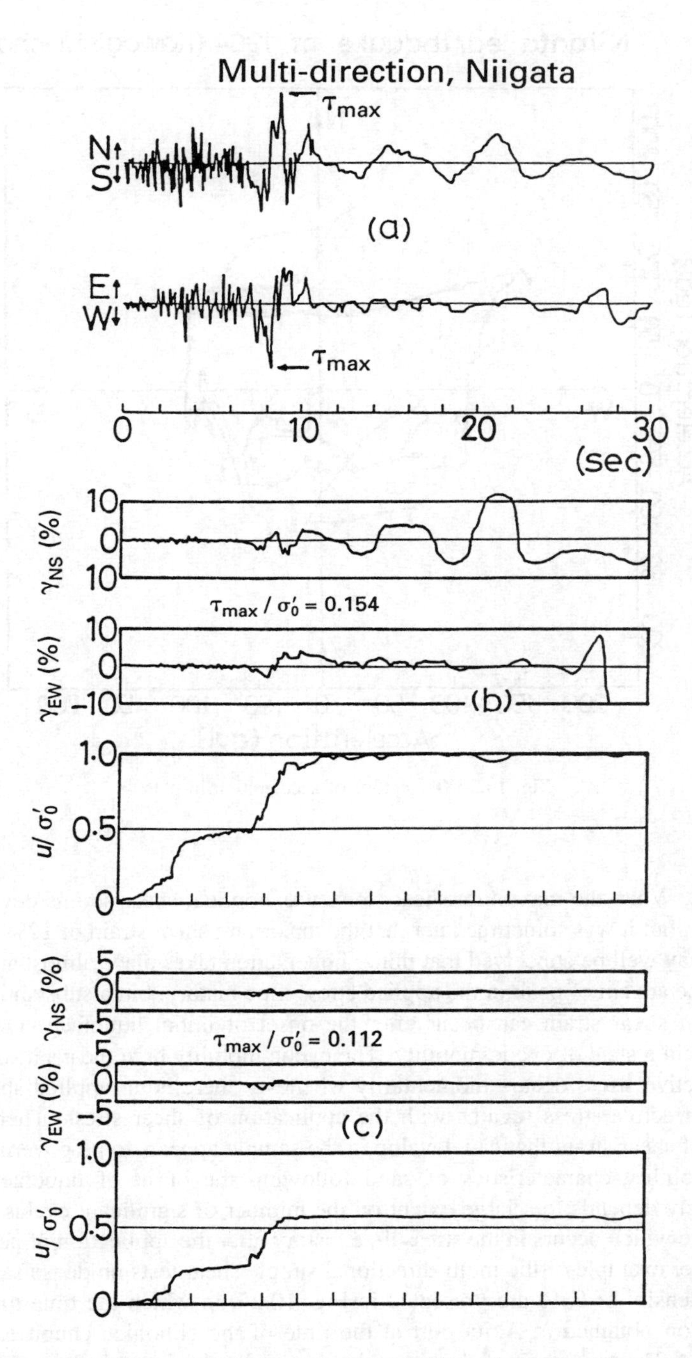

Fig. 10.23 Records of shear strain and pore pressure build-up in torsional tests employing the irregular time history of a load.

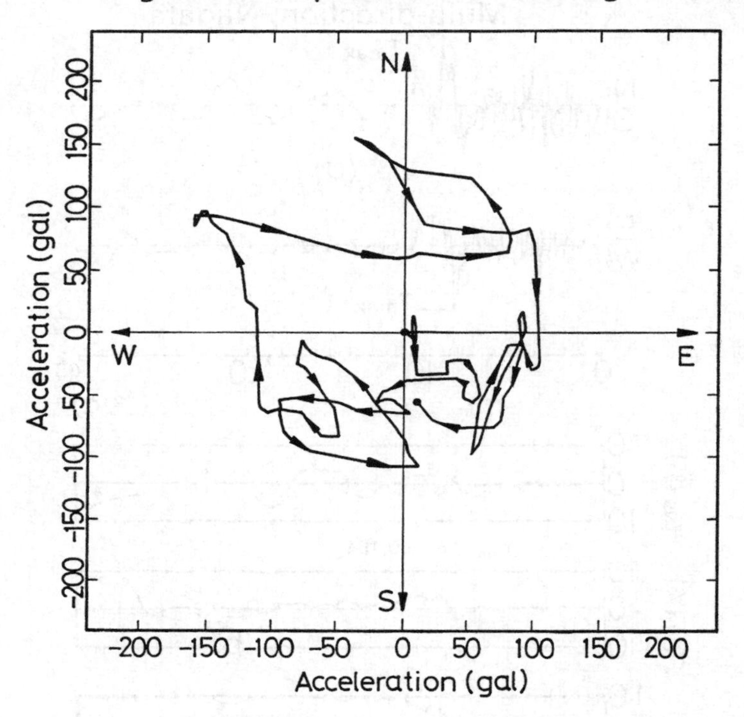

Fig. 10.24 Trajectory of acceleration in plan.

specimen. When the maximum shear stress was applied, shear strains developed to about 5%, but it was sometime later that the maximum shear strain of 12% occurred. Thus it may well be conceived that initial liquefaction takes place almost at the same time as the advent of peak in the applied stress time history, but a substantially large amount of shear strain can occur after the onset of initial liquefaction where the sample is in a state of cyclic mobility. The cyclic mobility here means a state where zero effective stress occurs momentarily whenever there is no applied shear stress but the effective stress regains with the application of shear stress. Therefore the amount of shear strain that can develop in the sample appears to be governed by the cyclic mobility characteristics of sand following the onset of liquefaction, and accordingly depends to a large extent on the number of significant cycles and their magnitude which occurs in the stress time history after the application of peak stress.

Another example of the multi-directional simple shear tests on dense sand with a relative density of 93% are presented in Fig. 10.25, in which the time histories of acceleration obtained at Akita port at the time of the Nihonkai-chubu earthquake were used. The trajectory of the time histories in the horizontal plane is shown in Fig. 10.26. The overall trend in the response of pore water pressure and shear strains

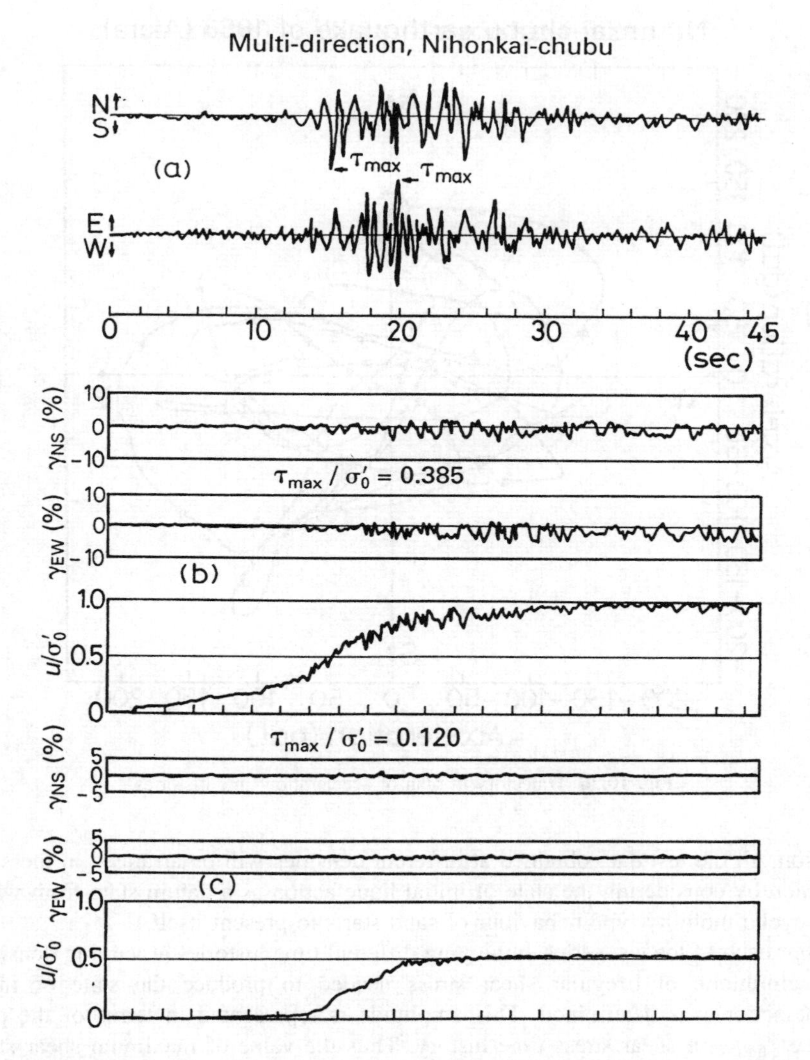

Fig. 10.25 Records of shear strain and pore pressure build-up in torsional tests employing the irregular time history of a load.

is similar to those observed in the loose sample of sand, except for the fact that the pore water pressure barely reached a value exactly equal to the initial confining stress. Although this is generally the case with increasing density of the sand, the attainment of a state of near-liquefaction with the pore water pressure becoming stationary in close proximity to the initial confining stress produced a state of sand behaviour resembling the initial liquefaction with the D.A. strain of 5% which is observed in the loose samples of sand under cyclic loading conditions. For this

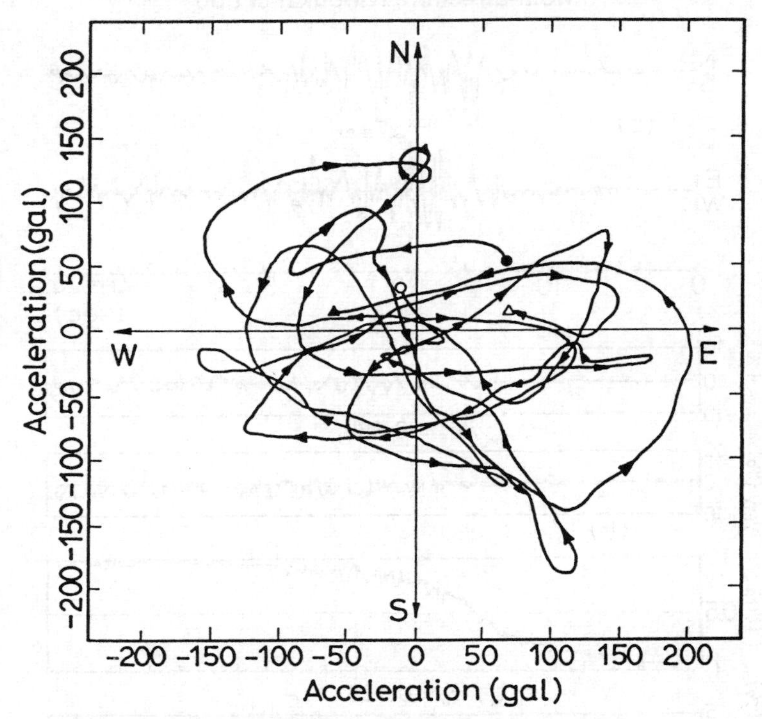

Fig. 10.26 Trajectory in plan of acceleration time histories.

reason, all the test data obtained at different densities will be arranged in the same fashion by considering the state of initial liquefaction as a datum state from which the cyclic mobility type behaviour of sand starts to present itself.

For each of the test series employing different time histories at varying densities, the amplitude of irregular shear stress needed to produce the state of initial liquefaction was determined. This amplitude is represented in terms of the peak value $\tau_{\max,\ell}$ in shear stress time history. Thus the value of maximum shear stress ratio $\tau_{max,\ell}/\sigma'_0$ was obtained for each of the test series. Note that the simple shear tests were run on isotropically consolidated samples.

On the other hand, tests with uniform loading in one direction only were conducted using the same simple shear test apparatus and the cyclic stress ratio causing different levels of shear strain amplitudes under 20 cycles of constant-amplitude cyclic stress application was determined for the sand specimen prepared under identical conditions. By comparing the results of these two types of tests, the coefficients for correction were established. The coefficient allowing for the effects of load irregularity in one direction alone will be denoted by C_2 and the coefficient C_5 will be used to represent the effects of multi-directionality in the seismic loading. Thus, the coefficient $C_2 C_5$ represents combined effects of load irregularity and

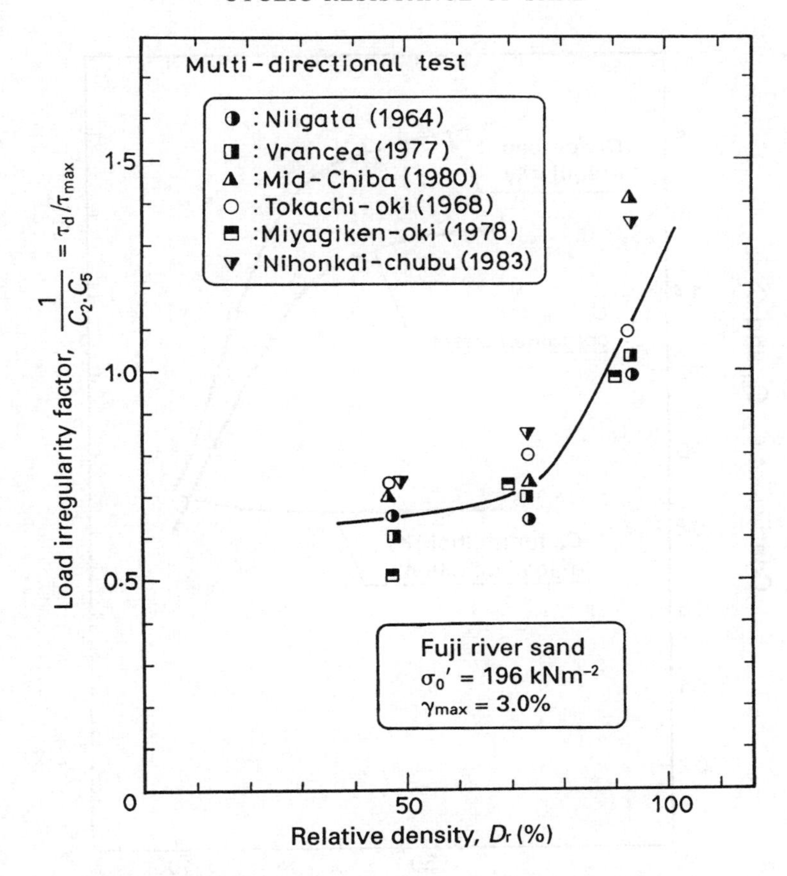

Fig. 10.27 Load irregularity factor as a function of relative density.

multi-directionality. Using these two coefficients, the maximum shear stress ratio $\tau_{\max,\ell}/\sigma'_0$ causing some specified level of shear strain, in multi-directional irregular loading can be correlated, as follows, with the cyclic stress ratio $(\tau_{d\ell}/\sigma'_0)_{20}$ inducing the same level of strain in 20 cycles of uniform loading in one direction:

$$\frac{\tau_{\max,\ell}}{\sigma'_0} = C_2\, C_5(\tau_{d\,\ell}/\sigma'_0)_{20} \tag{10.3}$$

where $\tau_{d\ell}$ denotes the amplitude of uniform cyclic shear stress causing the specified level of shear strain and $\tau_{\max,\ell}$ denotes the peak shear stress in any irregular time history of shear stress variation which is large enough to induce the specified shear strain.

The outcome of the simple shear tests on the Fuji river sand with different densities is summarized in Fig. 10.27 in which the inverse of combined coefficient $C_2\,C_5$ is plotted versus the relative density of the sand. The solid-line curve indicated

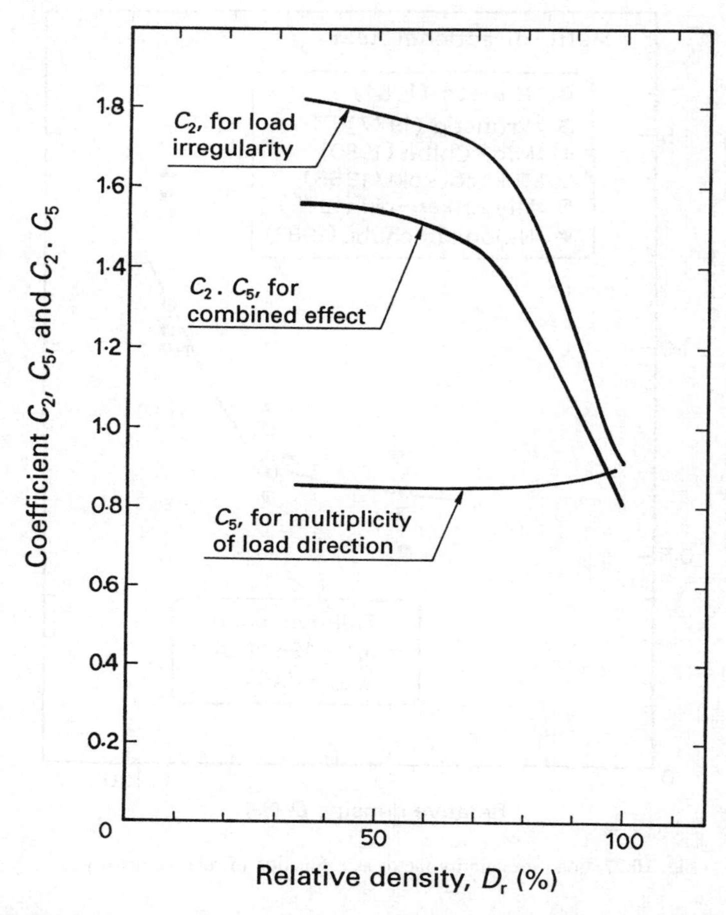

Fig. 10.28 Coefficient allowing for the effects of directionality and irregularity of load variation during earthquakes.

in the figure is an average over six suites of tests employing different time histories in the loading scheme. As a representative case, the condition inducing a single amplitude of simple shear strain of $\gamma_{max} = 3\%$ is presented in Fig. 10.27. The test results indicated that the same value of coefficient can hold valid as well for development of other shear strains. On the basis of the test results shown in Fig. 10.27, the combined coefficient C_2C_5 is demonstrated in Fig. 10.28, together with individual value of C_2 and C_5 which were estimated based on the results of other series of simple shear tests where irregular loads were employed in one direction alone.

The amplitude of simple shear stress $\tau_{d\ell}$ in uniform loading appearing in eqn (10.3) corresponds to a half of the single amplitude of axial stress $\tau_{d\ell}$ needed to cause liquefaction in the triaxial test. Therefore, by using the relation of eqn (10.2), eqn (10.3) is rewritten as

$$\frac{\tau_{\max,\ell}}{\sigma_v{}'} = C_1 C_2 C_5 \left(\frac{\sigma_{d\ell}}{2\sigma_0{}'}\right)_{20}. \tag{10.4}$$

On the other hand, the cyclic stress ratio defined as the amplitude of uniform cyclic shear stress τ_d divided by the effective vertical stress $(\tau_{d\ell}/\sigma_v{}')_{20}$ is related with $\tau_{\max,\ell}/\sigma_v{}'$ through

$$\frac{\tau_{\max,\ell}}{\sigma_v{}'} = C_2\, C_5 \left(\frac{\tau_{d\ell}}{\sigma_v{}'}\right)_{20}. \tag{10.5}$$

The outcome of the irregular loading tests summarized in Fig. 10.28 indicates that the coefficient $C_2\, C_5$ takes a value of about 1.55 for the sand with a relative density lower than about 70%. Thus if the K_0 value is assumed as 0.5, the coefficient C_1 takes a value of 1.5 from eqn (10.2). Introducing the above values in eqn (10.4) and (10.5), one obtains

$$\frac{\tau_{\max,\ell}}{\sigma_v{}'} \doteqdot \left(\frac{\sigma_{d\ell}}{2\sigma_0{}'}\right)_{20} \doteqdot \frac{1}{0.65}\left(\frac{\tau_{d\ell}}{\sigma_v{}'}\right)_{20}. \tag{10.6}$$

This indicates the relations among variously defined cyclic stress ratios at liquefaction or at the state of 5% D.A. strain development. While the above correlations were established for the amplitude of cyclic shear stress great enough to cause liquefaction, the same relations may be considered to hold valid for any amplitude of cyclic shear stress which is smaller or larger than that needed to produce a state of liquefaction. Thus the relations of eqn (10.6) are rewritten more generally as,

$$\frac{\tau_{\max,}}{\sigma_v{}'} \doteqdot \left(\frac{\sigma_d}{2\sigma_0{}'}\right)_{20} \doteqdot \frac{1}{0.65}\left(\frac{\tau_d}{\sigma_v{}'}\right)_{20}. \tag{10.7}$$

10.10 Effects of confining stress and initial shear stress on liquefaction resistance

All the cyclic shear tests ever performed on sands have indicated that there is a tendency for cyclic strength to decrease with increasing confining stress. Such an effect of confinement has a significant influence on problems such as the case where safety of foundation soils underlying high dams or tall buildings is to be evaluated. This effect was investigated by Rollins and Seed (1988), and Seed and Harder (1990) in terms of a correction factor K_σ which is defined as the cyclic stress ratio causing 5% D.A. strain in 20 cycles under any confining stress normalized to the corresponding value of cyclic stress ratio at an effective confining stress of 1 kgf/cm^2 = 0.1 MPa. The values of K_σ obtained by Seed and Harder (1990) from the

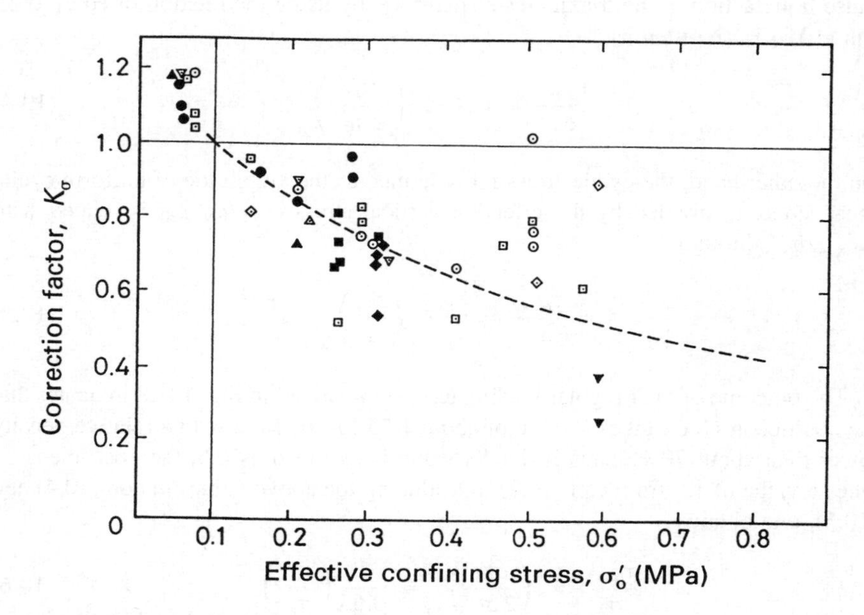

Fig. 10.29 Effects of confining stress on the cyclic strength (Seed and Harder, 1990).

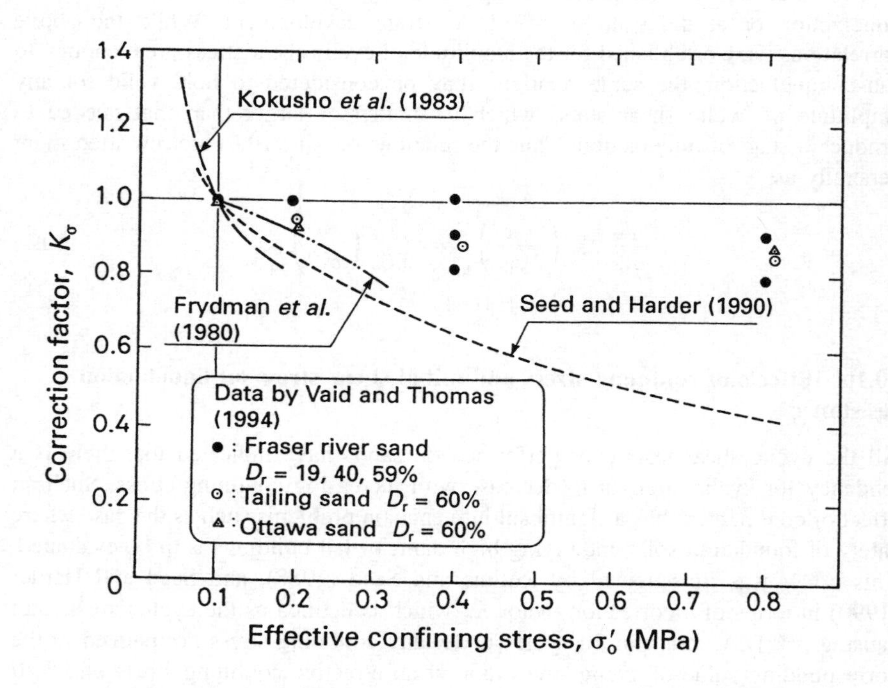

Fig. 10.30 Effects of confining stress on the cyclic strength of sands.

compilation of test data on materials of earth dams and clean sands are shown in Fig. 10.29 where it can be seen that the cyclic strength is reduced to about 40% of its value at $\sigma'_0 = 0.1$ MPa, if the confining stress becomes as great as 0.8 MPa. The data by Kokusho *et al.* (1983) and Frydman *et al.* (1980) are also compiled in the same fashion and shown in Fig. 10.30. Laboratory studies on reconstituted samples of a variety of sands by Vaid and Thomas (1994) are also shown in Fig. 10.30 where it may be seen that the correction factor K_σ could take a value around 0.4 at a high confining stress of $\sigma'_0 = 0.8$ MPa. Since the value of K_σ varies depend upon the type of soil, it is desirable to determined the K_σ value by tests for individual material under consideration.

As mentioned in Section 10.2.2, the presence of initial shear stress on the plane of cyclic shear stress application exerts a significant influence on the pore water pressure build-up and hence on the liquefaction characteristics of sand. Thus soil elements beneath the edge of a structure or within the body of earth dams are considered to exhibit a cyclic resistance which is different from that of a soil element in the free field under level ground conditions. The effects of the initial shear stress were also studied by Rollins and Seed (1988), and Seed and Harder (1990) who compiled cyclic triaxial test data on sands which were performed by applying initial shear stresses. In this type of test, the specimen is subjected initially to an axial stress which is greater than the lateral stress and then subjected to cyclic axial stress. These tests are equivalent to the ACT test described in Section 10.2.2 in that the cyclic shear stress is applied while keeping the specimen always subjected to the initial shear stress. The compilation of the test data by many workers indicated that for sands with a relative density greater than about 50%, the cyclic stress ratio required

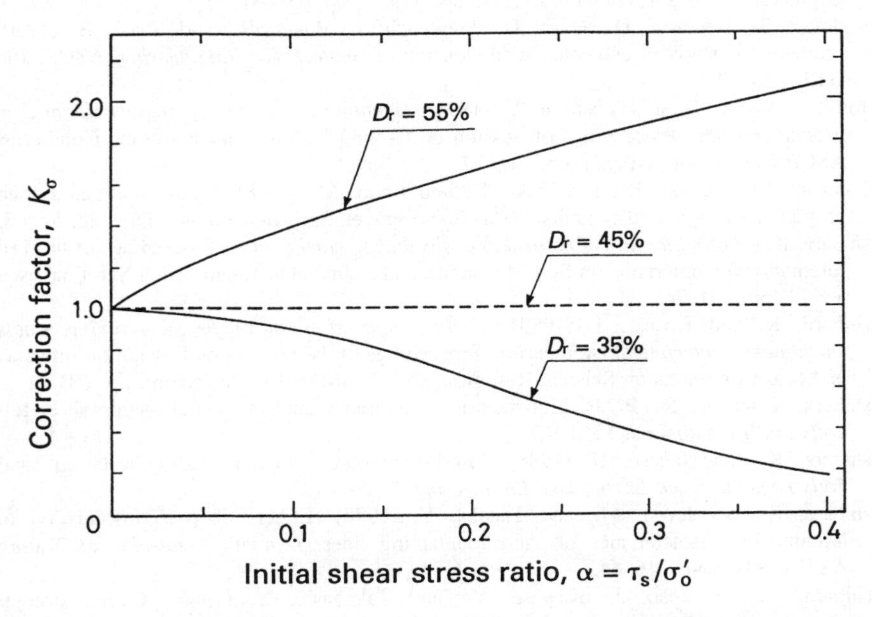

Fig. 10.31 Effects of initial shear stress on the cyclic strength of sands (Rolllins and Seed, 1990).

to cause 5% D.A. strain tends to increase as the initial shear stress increases. However, the opposite tendency was found to be true for loose sands: the cyclic strength tends to decrease with increasing initial shear stress. In summarizing these data, the ratio of the initial shear stress τ_s divided by the effective confining stress $\sigma_0{}'$ was taken as a parameter α. The effect of this alpha parameter on the cyclic strength was expressed in terms of a correction factor K_α which was defined as the cyclic strength under any initial shear stress normalized to the corresponding value without the initial shear stress.

The outcome of such data compilation by Rollins and Seed (1988) is shown in Fig. 10.31 where the correction factor K_α is plotted versus the initial stress ratio α. The two curves in this figure are conservative average chosen for the dense and loose sands. For sands with other relative densities, the relation between K_α and α may be inferred by interpolation. At the current state of the art, the correlation between K_α and α is not firmly established because of the scatter in the data, but a new interpretation as conceived by Pillai (1991) may be helpful to firm up the correlation. By using the chart as shown in Fig. 10.31, the corrected cyclic strength for sand with a given alpha value and relative density can be determined by multiplying the cyclic strength value under level ground conditions by the correction factor K_α.

References

Finn, W.D.L. (1981). *Liquefaction potential: developments since 1976*. Proceedings of the First International Conference on Recent Advances in Geotechnical Earthquake Engineering and Soil Dynamics, St. Louis, Vol. 2, pp. 655–81.

Frydman, S., Hendron, D., Horn, H., Steinbach, J., Baker, R., and Shaal, B. (1980). Liquefaction study of cemented sand. *Journal of Geotechnical Engineering*, ASCE, **106**, GT3, 275–97.

Goto, S., Nishio, S., and Yoshimi, Y. (1994). *Dynamic properties of gravels sampled by ground freezing*. Proceedings of Session on Ground Failures under Seismic Conditions, ASCE Convention, Atlanta, pp. 141–57.

Hatanaka, M., Suzuki, K., Kawasaki, T., and Endo, M. (1988). Cyclic undrained shear properties of high quality undisturbed Tokyo gravel. *Soils and Foundations*, **28**, 57–68.

Ishihara, K. (1985). *Stability of natural deposits during earthquakes*. Proceedings of the 11th International Conference on Soil Mechanics and Foundation Engineering, San Francisco, Vol. 1, pp. 321–76.

Ishihara, K. and Koseki, J. (1989). *Cyclic shear strength of fines-containing sands, earthquake geotechnical engineering*. Proceedings of the Discussion Session on Influence of Local Conditions on Seismic Response, 12th ICSMFE Rio de Janeiro, pp. 101–6.

Ishihara, K. and Li, S. (1972). Liquefaction of saturated sand in triaxial torsion shear test, *Soils and Foundations*, **12**, 19–39.

Ishihara, K. and Nagase, H. (1988). Multi-directional irregular loading tests on sand. *Soils Dynamics and Earthquake Engineering*, 7, 201–12.

Ishihara, K., Sodekawa, M., and Tanaka, Y. (1978). Effects of overconsolidation on liquefaction characteristics of sands containing fines. *Dynamic Geotechnical Testing*, ASTM, STP 654, 246–64.

Ishihara, K., Troncoso, J., Kawase, Y., and Takahashi, Y. (1980). Cyclic strength characteristics of tailings materials. *Soils and Foundations*, **20**, 127–42.

Kokusho, T. and Tanaka, Y. (1994). *Dynamic properties of gravel layers investigated by in situ freezing sampling*. Proceedings of Session on Ground Failures under Seismic Condition, ASCE Convention, Atlanta, pp. 121–40.

Kokusho, T., Yoshida, Y., Nishi, K., and Esashi, Y. (1983). *Evaluation of seismic stability of dense sand layer (part 1) - dynamic strength characteristics of dense sand*. Report 383025. Electric Power Central Research Institute, Japan (in Japanese)

Kokusho, T., Tanaka, Y., Kudo, K., and Kawai, T. (1995). *Liquefaction case study of volcanic gravel layer during the 1993 Hokkaido–Nanseioki earthquake*. Proceedings of the 3rd International Conference on Recent Advances in Soil Dynamics and Geotechnical Earthquake Engineering, Vol. 1. St. Louis.

Konno, T., Hatanaka, M., Ishihara, K., Ibe, Y., and Iizuka, S. (1994). Gravelly soil properties evaluation by large scale in situ cyclic shear tests. Proceedings of Session on Ground Failures under Seismic Conditions, ASCE Convention, Atlanta, pp. 177–200.

Ladd, R.S. (1974). Specimen preparation and liquefaction of sands. *Journal of ASCE*, 100, GT10, 1180–4.

Lee, K.L. and Seed, H.B. (1967). Cyclic stress condition causing liquefaction of sand. *Journal of Soils Mechanics and Foundation Engineering*, ASCE, **93**, SM1, 47–70.

Mulilis, J.P., Seed, H.B., Chan, C.K., Mitchell, J.K., and Arulanandan, K. (1977). Effects of sample preparation on sand liquefaction. *Journal of ASCE*, 103, GT2, 91–108.

Pillai, V.S. (1991). *Liquefaction analysis of sands: some interpretation of Seed's K_α and K_σ correction factors using steady state concept*. Proceedings of the 2nd International Conference on recent Advances in Geotechnical Earthquake Engineering and Soil Dynamics, St. Louis, Vol. 1, pp. 579–87.

Rollins, K.M. and Seed, H.B. (1988). Influence of buildings on potential liquefaction damage. *Journal of Geotechnical Engineering*, ASCE, **116**, GT2, 165–85.

Seed, H.B. (1979). Soil liquefaction and cyclic mobility evaluation for level ground during earthquakes. *Journal of ASCE*. **105**, GT2, 201–55.

Seed, H.B. and Lee, K.L. (1966). Liquefaction of saturated sands during cyclic loading. *Journal of ASCE*, **92**, SM6, 105–34.

Seed, R.B. and Harder, L.F. (1990). *SPT-based analysis of cyclic pore pressure generation and undrained residual strength*. Proceedings of the B. Seed Memorial Symposium, Vol. 2, 351–76.

Silver, M.L., Chan, C.K., Ladd, R.S., Lee, K.L., Tiedemann, D.A., Townsend, F.C., Valera, J.E., and Wilson, J.H. (1976). Cyclic triaxial strength of standard test sand. *Journal of ASCE*, **102**, GT5, 511–23.

Tanaka, Y., Kokusho, K., Kudo, K., and Yoshida, Y. (1991). *Dynamic strength of gravelly soils and its relation to the penetration resistance*. Proceedings of the 2nd International Conference on Recent Advances in Geotechnical Earthquake Engineering and Soil Dynamics, St. Louis, Vol. 1, pp. 399–406.

Tatsuoka, F., Toki, S., Miura, S., Kato, H., Okamoto, M., Yamada, S., Yasuda, S., and Tanizawa, F. (1986*a*). Some factors affecting cyclic undrained triaxial strength of sand. *Soils and Foundations*, **26**, 99–116.

Tatsuoka, F., Ochi, K., Fujii, S., and Okamoto, M. (1986*b*). Cyclic undrained triaxial and torsional shear strength of sands for different sample preparation methods. *Soils and Foundations*, **26**, 23–41.

Toki, S., Tatsuoka, F., Miura, S., Yoshimi, Y., Yasuda, S., and Makihara, Y. (1986). Cyclic undrained triaxial strength of sand by a cooperative test program. *Soils and Foundations*, **26**, 117–28.

Vaid, Y.P. and Thomas, J. (1994). *Post liquefaction behaviour of sand*. Proceedings of the 13th International Conference on Soil Mechanics and Foundation Engineering, New Delhi, Vol. 1, 1303–10.

Wong, R.T., Seed, H.B. and Chan, C.K. (1975). Cyclic loading liquefaction of gravelly soils. *Journal of Geotechnical Engineering*, ASCE, GT6, 571–83.

Yoshimi, Y., Richart, F.E., Prakash, S., Balkan, D.D., and Ilyichev, V.A. (1977). *Soil Dynamics and its application to foundation engineering*. Proceedings of the 9th International Conference on Soil Mechanics and Foundation Engineering, Tokyo, Vol. 2, pp. 605–50.

Yoshimi, Y., Tokimatsu, K., Kaneko, O., and Makihara, Y. (1984). Undrained cyclic shear strength of a Niigata sand. *Soils and Foundation*, **24**, 131–45.

Yoshimi, Y., Tokimatsu, K., and Hosaka, Y. (1989). Evaluation of liquefaction resistance of clean sands based on high-quality undisturbed samples. *Soils and Foundations*, **29**, 93–104.

11

SAND BEHAVIOUR UNDER MONOTONIC LOADING

11.1 Flow and non-flow in undrained sand samples

A deposit of sand is composed of an assemblage of particles in equilibrium where inter-granular forces are transmitted through points of contact. When shear stress is applied, the resulting deformation is always accompanied by a volume change which is known as dilatancy. This shear-induced volume change accrues as a result of two competing modes of particle movement, namely, slip-down and roll-over (Dafalias, 1993). The slip-down movement of grains tends to reduce the volume by repacking the sand aggregate into a denser state. This mechanism is activated predominantly in loose deposits of sand. The roll-over mechanism tends to increase the volume which is characteristic in the behaviour of dense sand. When the slip down takes place, particles are filling gaps in the void and not moving largely in the direction of shearing. Therefore the slip-down movement can occur rather easily without mobilizing a large amount of shear strain. It is for this reason that the volume reduction is generally observed at an early stage of loading in the tests on sands with a wide range of density. On the contrary, a larger movement is always required for particles to roll over neighbouring ones and hence the volume increase or dilation is generally induced at a later stage of shear stress application where the sand is largely deformed.

The two mechanisms working more or less simultaneously are manifested in the stress–strain behaviour observed in the laboratory test. Figure 11.1 illustrates three different types of stress–strain relations obtained from undrained shear tests on saturated samples of sand. When the density is large, the sand tends to exhibit strain-hardening behaviour where the shear stress always goes up with increasing shear strain. If this sample is sheared in the drained condition, the dilation will take place reflecting the roll-over mechanism. The dense sand in such a state is called *strain-hardening type* and referred to as being *dilative or non-flow type*. When the density is low, the strain-softening behaviour is exhibited in the sample with a drop in shear stress accompanied by an unlimitedly large strain. If subjected to shear in the drained condition, this sample will exhibit contraction in volume indicating predominant occurrence of the slip-down mechanism within the sample. Thus loose sand in such a state is called *strain-softening type* and referred to as being *contractive or flow type*. When the density is moderate, the sand first exhibits the strain-softening at moderate shear strains but it starts to show the strain-hardening behaviour as the strain increases. In a drained shear, this sample will show a volume contraction first and then dilation with increasing shear strain. This is the reflection of the slip-down mechanism working at moderate shear strains, followed by

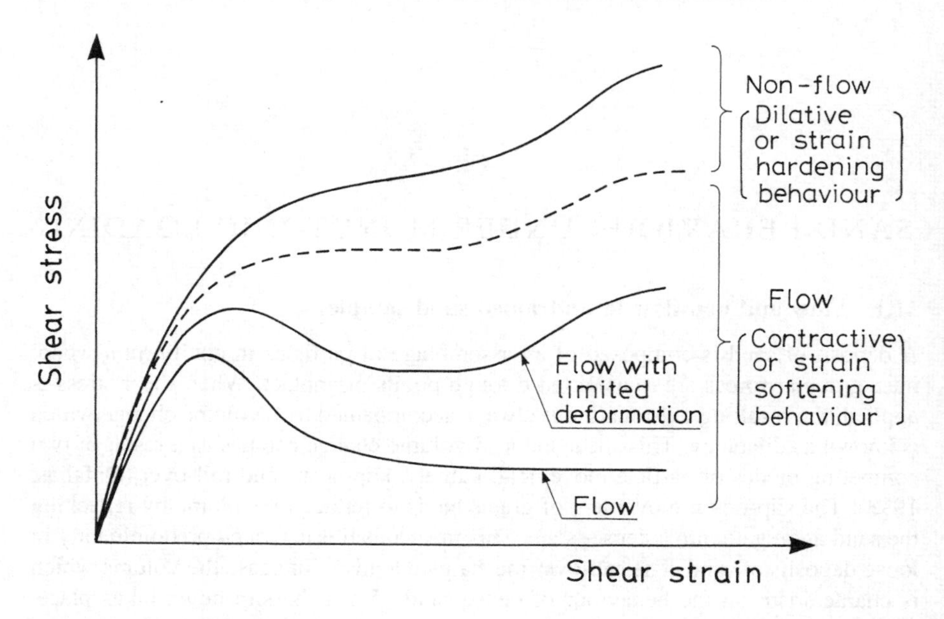

Fig. 11.1 Classification of undrained behaviour of sandy soils based on contractiveness and dilativeness.

activation of the roll-over mechanism at large strains. The behaviour of sand as above is called *flow type with limited deformation*. In what follows, the flow type and the flow type with limited deformation will be merged into one group which is jointly referred to as *flow type*.

The undrained behaviour of sand following seismic shaking is expected not to be influenced by prior application of cyclic loads. Evidence in support of this can be provided by two sets of undrained triaxial compression test results on a loose specimen of Toyoura sand, as shown in Figs 11.2 and 11.3 (Ishihara *et al.* 1991). In each set of tests, the first sample was subjected to an initial shear stress under drained condition and then loaded undrained cyclically, followed by monotonic loading until an axial strain of about 20% developed. The second sample was subjected to the same initial shear stress, and then sheared undrained monotonically under strain-controlled conditions. The results of one of such sets of tests are displayed in Fig. 11.2 where the samples with a relative density of 16% were consolidated to a high initial confining stress of $\sigma'_0 = 0.1$ MPa. It can be seen that, over the range of moderate to large strains, the stress–strain curve and the stress path for the monotonic loading are almost coincident with those obtained for the cyclic to monotonic loading. The results of another set of tests on samples consolidated to a lower confining stress of $\sigma'_0 = 0.02$ MPa are shown in Fig. 11.3 where the coincidence is observed as well between these two types of tests. Thus, it would be reasonable to assume that the soil behaviour following the seismic loading can be well understood by investigating the deformation characteristics which is manifested

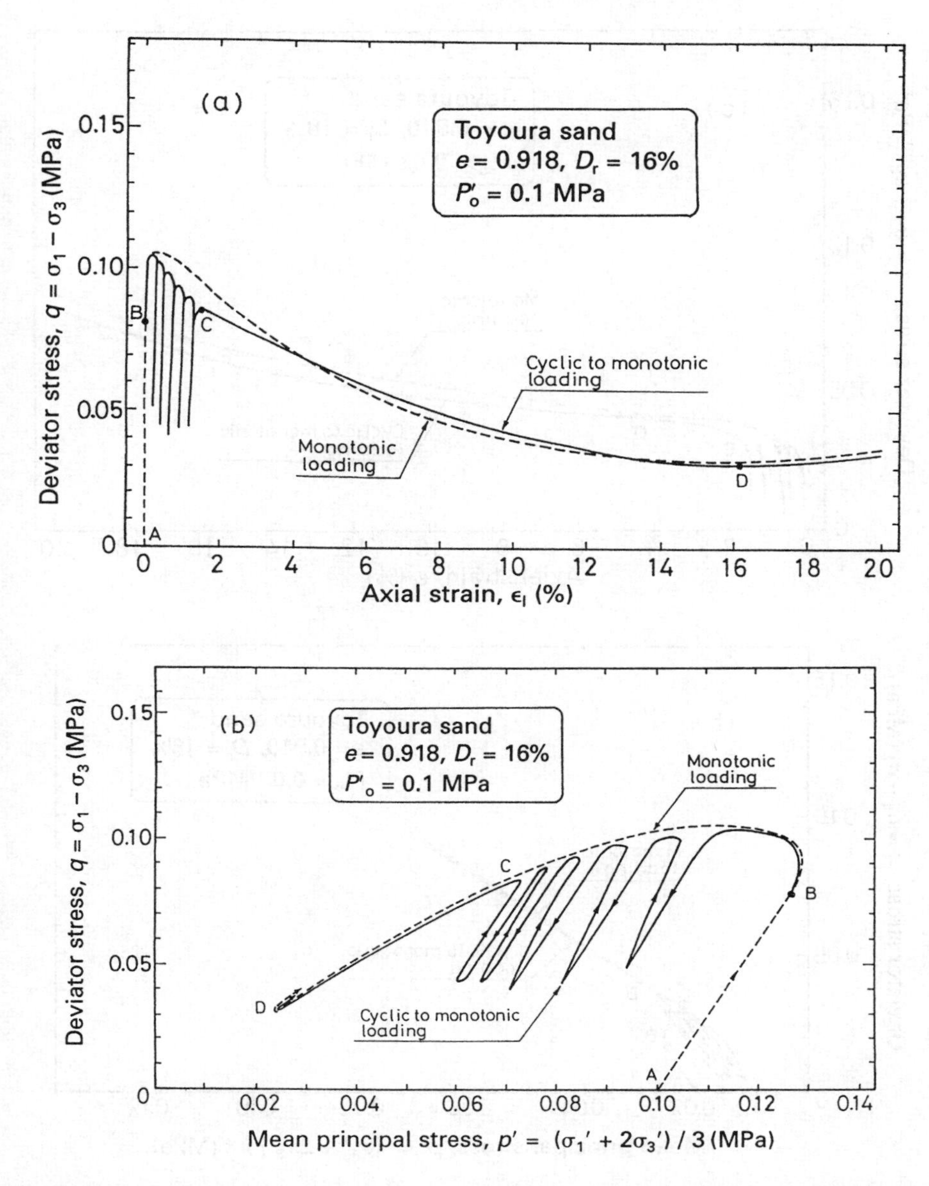

Fig. 11.2 Undrained behaviour of Toyoura sand under monotonic or cyclic to monotonic loadings conditions.

under the monotonic loading conditions. It is on this ground that the description of soil behaviour is made in what follows based on the results of the monotonic loading tests (Ishihara, 1993).

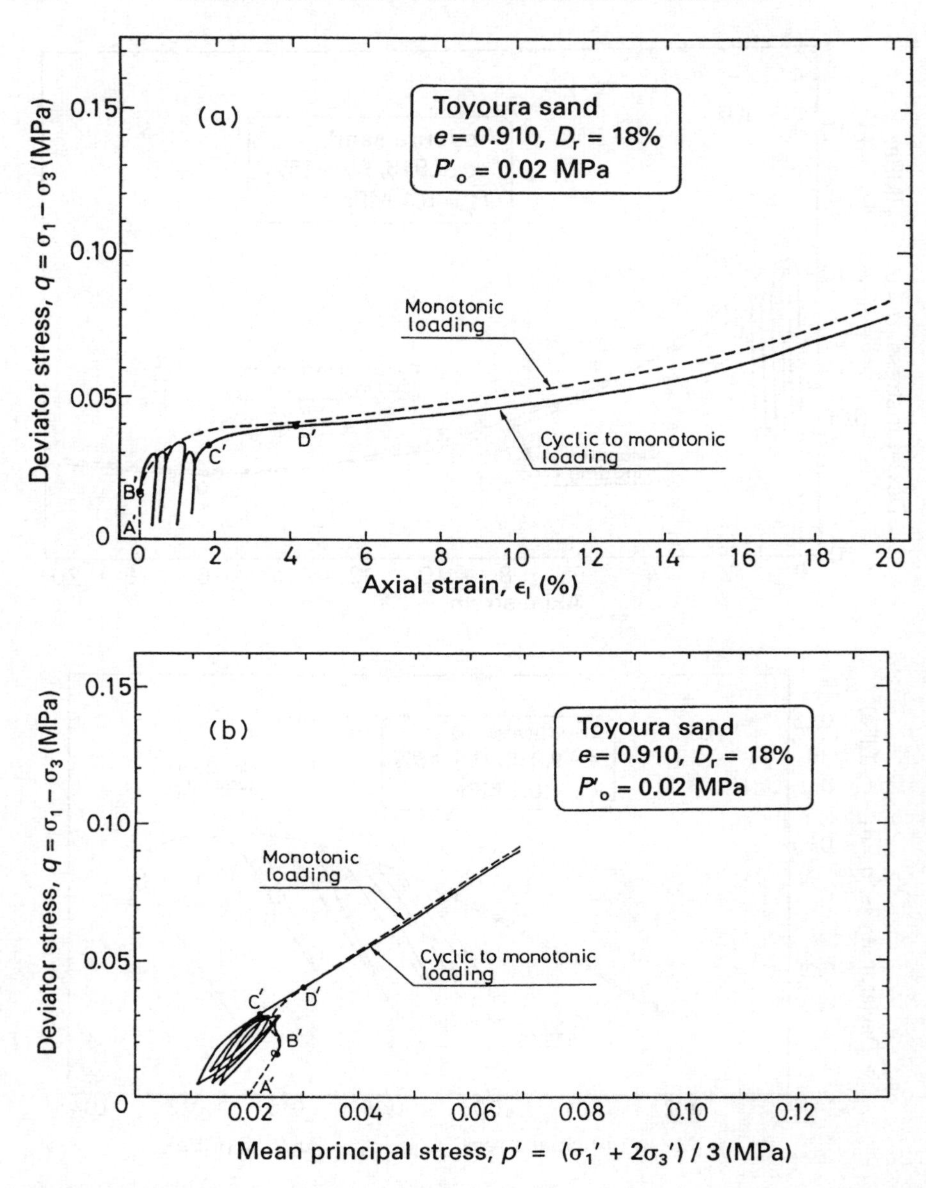

Fig. 11.3 Undrained behaviour of Toyoura sand under monotonic or cyclic to monotonic loadings conditions.

11.2 Compression characteristics of sand and the method of sample preparation

It has been shown that the behaviour of sand is influenced notably by the fabric formed during its deposition. To demonstrate this effect, samples were prepared in

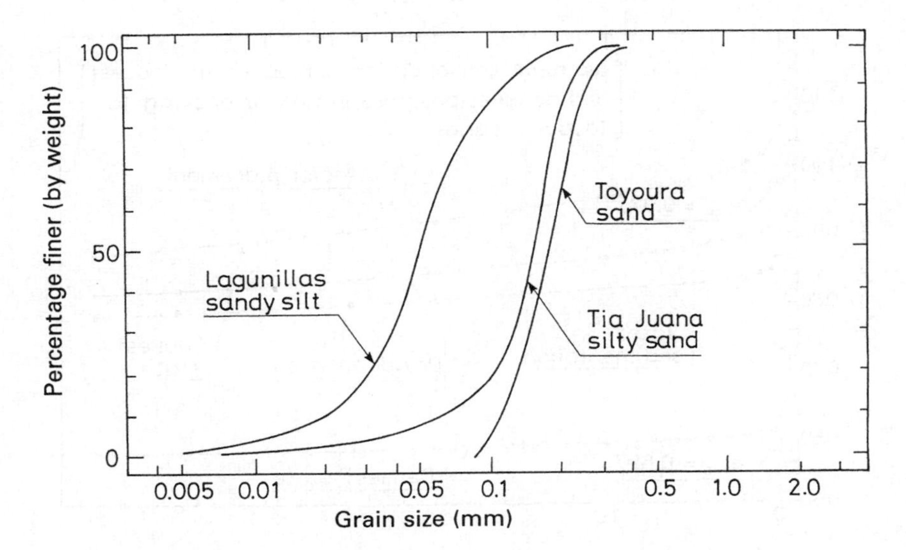

Fig. 11.4 Grain size distribution curves for sandy soils used in the tests.

the triaxial test device by three different methods. They will be referred to as the method of moist placement, dry deposition, and water sedimentation. Detailed procedures of the three sample preparation methods are described in Appendix A. The Japanese standard sand called Toyoura sand was used in this test. It consists of subangular particles and has a means diameter of D_{50} = 0.17 mm and a uniformity coefficient of U_c = 1.7, as seen in the grain size distribution curve in Fig. 11.4.

Effects of the fabric seem to be manifested most conspicuously in the form of void ratio of the samples prepared by different methods. Therefore, it is of interest, first of all, to see what would be the highest void ratio that can be produced by each of the three methods of sample preparation. The highest void ratio at an initial stage, attained by the moist placement method was found to be 1.03 for Toyoura sand as shown in Fig. 11.5. The relationship between the void ratio and effective confining stress during the isotropic consolidation is also shown in Fig. 11.5, where it can be seen that the highest void ratio of the sample by this method is still higher at the start of consolidation than the value of the maximum void ratio of e_{max} = 0.977 which is determined by the method of Japanese Society of Soil Mechanics and Foundation Engineering (JSSMFE). The consolidation curves of the loosest possible samples obtained by the dry deposition and water sedimentation are also displayed in Fig. 11.5. It can be seen that the void ratios attained are substantially lower than that obtained by the method of moist placement. The results of isotropic consolidation tests on the sample prepared by moist placement but compacted to the densest possible state by tamping in the forming mould are demonstrated in Fig. 11.5. It can be seen that the void ratio at the start of consolidation is only slightly larger than the minimum void ratio of e_{min} = 0.597 determined by the JSSMFE method. It was observed that the change in void ratio of the densest sample during isotropic

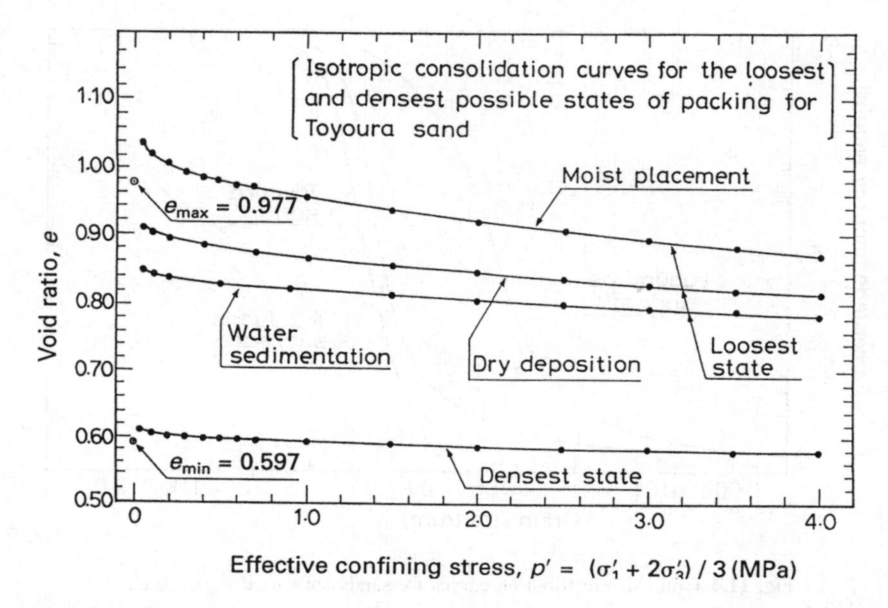

Fig. 11.5 Isotropic consolidation curves of samples of Toyoura sand prepared by three methods.

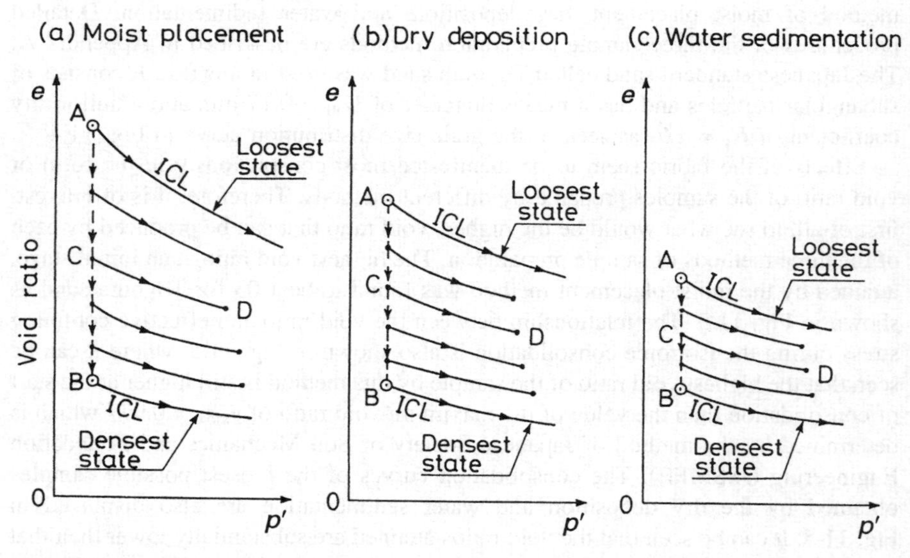

Fig. 11.6 Consolidation characteristics of samples prepared by: (a) moist placement; (b) dry deposition; (c) water sedimentation.

consolidation is generally very small over the wide range of the mean principal stress used in the test. Similar consolidation curves were obtained also for the densest possible samples prepared by the method of dry deposition and water sedimentation. Thus it can be mentioned that, for the densest state of sand, the isotropic consolidation curve is determined almost uniquely, no matter what method is employed for preparing the test sample. For an intermediate state between the maximum and minimum void ratio by any means of preparation, samples can be formed by controlling the tapping energy that is applied initially to the forming mould during sample preparation. Thus it may be mentioned that, by means of any method of sample preparation, any initial state having pre-assigned void ratio and confining stress could be produced in the sample by controlling the tapping energy during sand preparation, but the widest range in void ratio that can be attained is by means of moist placement while the method of water sedimentation would result in the smallest range in the attainable void ratio. This fact is illustrated schematically in Fig. 11.6.

11.3 Steady state of sand

The samples of Toyoura sand prepared by the method of moist placement have been shown to produce void ratios varying in the widest range. Thus both contractive and dilative characteristics or strain-softening and strain-hardening behaviour can be observed in such samples during subsequent loading. In order to examine the deformation behaviour of sand from a broader perspective, it is therefore considered highly desirable to prepare all the test samples by this procedure. Thus all the undrained triaxial compression tests described herein are conducted on samples of Toyoura sand prepared by the moist placement method at relative densities varying from 7% to 65%. The following is the outcome of the tests conducted in this vein.

The results of a series of undrained triaxial compression tests on loose samples of Toyoura sand with a relative density of 16% are presented in Fig. 11.7. It can be seen that the peak in stress–strain curves appears eminently when the initial confining stress is large, but the peak tends to dwindle as the initial confining stress becomes smaller. It is also to be noted that, despite the large difference in the stress-strain behaviour at an early stage of loading, the samples tend to exhibit an almost identical behaviour at a later stage of loading where the developed axial strain becomes as large as 25%. In this largely deformed state, the deviator stress is seen to stay at an approximately constant value of 80 kPa with an effective confining stress of about 60 kPa. The state of the sand deforming continually, keeping the volume constant, under a constant shear stress and confining stress, is called the steady-state by Castro (1975) and Castro and Poulos (1977). The shear stress of a sand mobilized at the steady state has been called the steady-state strength or residual strength. One of the characteristic features observed conspicuously in this test series is the fact that the stress path goes upwards to the right after the shear stress has reached a minimum value. This state of minimum shear stress corresponds to what is called the state of phase transformation (Ishihara *et al.* 1975), because it defines a transient state in which the change from contractive to dilative behaviour occurs in the sand. In a sample much looser than D_r = 16%, the dilative behaviour never shows up and the state of phase transformation coincides with the steady-state.

The results of another series of tests on somewhat denser samples prepared by the

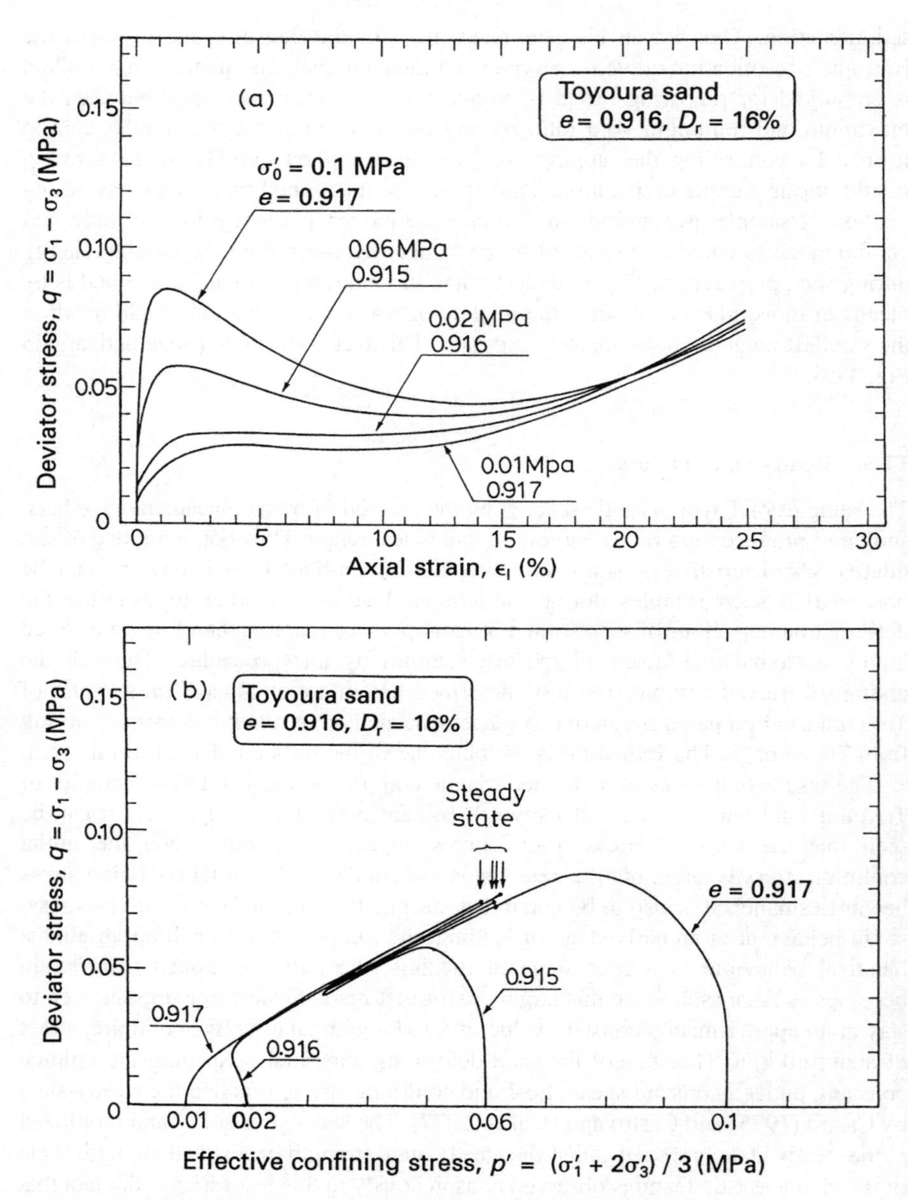

Fig. 11.7 Undrained behaviour of loose samples of Toyoura sand.

method of moist placement at a relative density of 38% are presented in Fig. 11.8, where the same tendency is observed in the overall behaviour with respect to the influence of the initial confining stress on the stress-strain and pore pressure response.

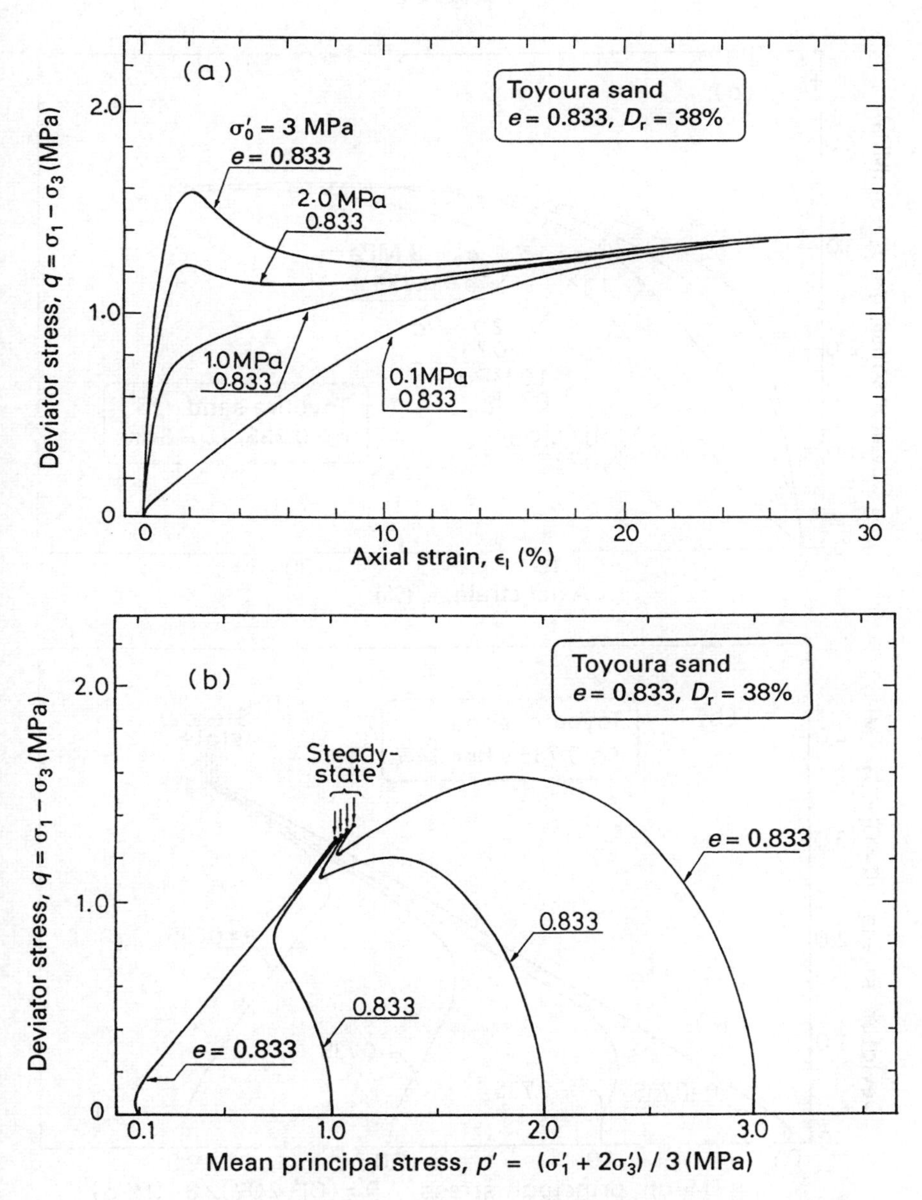

Fig. 11.8 Undrained behaviour of a medium loose sample of Toyoura sand.

The results of still other tests on dense samples with a relative density of $D_r = 64\%$ are presented in Fig. 11.9, where it may be seen that the steady state was attained at an effective confining stress of $p' = 2.8$ MPa and a deviator stress of $q = 3.6$ MPa.

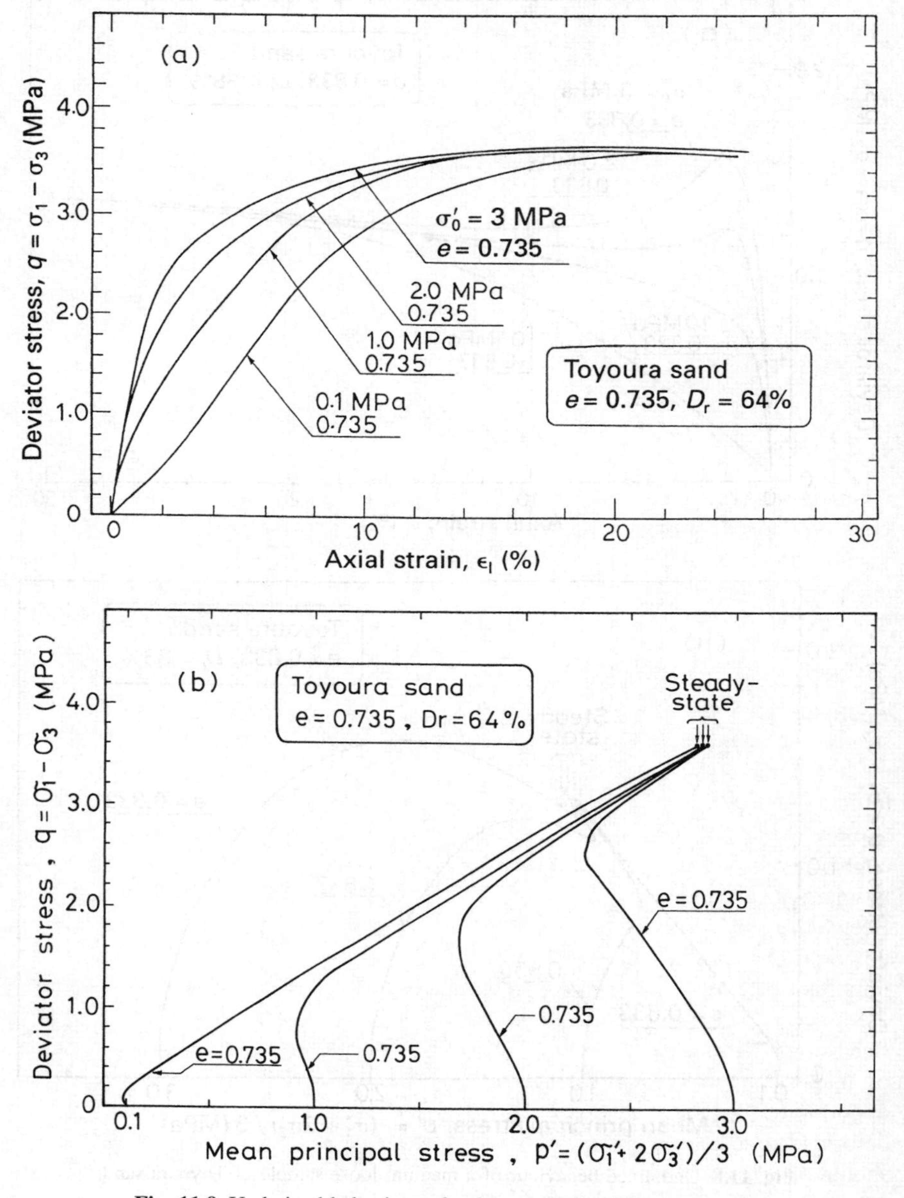

Fig. 11.9 Undrained behaviour of a dense sample of Toyoura sand.

As indicated by Castro and Poulos (1977) and shown as well in the results of tests herein described, the state of stress of sand at a steady-state deformation is determined uniquely by the void ratio alone, and for the sand sheared undrained, the pore water pressure increases or decreases, depending upon the initially applied confining stress, so as to bring the effective confining stress to a unique value which is inherent to the void ratio. Thus it becomes possible to plot the confining stress at the steady state versus the void ratio for the data sets demonstrated above and to draw a line which is generally referred to as the steady-state line. If the deviator stress q is plotted versus the effective confining stress p' both at the steady state, an angle of interparticle friction ϕ_{ss} is determined through the following relation,

$$M_s = \left(\frac{q}{p'}\right)_{ss} = \frac{6 \sin \phi_{ss}}{3 - \sin \phi_{ss}} \tag{11.1}$$

where M_s is the ratio of q and p' at the steady state. For Toyoura sand the angle of interparticle friction was determined to be $\phi_{ss} = 31°$ corresponding to $M_s = 1.24$.

11.4 Quasi-steady state

Observation of the general behaviour described above indicates that, if the density is great or when the initial confining stress is sufficiently low, the sand tends to exhibit dilative or strain-hardening characteristics with the shear stress rising with increasing shear strain until the steady state has been reached at the end. It is to be noticed that, at this ultimate state, the shear stress attains its maximum and this value may be taken as the strength of the sand. However, when the sand is loose and is subjected to a large confining stress, it tends to deform fairly largely at the beginning exhibiting contractive behaviour, and then starts to dilate reaching the steady state at the end. An example of such a case for Toyoura sand prepared by the method of moist placement is shown in Fig. 11.10. It can be seen that, for the case of initial confining stress of $\sigma_0 = 0.5$ MPa, there occurs a temporary drop in the shear stress accompanied by a large strain whereupon the sand changes its behaviour from contractive to dilative. A similar drop in shear stress has been observed and reported by many investigators (Castro, 1975; Hanzawa, 1980; Mohamad and Dobry, 1986; Been *et al.* 1991; Konrad, 1990 *a,b*; Vaid *et al.* 1990; Georgiannou *et al.* 1991), and this type of behaviour has been called *flow with limited deformation*. One of the important bearings of this state is that the shear stress mobilized at this moderately deformed condition is definitely smaller than the stress mobilized at the ultimate steady state with much larger strains. Therefore, when the residual strength becomes of major concern in relation to some practical problems, the presence of two opportunities for the sand to exhibit different levels of strength both at fairly largely deformed conditions will pose a serious question as to which of these two strengths should be taken as the residual strength. The answer to this question will be different depending upon individual circumstances encountered in practice. It is to be noted, however, that most of the investigators have adopted the definition of the steady state by drawing attention to this intermediate stage where the minimum strength is encountered. It would be necessary, therefore, to quantitatively revisit the characteristic feature of this state of minimum strength. It should be mentioned here that the state of minimum strength does coincide with what is termed the point

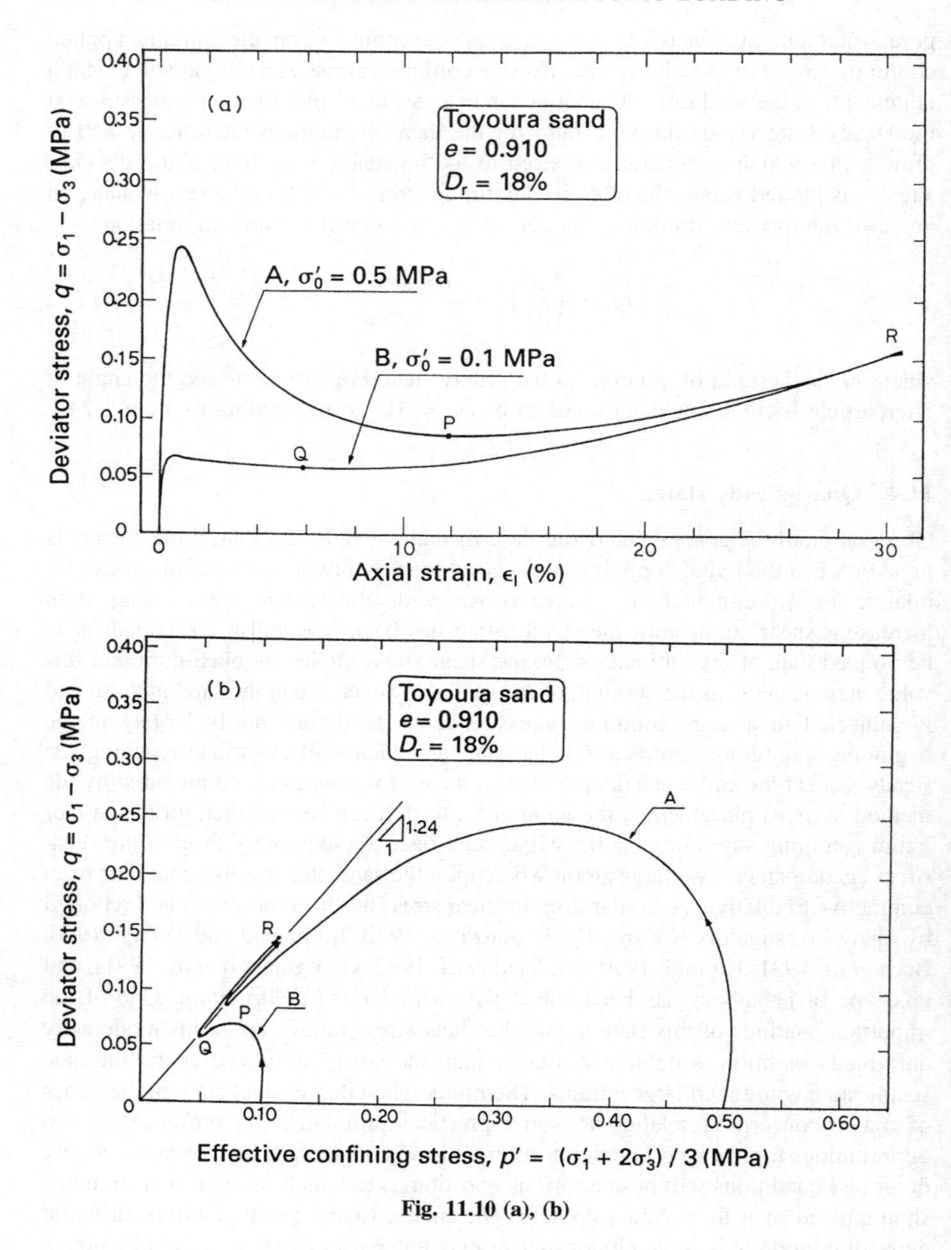

Fig. 11.10 (a), (b)

of phase transformation, although it has been defined (Ishihara *et al.* 1975) to imply a temporary state of transition from contractive to dilative behaviour of sand in a broader sense, irrespective of whether or not it involves a temporary drop in shear stress. What is at stake now is the particular case of phase transformation where a

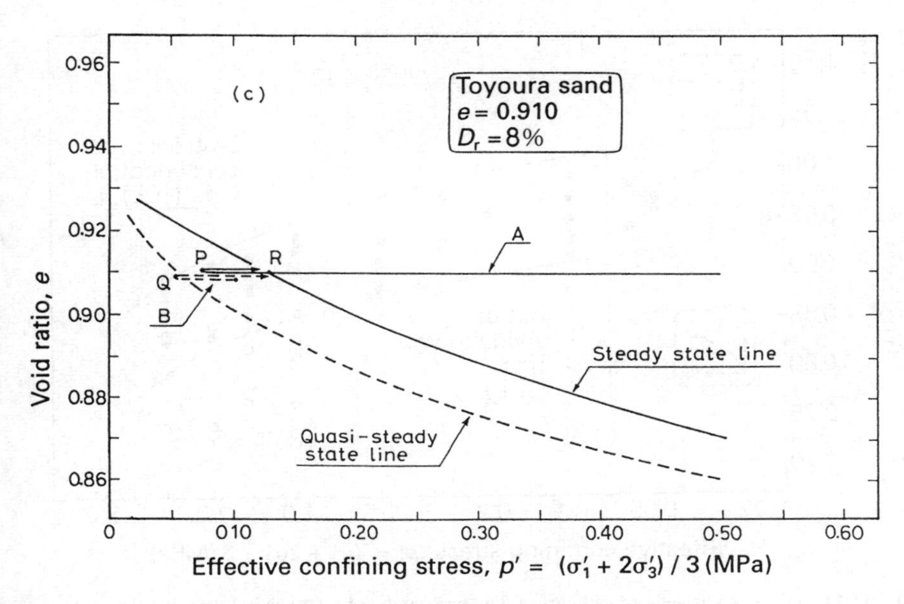

Fig. 11.10 (c) Determination of the quasi-steady state (QSS).

temporary drop in shear stress takes place over a limited range of shear strains, as exemplified by the test data shown in Fig. 11.10. Such a case was termed the *quasi-steady state* by Alarcon-Guzman *et al.* (1988), Been *et al.* (1991), and Vaid *et al.* (1990), as opposed to the conventionally defined steady state which is reached at larger shear strains.

In the following, the term quasi-steady state (QSS) will be used to signify the state of minimum shear stress as indicated above. It has been recognized from a number of laboratory tests that the occurrence or non-occurrence of the temporary drop in shear stress as above is governed by the void ratio and confining stress at the time of consolidation, and it could occur only in loose samples when sheared from large initial confining stresses. Thus it is possible to distinguish between two classes of initial state of consolidation where the temporary drop in shear stress could or could not occur in the samples upon subsequent application of shear stress. Shown in Fig. 11.11 is the e–p' diagram plotting all the test results differentiating between the above two behaviours for the Toyoura sand prepared by moist placement. The solid circles indicate the case where the quasi-steady state was observed with the minimum strength and open circles designate the conditions without such a decrease in shear stress. Then it becomes possible to draw a line of demarcation through the date points separating the initial conditions with or without the occurrence of minimum strength in the subsequent stage of undrained loading. Such a boundary line is indicated in Fig. 11.11 and will be called the *initial dividing line* (IDL). It should be kept in mind that this line is not a curve projected on the e–p' plane but lies on this plane as does the isotropic consolidation line (ICL).

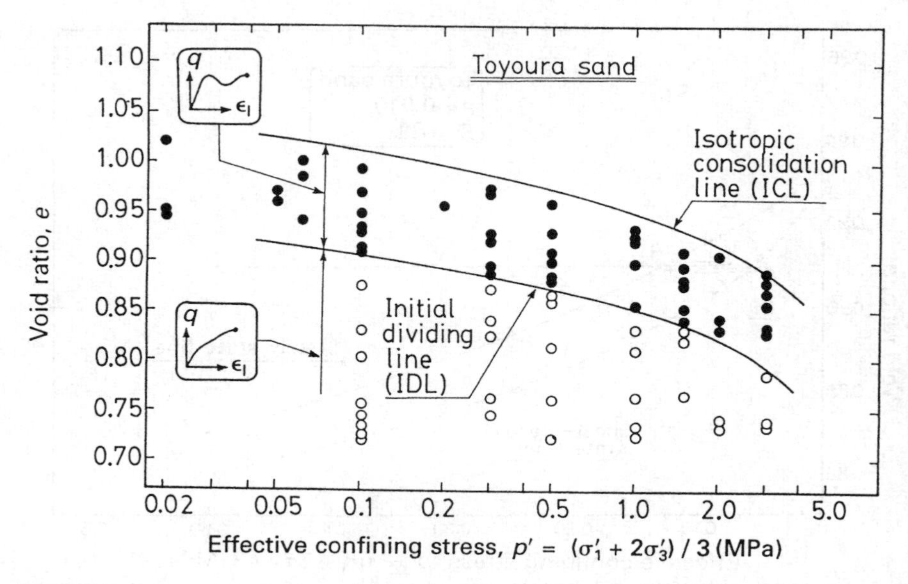

Fig. 11.11 IDL for moist-placed samples of Toyoura sand: solid circles indicate that the QSS was observed with the minimum strength; open circles denote conditions without such a decrease in shear stress.

When a sample is sheared from an initial state of stress located above the initial dividing line (IDL) in the e–p' diagram, the sample is shown to exhibit a temporary drop in shear stress over a certain range of shear strains but to gain strength upon further straining. If this minimum shear stress is to be taken as the residual strength used in stability analyses, it would be necessary to establish a rule to specify this condition. This can be done by locating a stress point in the p–q diagram where the quasi-steady state or minimum strength occurs such as the points P and Q shown in Fig. 11.10(a) and (b). If the effective confining stress at this stage is plotted on the e–p' diagram, points are obtained such as P and Q in Fig. 11.10(c). A number of points of the quasi-steady state were obtained in this manner from a majority of test data with varying void ratios. The outcome of such data compilation on Toyoura sand is displayed in Fig. 11.12, where the initial state at consolidation is indicated by open squares and solid circles indicate the points of quasi-steady state. It may be seen that there are some scatters in the data points. This scattering can be taken for granted, however, in view of the nature of quasi-steady state being defined as a special case of the state of phase transformation. It is to be remembered that the state of phase transformation changes in general depending upon the magnitude of initial confining stress and accordingly there are reasons that the quasi-steady state can not be precisely defined. It should be noted, however, that the state of phase transformation is least affected by the initial confining stress, if the sample is sheared from the initial state above the initial dividing line (IDL), as compared to other cases

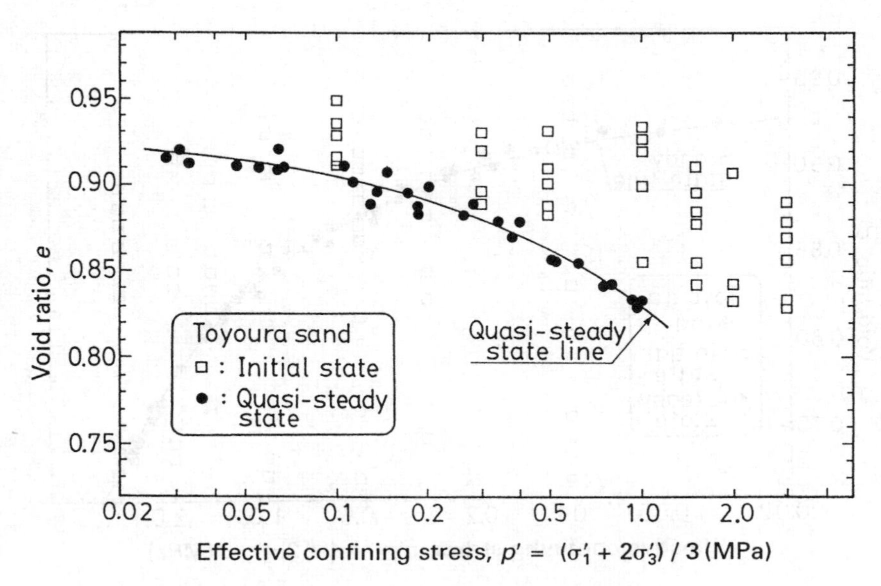

Fig. 11.12 QSSL for moist-placed samples of Toyoura sand.

in which the initial state of the sample lies below IDL. Thus it would be possible to draw a line representing an average condition of the quasi-steady state in the e–p' diagram as shown in Fig. 11.12 and to use it as a convenient reference line to facilitate the evaluation of the minimum residual strength. The average line will be referred to as the quasi-steady state line (QSSL).

In the same fashion, the steady state at large strains is determined by locating a point in the stress path such as point R in Fig. 11.10(b). The steady-state line established from the same data set is indicated in Fig. 11.13, where an additional set of data obtained by tests from below the initial dividing line is also included. It may be noted that there is less scatter in the data points to determine the steady state line (SSL). The steady state in undrained loading as above was shown to be coincident with the steady state attained in drained application of shear stress (Verdugo, 1992). The QSSL and SSL established above are demonstrated in Fig. 11.14 together with the initial consolidation line (ICL) and the initial dividing line (IDL). It is to be noticed that the QSSL and SSL do not actually lie on the e–p' plane but are the lines projected on that plane, whereas the IDL and ICL are the lines existing on the e–p' plane. As noted in Fig. 11.14, the QSSL always lies below the SSL in the e–p' diagram but they become closer to each other as the initial confining stress becomes smaller. It is to be noted that the QSSL becomes significant only when shear stress application is started from the initial state of void ratio located above the initial dividing line (IDL) in the e–p' diagram. When the loading is started from the initial state below the IDL, the QSSL does not bear any practical significance and the residual strength should be evaluated based on the SSL. The basic concept of the

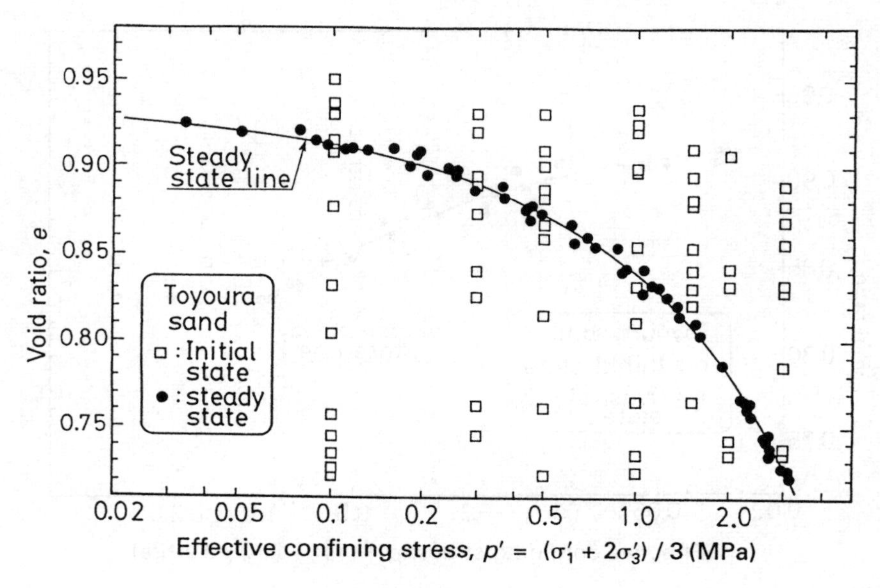

Fig. 11.13 SSL of Toyoura sand.

quasi-steady state envisaged for sand is essentially the same as the critical state defined for clay (Schofield and Wroth, 1968; Poorooshasb, 1989).

11.5 Quasi-steady state of silty sands

The behaviour of clean sands as represented by Toyoura sand has been discussed in detail in the foregoing sections. In reality, however, sand deposits contain, more often than not, some amount of fines consisting of non-plastic to low plasticity silts or clays. With some fines, the sand is generally deposited with high void ratios and tends to be contractive, whatever may be the method of deposition. Accordingly, by any means of sample preparation described above, samples with highly contractive behaviour can be formed in the laboratory. In view of many instances of actual flow failures during earthquakes having occurred in silty sand or sandy silt deposits, the potential for flow failure is considered to be much higher for a dirty sand than for a clean sand. To investigate the deleterious effects of fines on the deformation behaviour, a series of undrained triaxial tests were performed on silty sand and sandy silt samples. Two methods of sample preparation, the dry deposition and water sedimentation techniques, were mainly employed to form specimens for triaxial testing. The method of moist placement was not always used for siltier materials because of its difficulty in forming consistently good samples. The results of the tests and their implication for practical usage will be described in the following.

The silty sand tested was obtained from an alluvial deposit at Tia Juana, Venezuela. The gradation curve of this soil is shown in Fig. 11.4. The specific

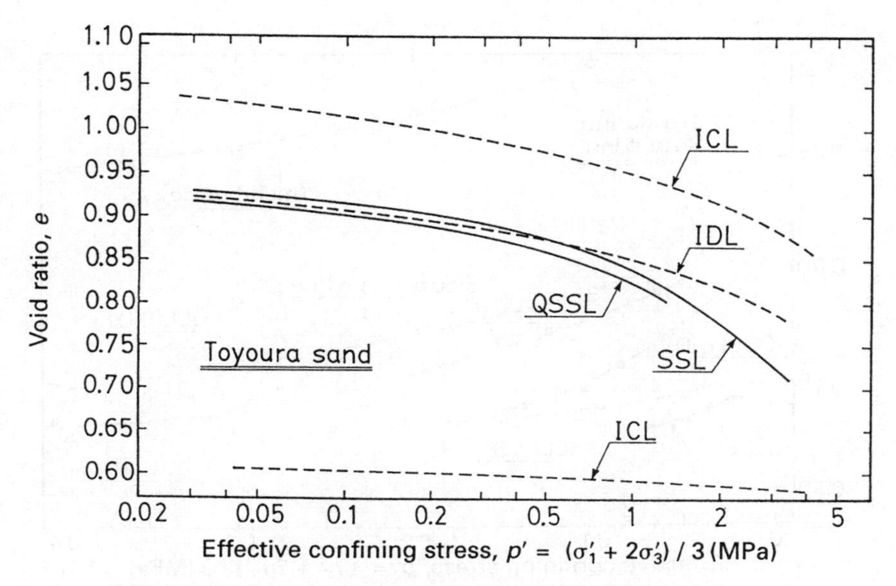

Fig. 11.14 Characteristic lines of Toyoura sand in a e–log p' diagram.

gravity is 2.68 and is non-plastic. The maximum and minimum void ratios determined by the method of JSSMFE are 1.099 and 0.620, respectively.

In the first series of tests using the method of dry deposition, the specimens formed at its highest possible void ratio were consolidated to various effective confining stresses. The consolidation curve obtained in this test series is shown by a dashed line and denoted by ICL (D.D.) in Fig. 11.15. Undrained shear stress application was started from the state of the samples lying on this isotropic consolidation line (ICL). The individual states of the samples tested are indicated by an open square. The stress–strain relations and stress paths obtained from this test series are presented in Fig. 11.16, where it may be seen that the ultimate steady state is reached at a stage where the shear strains grow to values as high as 30%. Individual states of the samples at QSSL and SSL are indicated by full squares and half-filled squares, respectively, in the e–p' diagram of Fig. 11.15.

Samples formed with the highest possible void ratio using the method of water sedimentation were also tested in the same fashion in a second series of tests. The isotropic consolidation curve obtained for the water-sedimented samples is shown in Fig. 11.15 by open triangles and denoted by ICL (W.S.). The stress–strain curves and stress paths obtained in this test series are demonstrated in Fig. 11.17, where it can be clearly seen that the quasi-steady state is produced at low levels of shear stress over the range of strains between 1 and 7%, whereas the ultimate steady state is encountered at larger shear stresses in the shear strain range exceeding 25%. Conditions of individual samples at the QSS and steady state are indicated by filled triangles and half-filled triangles, respectively, in the e–p' plot of Fig. 11.15.

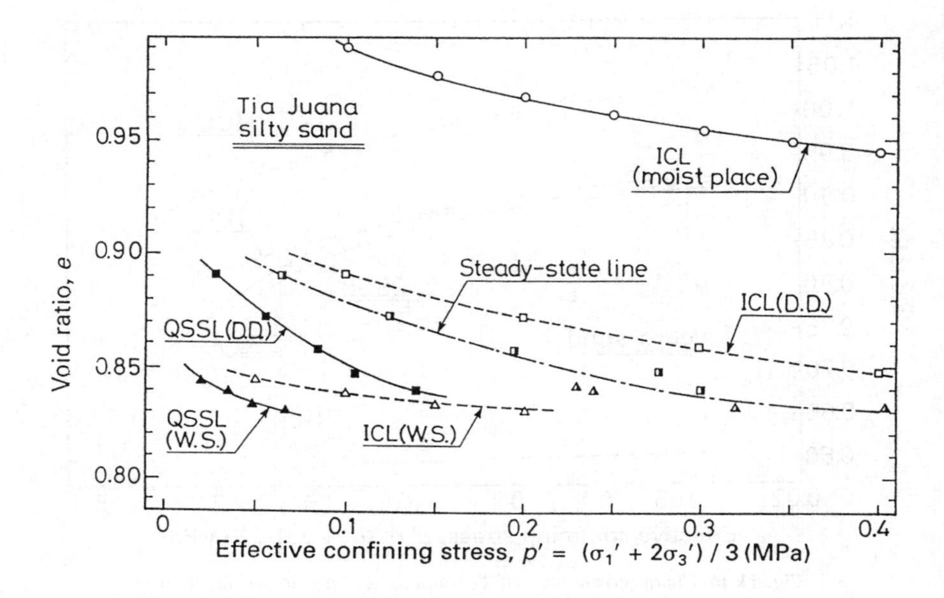

Fig. 11.15 SSL and QSSL for samples of Tia Juana silty sand.

A sandy silt material obtained from an alluvial deposit at Lagunillas, Venezuela was also tested in the same fashion by preparing samples by the method of dry deposition and water sedimentation. The gradation curve is given in Fig. 11.4, and the specific gravity and plasticity index are 2.68 and 4.0, respectively. The maximum and minimum void ratios nominally determined by JSSMFE method are 1.389 and 0.766, respectively. The consolidation curves for the samples prepared by the respective method to their highest possible void ratios are displayed by dashed lines in Fig. 11.18. The stress–strain relations and stress paths obtained for the samples of Lagunillas sandy silt are not presented here. The QSS and steady state obtained in this series of tests are indicated in the e–p' diagram of Fig. 11.18 by the same symbols and notations as used in Fig. 11.15 for Tia Juana silty sands. The conditions of individual samples at QSS and steady state show that, unlike the case of Tia Juana soil, the QSSL is almost coincident with the SSL for Lagunillas sandy silt.

The observation of the test data compiled in Figs 11.15 and 11.18 indicates several features of interest regarding the effects of fabrics as created by different methods of sample preparation. Firstly, it can be seen that the two methods of sample preparation employed in the present test scheme produce distinctly different isotropic consolidation lines. By the dry deposition method, a fabric with much higher void ratios was formed as compared to the fabric created by the method of water sedimentation under the same consolidation pressure. In accord with this, the QSSL is also distinctly different between the two groups of samples prepared by the

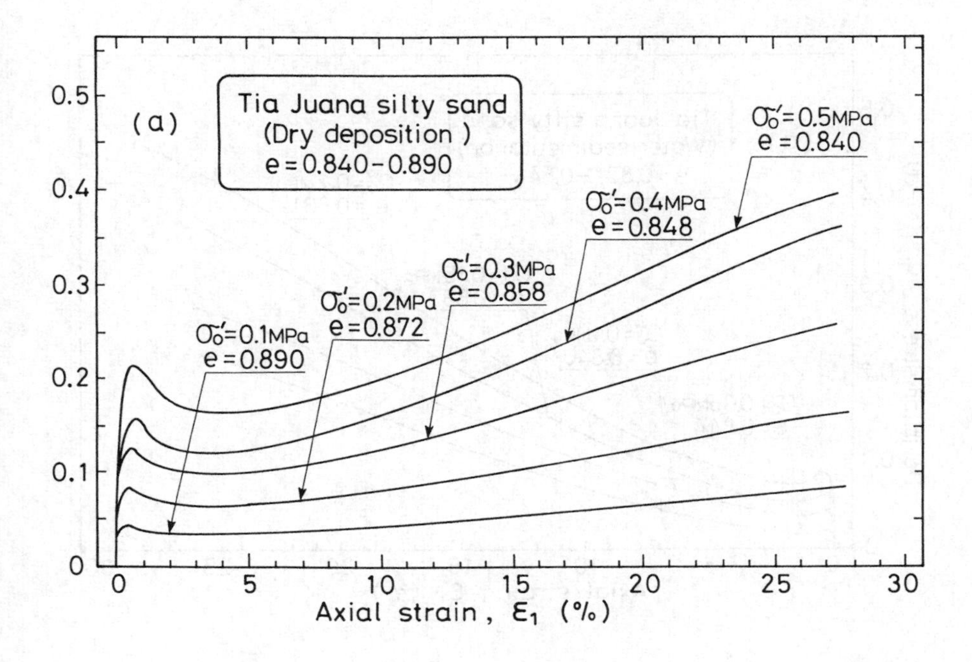

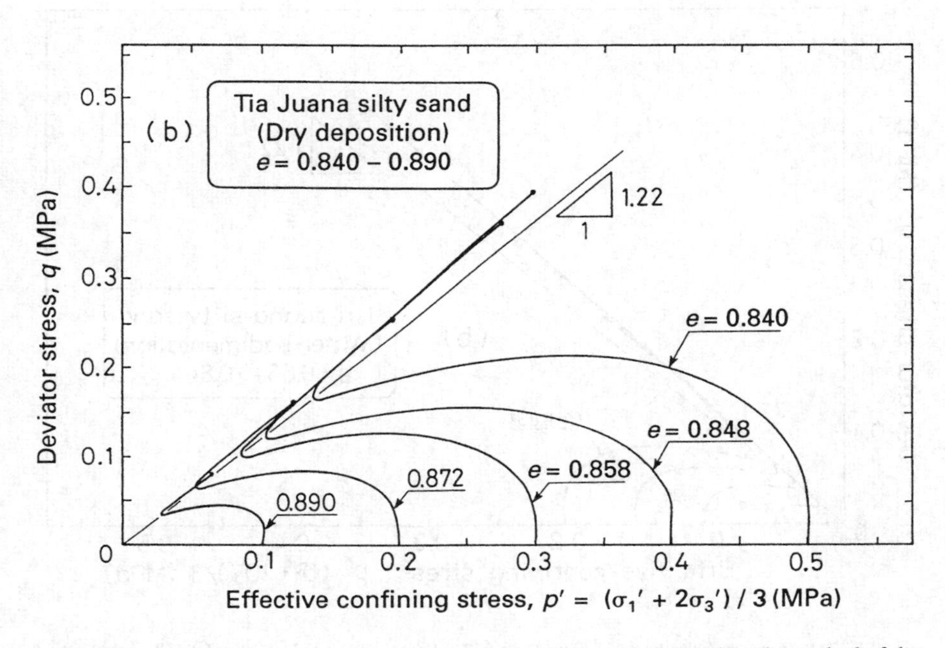

Fig. 11.16 Undrained behaviour of samples of Tia Juana silty and prepared by the method of dry deposition.

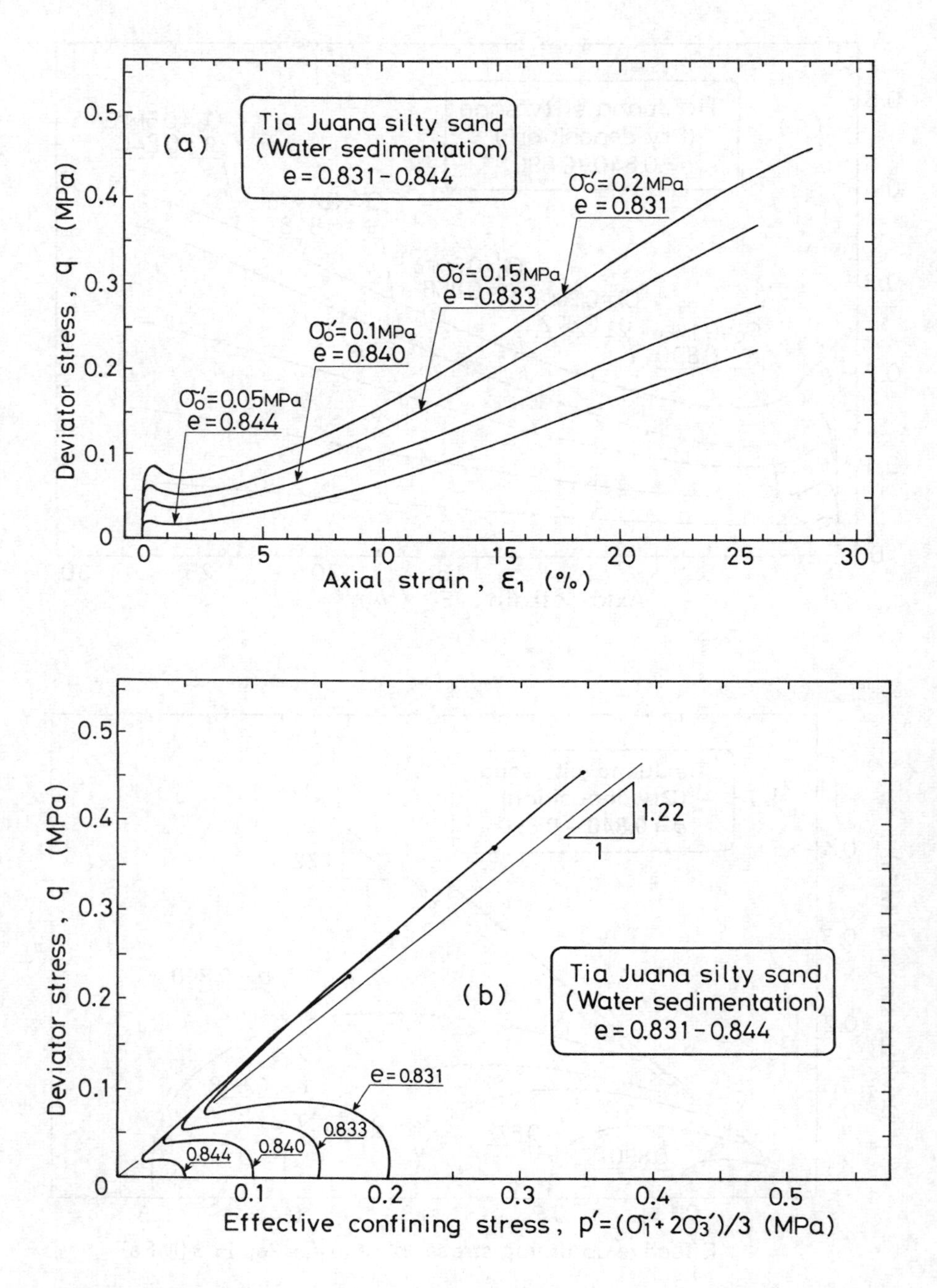

Fig. 11.17 Undrained behaviour of samples of Tia Juana silty sand prepared by the method of water sedimentation.

different methods. As far as the SSL is concerned, however, the two groups of samples are shown to yield an all but unique single line. Thus, it may be mentioned conclusively that the quasi-steady state is profoundly influenced by the fabrics formed by different modes of sand deposition which are represented by the different ICLs inherent to each method of sample preparation, whereas the ultimate steady state is established uniquely for a given sand independently of the mode of sand deposition. The non-uniqueness of the quasi-steady state can be understood with good reasons, if one is reminded of the state of the sand being deformed in the moderate strain range still preserving its inherent fabric structure formed during the process of deposition. When the sand is deformed largely, the remnants of initially formed fabric structure are completely erased and consequently the ultimate steady state becomes unaffected by the mode of sand deposition. For reference sake, an ICL produced by the moist placement method is also shown in Figs 11.15 and 11.18. It may be seen that this method produces another type of highly porous fabric structure in the same silty sand. Thus, it may be mentioned in general that the method of moist placement could create probably the highest-level porous structure, whereas a moderately porous structure and the lowest-level porous fabric are produced in the laboratory sample by means of the dry deposition and water sedimentation, respectively.

Another feature of interest worthy of note is the fact that the void ratio in the initial fabric by dry deposition is higher than the void ratio created by the method of water sedimentation. Accordingly, as is apparent from the stress paths in Figs 11.16 and 11.17, the samples by dry deposition were driven to the steady state by generating so great an amount of pore water pressure that the remaining confining

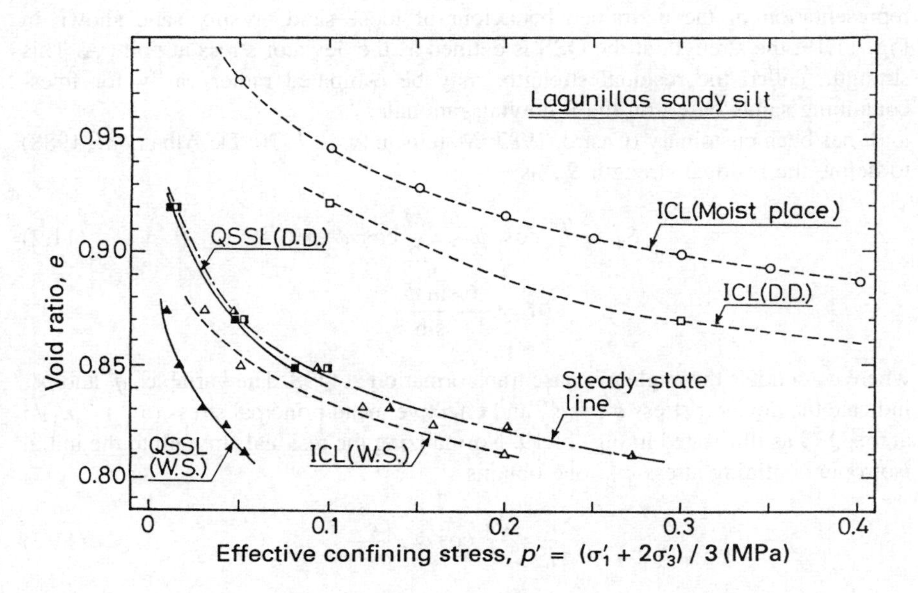

Fig. 11.18 SSL and QSSL for samples of Lagunillas sandy silt.

stress at the ultimate steady state is smaller than the initial value of confining stress. In contrast to this, a moderate amount of pore water pressure generation is sufficient to take the water-sedimented samples to the quasi-steady state and consequently the ultimate steady state is attained with the effective confining stress in excess of its initial value at the time of consolidation. This tendency is more notably observed in the sandy silt of Lagunillas as compared to that of silty sand from Tia Juana. Worthy of note is the fact that the QSS and steady state of dry-deposited samples happen to coincide for the sandy silt from Lagunillas.

Lastly, it is noted in Figs 11.16(b) and 11.17(b) that the angle of phase transformation ϕ_s at the QSS is the same for the two kinds of specimens prepared by the method of dry deposition and water sedimentation. This observation has been proven to be valid for other clean sands and silty sands tested. Thus it can be mentioned that the angle of phase transformation is determined uniquely irrespective of fabrics formed under different modes of deposition.

11.6 Residual strength of fines-containing sand

The outcome of the tests shown in Figs 11.16 and 11.17 indicate that for silty sand there is a quasi-steady state over the range of shear strain between 3 and 10%, where a moderately large deformation can develop while keeping the magnitude of deviator stress at the lowest level. If soils in field deposits are brought to such a state as a result of seismic load application, there will occur an intolerable amount of lateral displacements causing serious damage to embankments or structures lying on such deposits. Therefore the strength of the soils mobilized at the quasi-steady state has an important ramification for engineering practice. In the schematic representation of the undrained behaviour of loose sand or silty sand shown in Fig. 11.19, the strength at the QSS is defined as the deviator stress at point A. This strength, called the residual strength, may be estimated rather easily for fines-containing sands based on the following rationale.

It has been customary (Castro, 1987; Marcuson *et al.* 1990; De Alba *et al.* 1988) to define the residual strength S_{us} as

$$S_{us} = \frac{q_s}{2} \cos \phi_s = \frac{M}{2} \cos \phi_s . p'_s \qquad (11.2)$$

$$M = \frac{6 \sin \phi_s}{3 - \sin \phi_s}$$

where ϕ_s denotes the angle of phase transformation at QSS. The variables q_s and p'_s indicate the deviator stress $\sigma'_1 + \sigma'_3$ and effective mean principal stress $(\sigma'_1 + 2\sigma'_3)/3$ at the QSS as illustrated in Fig. 11.19. Normalizing the residual strength to the initial isotropic confining stress σ'_0, one obtains

$$\frac{S_{us}}{\sigma'_0} = \frac{M}{2} \cos \phi_s . \frac{p_s{}'}{p_c{}'} \qquad (11.3)$$

where p'_c is used instead of σ'_0 when referring to the point in the stress path. The

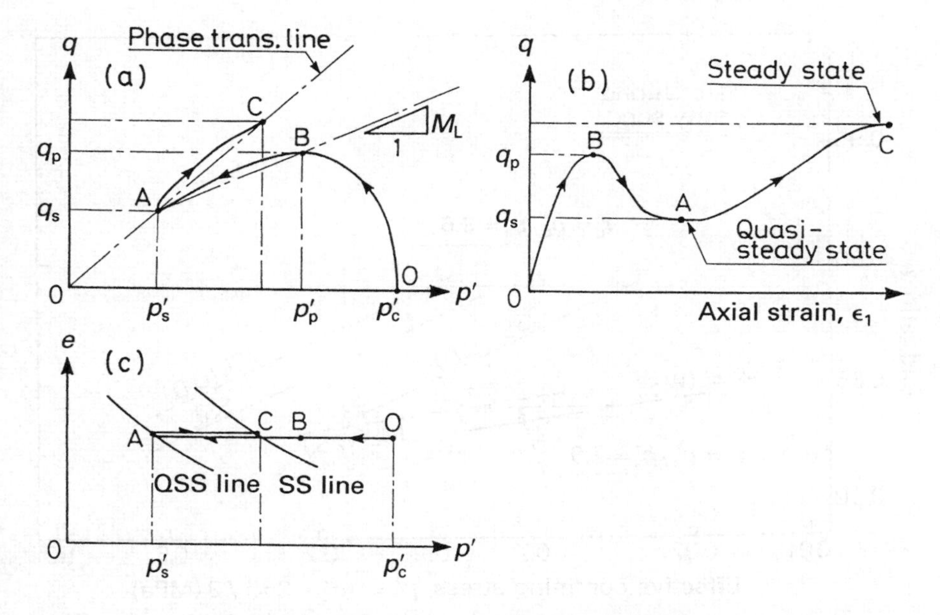

Fig. 11.19 Characteristics of undrained behaviour of loose sand.

ratio defined above is called the *normalized residual strength*. At this stage, it is convenient to introduce a parameter r_c defined as,

$$r_c = \frac{p_c{}'}{p_s{}'}. \tag{11.4}$$

This ratio between the effective confining stress at the initial state and at the quasi-steady state is referred to as the *initial state ratio*. Then, eqn (11.3) is rewritten as

$$\frac{S_{us}}{\sigma_0{}'} = \frac{M}{2} \cos \phi_s . \frac{1}{r_c}. \tag{11.5}$$

If the test data in Figs. 11.15 and 11.18 are replotted in the diagram of the void ratio against the logarithm of confining stress, the condition of initial isotropic consolidation and the quasi-steady state may be represented approximately by two straight lines. The ICL and QSSL thus obtained for Tia Juana sand and Lagunillas sandy silt are displayed in Figs. 11.20 and 11.21. It is important to remember that the value of r_c ought to remain constant over the range of effective confining stress being considered. In the plot of test data shown in Figs. 11.20 and 11.21, one can recognize that, for both Tia Juana silty sand and Lagunillas sandy silt, the slope of the QSSL is parallel to that of the ICL for each set of the test data obtained using two methods of sample preparation. Thus there are reasons for the concept of the initial state ratio to be applied for the test data shown above. The values of r_c directly read

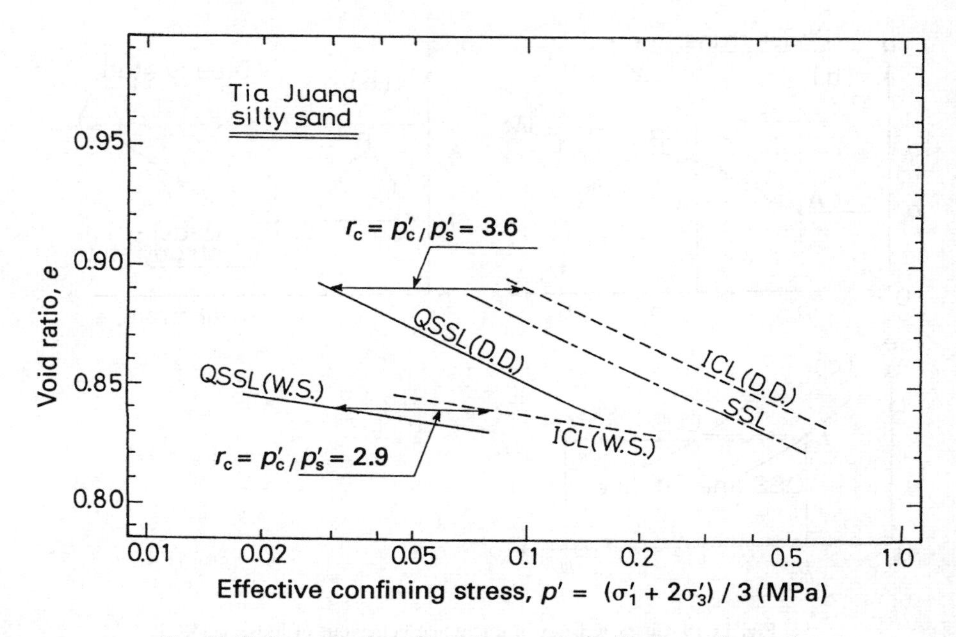

Fig. 11.20 Determination of initial state ratio for Tia Juana silty sand.

off from Figs. 11.20 and 11.21 are shown in Table 11.1. The angle of phase transformation at the quasi-steady state ϕ_s was read off from the stress path plots and also shown in Table 11.1. Thus, the value of $S_{us}/\sigma_0{}'$ was obtained using the relation of eqn (11.5) and shown in Table 11.1.

On the other hand, it is possible to directly read off the minimum value of shear stress from the stress–strain curves shown in Figs. 11.16 and 11.17. This minimum value that is the deviator stress q_s at QSS is used in eqn (11.2) to obtain the residual strength S_{us}. The values of S_{us} thus obtained are plotted in Fig. 11.22 versus the confining stress at the time of consolidation. Similar data compilation was also made for the sandy silt from Lagunillas and shown in Fig. 11.23. It may be seen in these figures that the residual strength at QSS tends to increase in proportion to the initial confining stress. Thus the relationship between S_{us} and σ'_0 can be represented approximately by a straight line as accordingly indicated in the figures. The slope of this straight line gives the value of the normalized residual strength. It is self-evident that the value of S_{us}/σ'_0 thus obtained for Tia Juana silty sand and Lagunillas sandy silt is coincident with the corresponding value of normalized residual strength listed in Table 11.1 which has been obtained through the concept of the initial state ratio using the relation of eqn (11.4).

The test data arranged in Figs. 11.22 and 11.23 indicate that the normalized residual strength of a given material tends to vary to some extent depending upon the fabric formed by different modes of deposition.

Table 11.1 Parameters characterizing undrained stress paths of fines-containing sand

Soil type	Tia Juana silty sand		Lagunillas sandy silt	
Method of sample preparation	D.D.*	W.S.**	D.D*	W.S.**
ϕ_s degrees	30.5	30.5	31	31
Initial state index, $r_c = p'_c/p'_s$	3.6	2.9	6.2	4.0
$\frac{S_{us}}{\sigma_0{}'} = \frac{\cos\phi_s}{2}\frac{M}{r_c}$	0.146	0.181	0.086	0.134

* Dry deposition

** Water sedimentation

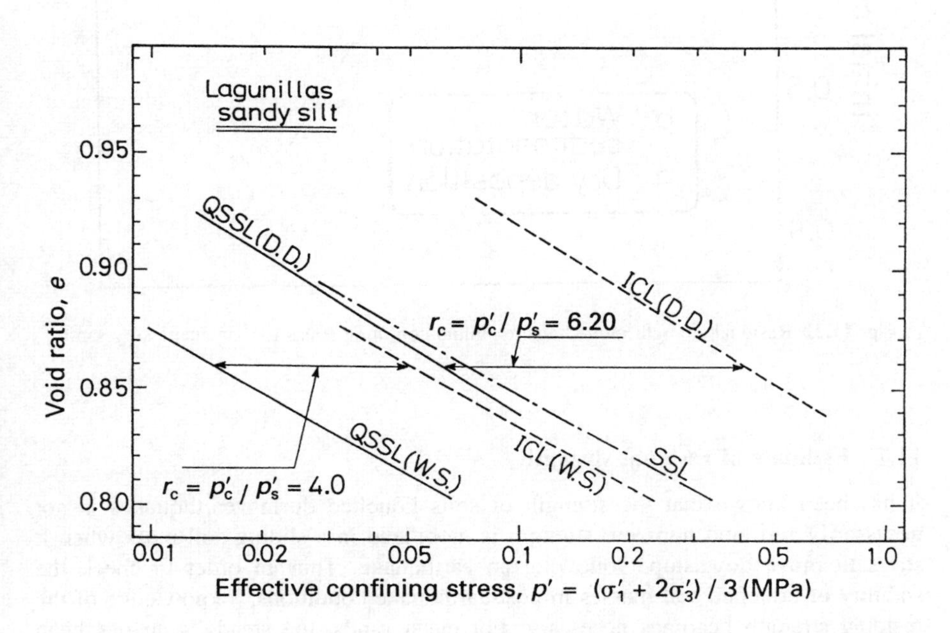

Fig. 11.21 Determination of initial state ratio for Lagunillas sandy silt.

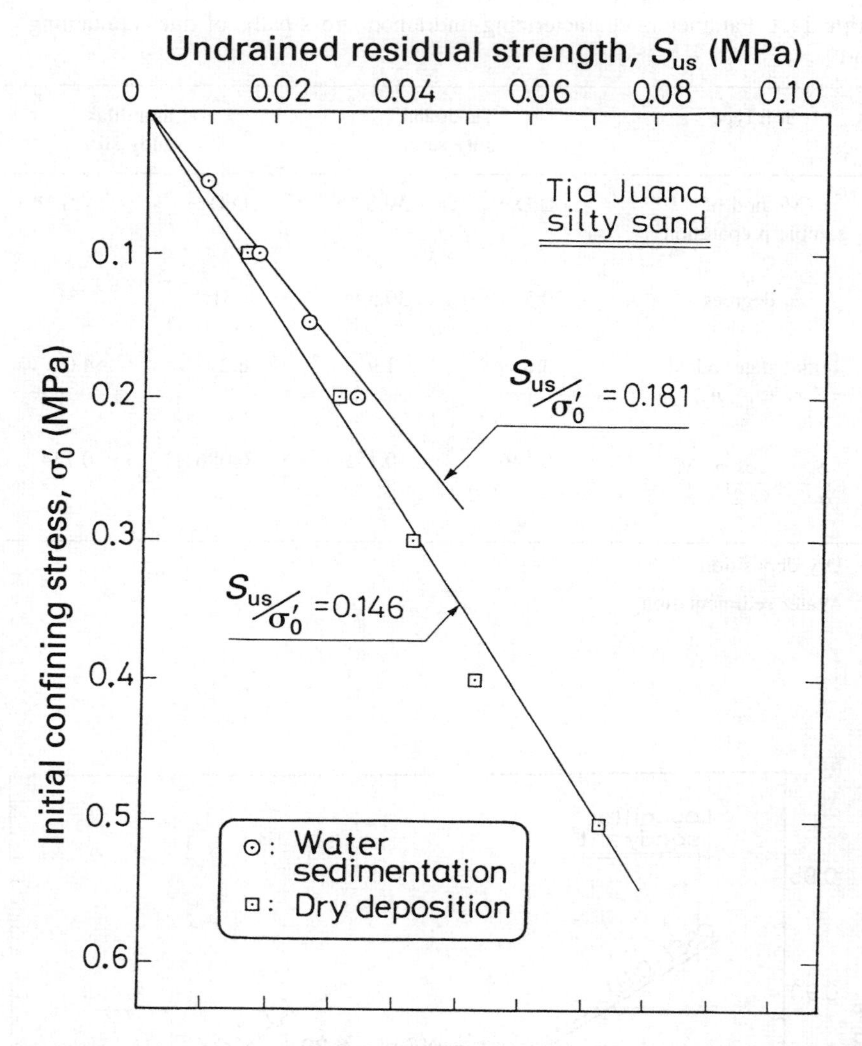

Fig. 11.22 Residual strength plotted against initial confining stress for Tia Juana silty sand.

11.7 Estimate of residual strength

It has been known that the strength of soils liquefied during earthquakes is not necessarily zero and non-zero strength is mobilized in a sliding soil mass when it starts to move downslope following an earthquake. Thus in order to check the stability of liquefied soil masses in post-earthquake conditions, a knowledge of the residual strength becomes necessary. For clean sands, the steady state has been known to be governed solely by the void ratio and attempts have been made to

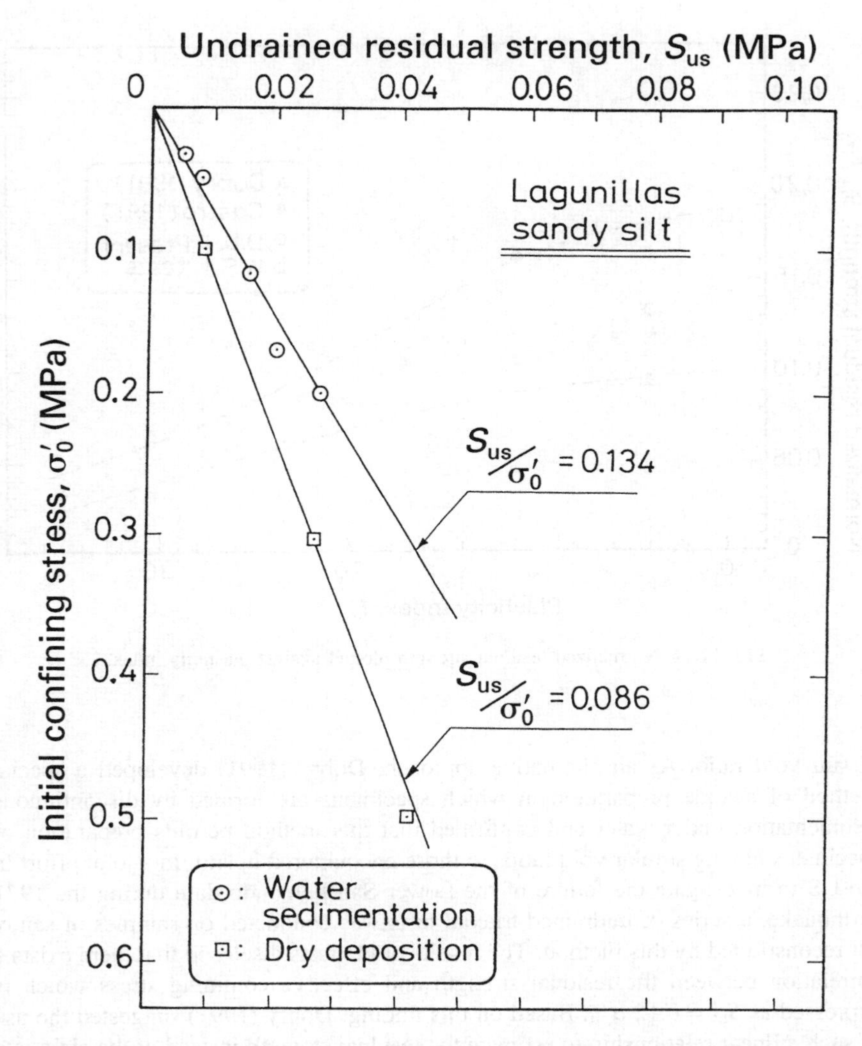

Fig. 11.23 Residual strength plotted against initial confining stress for Lagunillas sandy silt.

estimate the in situ void ratio by means of undisturbed sampling (Poulos *et al.* 1985a). It has also been pointed out that the undrained residual strength is so sensitive to minute variation of the void ratio that it is generally a difficult task to estimate the residual strength with a reasonable degree of accuracy. On the other hand, it was pointed out by Baziar and Dobry (1991) that most in situ soils in alluvial or reclaimed deposits which are exposed to the potential of liquefaction during earthquakes are composed of sands containing 10 to 80% fines and for this type of soil, there is not necessarily a strong need for accurately determining the

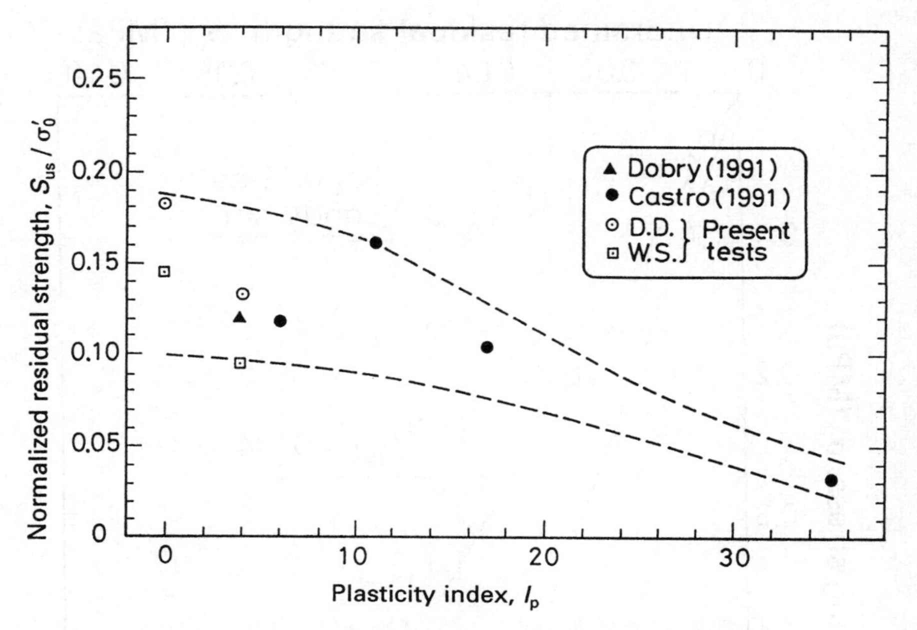

Fig. 11.24 Normalized residual strength plotted against plasticity index.

in situ void ratio. As an alternative approach, Dobry (1991) developed a special method of sample preparation in which specimens are formed by discontinuous sedimentation under water and confirmed that this method permits preparation of specimens having similar void ratios to those encountered in situ. In a joint effort in the US to investigate the failure of the Lower San Fernando dam during the 1971 earthquake, a series of undrained triaxial tests was conducted on samples of sandy silt reconstituted by this method. The results of the tests disclosed that there exists a correlation between the residual strength and effective confining stress which is expressed as $S_{us} = 0.12\ \sigma'_0$. Based on this finding, Dobry (1991) suggested the use of such a linear relationship to estimate the residual strength instead of the elaborate in situ measurements of void ratio. The above relation is shown in Fig. 11.24 in terms of S_{us}/σ'_0 plotted versus the plasticity index of the soil.

Mine tailings is another kind of silty sand susceptible to liquefaction for which evaluation of residual strength has an important bearing on appraisal of the degree of post-earthquake stability. Vane test was used by Poulos *et al.* (1985b) to investigate this aspect for aluminum tailings ($I_p = 35$) consolidated slurry in a box in the laboratory. A small vane 2.0 cm in diameter and 4.0 cm in height was rotated fast enough to keep the shearing undrained. The result of tests showed a correlation, $S_{us} = 0.022\ \sigma'_v$. It may be assumed that the deposit normally consolidated from slurry had a value of $K_0 = 0.5$. Therefore, using the relation, $\sigma'_0 = (1 + 2K_0)\,\sigma'_v/3$, the value of σ'_0 is estimated to have been $\sigma'_0 = 2\sigma'_v/3$. Thus, the normalized residual strength may be obtained as $S_{us} = 0.033\ \sigma'_0$, as displayed in Fig. 11.24. Similar tests using the vane were conducted by Castro and Troncoso (1989) on samples of slimes obtained from test pits excavated at three tailings dams sites in Chile. The results of their tests are also presented in Fig. 11.24. The normalized residual strength obtained in Figs. 11.22 and 11.23 for the Tia Juana silty sand and sandy silt from Lagunillas is also shown in the diagram of Fig. 11.24. It may be seen that the normalized undrained residual strength has a tendency to decrease with increasing plasticity index of soils. Generally speaking, the consolidation characteristics can not be uniquely established for sands containing non-plastic to low plasticity fines. As is apparently the case with the soils from Tia Juana and Lagunillas shown in Figs. 11.15 and 11.18, the consolidation characteristics are governed to a great extent by conditions under which the soils are placed and consolidated. Because of the multiplicity of the consolidation curves inherent to each fabric formed by different modes of soil deposition, the quasi-steady state can not be determined uniquely for a given soil. In view of this observation, it appears reasonable to observe so much scatter in the plot of normalized residual strength shown in Fig. 11.24 particularly for soils having a low plasticity index.

11.8 Effects of the fabric on residual strength

The deformation characteristics of Toyoura sand discussed in Section 11.4 was based on the observed response of samples prepared by the method of moist placement. To established a general framework for understanding sand behaviour covering a wide range of void ratio, the observed response of the moist-placed samples was instructive and useful. However, the fabric of samples formed by this method may not represent all depositional states of sand existing in the field. Thus it is worthwhile studying the behaviour of samples prepared by other methods of preparation.

The results of undrained triaxial compression tests on specimens prepared by the method of dry deposition are shown in Fig. 11.25 where the void ratio is plotted versus the effective confining stress at the initial state and at the quasi-steady state. Also shown by the dotted area is the range of initial state of moist-placed specimens which exhibit contractive behaviour upon application of shear stress. This range is quoted from the diagram in Fig. 11.11. The QSSL for the moist-placed specimens is also quoted from Fig. 11.11 and shown in Fig. 11.25. It is to be noted that by the method of dry deposition the contractive behaviour is encountered only when the specimen is prepared at its loosest possible state. If such a specimen is compacted initially to produce a slightly dense state, its behaviour becomes dilative when sheared undrained. Therefore the initial consolidation line (ICL) corresponding to the loosest possible state does coincide with the initial dividing line (IDL), as accordingly indicated in Fig. 11.25.

To illustrate the effects of fabric, the characteristics lines displayed in Fig. 11.25 are rewritten in Fig. 11.26. Suppose two specimens, one by moist placement and the other by dry deposition, are prepared at their loosest states and consolidated to have

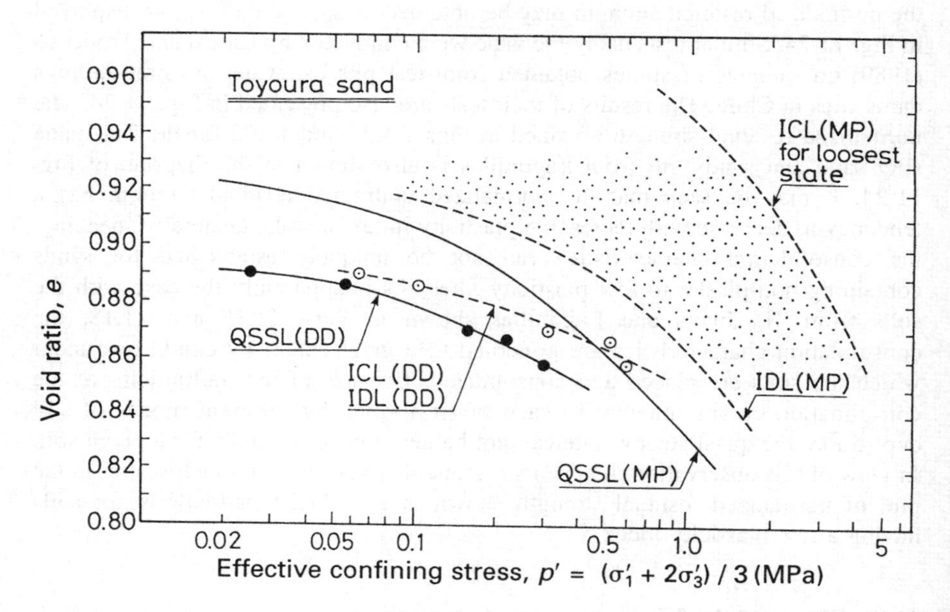

Fig. 11.25 Initial consolidation line (ICL) and quasi-steady state line (QSSL) for the loosest states of sand prepared by two different methods.

an identical void ratio, say, $e = 0.88$, as indicated by points A_1 and A_2 in Fig. 11.26. Needless to say, the moist-placed specimen must be consolidated to a confining stress which is higher than that of the dry-deposited specimen. Upon being sheared undrained, each specimen produces the quasi-steady state at points B_1 and B_2 thereby exhibiting the minimum strength. The value of residual strength itself is large for the moist-placed specimen because of its effective confining stress at QSS being higher than that at point B_2 for the dry-deposited specimen. However the horizontal distance in log between points A_1 and B_1 is much larger than the distance between A_2 and B_2. Since this distance is taken as the initial state ratio defined by eqn (11.4), its value is $r_c \doteqdot 10$ for the moist-placed specimen and $r_c \doteqdot 2.0$ for the dry-deposited specimen. Consequently, the normalized residual strength defined by eqn (11.5) becomes 5 times greater for the dry-deposited specimen as compared to the moist-placed specimen. Note that what is essentially important in the present discussion is not the residual strength but the normalized residual strength. Generally speaking, the values of M and $\cos \phi_s$ in eqn (11.5) remained unchanged for varying fabrics and therefore the normalized residual strength is governed by the initial state ratio. From the considerations as above, it becomes clear that, even though the void ratio is the same, the moist-placed specimen exhibits much less normalized residual strength than does the dry-deposited specimen. This is considered as a manifestation of the effect of the fabric created by the different methods of deposition in preparing the test specimens.

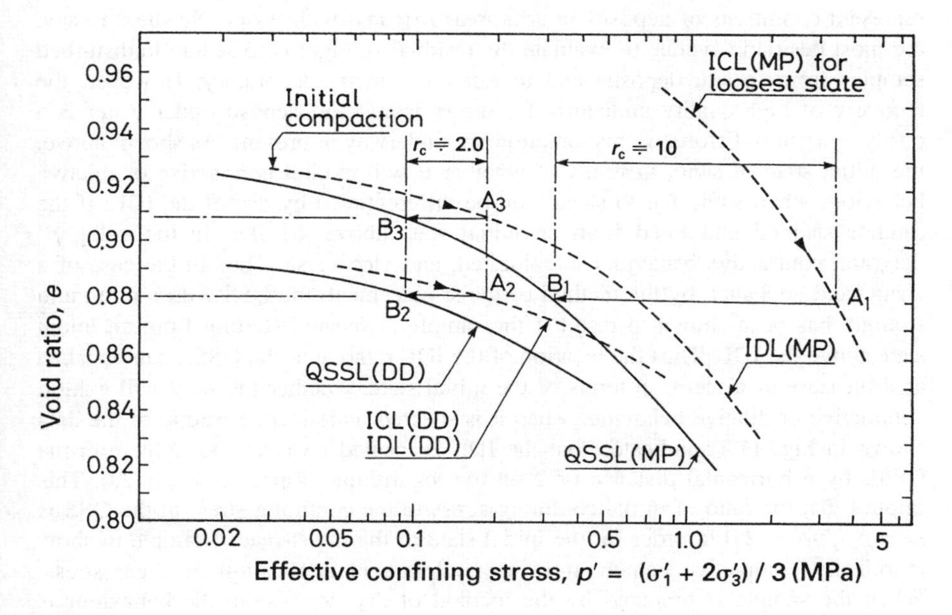

Fig. 11.26 Illustration of the implication of the r_c value.

Another feature of the fabric effect is observed by comparing the behaviour of two differently prepared specimens which are sheared undrained starting from the initial states at A_2 and A_3 in Fig. 11.26. In this case, the moist-placed specimen is compacted to some extent initially so that the consolidation can take place following the IDL. Upon being sheared, this specimen exhibits the minimum strength at a quasi-state which is encountered at point B_3. The other specimen prepared by the method of dry deposition is sheared undrained starting from the initial point A_2. This specimen attains the quasi-steady state at point B_2. It is to be noted here that the two specimens as above have the same value of the initial state ratio of $r_c \doteqdot 2.0$. Consequently, the normalized residual strength becomes identical, although the void ratio is different because of the difference in the fabric of the two specimens formed by different preparation methods. This observation again provides evidence to show the importance of the fabric effects in governing the behaviour of sand.

At present it is not possible to quantify the effects of fabric in terms of appropriate parameters that can be used in engineering practice. It would be desirable to conduct tests in the laboratory on specimens prepared by a method which is deemed to replicate depositional conditions in the field. Roughly speaking, the moist placement method may simulate man-made fills on land which is later saturated with water seeping into the sand fills. The method of water sedimentation may duplicate the mode of deposition in hydraulic fills and fluvial sands which are sedimented in underwater environments. The method of dry deposition appears to represent conditions of deposits in arid areas irrigated with water. Needless to say, the most desirable avenue to evaluate the residual strength is to secure undisturbed samples from in situ deposits and to test them in the laboratory. However, the recovery of high-quality undisturbed samples from sand deposit under water is a costly operation. Efforts in this direction are underway at present. As shown above, the initial state of sand, in terms of whether it will exhibit contractive or dilative behaviour when subjected to shear, can be differentiated by use of the IDL. If the sand is sheared undrained from an initial state above the IDL in the e–log p' diagram, contractive behaviour is exhibited, and vice versa. Thus in the case of a clean sand deposited by the method of moist placement, the QSS with a minimum strength has been shown to result if the sample is sheared starting from an initial state above the IDL. Thus the location of the IDL relative to the QSSL can be taken as a measure to discern, in terms of the initial state, whether the sand will exhibit contractive or dilative behaviour when it is sheared undrained. Scrutiny of the data shown in Fig. 11.12 indicated that the IDL is located away to the right from the QSSL by a horizontal distance of 2 on the logarithmic scale: i.e. $r_c \doteqdot 2.0$. This implies that the ratio of initial confining stress to the confining stress at the QSS is $r_c = p_c'/p_s' = 2.0$ in order for the initial state of the moist-placed sample to show marginally contractive behaviour in any subsequent application of shear stress. When the sample is prepared by the method of dry deposition, the behaviour is shown also to be just on the border between dilative and contractive behaviour, with a value of $r_c = 2.0$ as read directly from the data in Fig. 11.26. In summary, it can be concluded that whatever the fabric for sands formed by various modes of deposition, if a deposit shows contractive behaviour in subsequent loading, the deposit should be subjected to an initial confining stress at least twice as great as the confining stress at the QSS.

11.9 Effects of the deformation mode on residual strength

The residual strength or strength at QSS discussed in the foregoing sections has all been based on the data obtained from triaxial compression tests. However, a question arises as to whether or not the same consequence is obtained from other types of shearing tests. To examine this aspect, two series of undrained tests, one in triaxial compression and the other in triaxial extension, were conducted at the University of Tokyo on specimens of several sands all prepared by the method of dry deposition. As a result, the normalized residual strength for the triaxial compression tests was shown to be 2–7 times greater for two of the sands as compared to the strength tested in the triaxial extension. The other two of the sands tested showed much more pronounced decrease in residual strength in the triaxial extension. Observation in the same vein is reported by Vaid *et al.* (1990) and Vaid and Thomas (1994). While accuracy of the data still remains to be examined for the triaxial extension test, it appears likely that the normalized residual strength is, by and large, smaller in the triaxial extension mode than that in the mode of triaxial compression.

To examine the residual strength in the simple shear or torsional mode of deformation, a series of undrained torsional tests was conducted on hollow

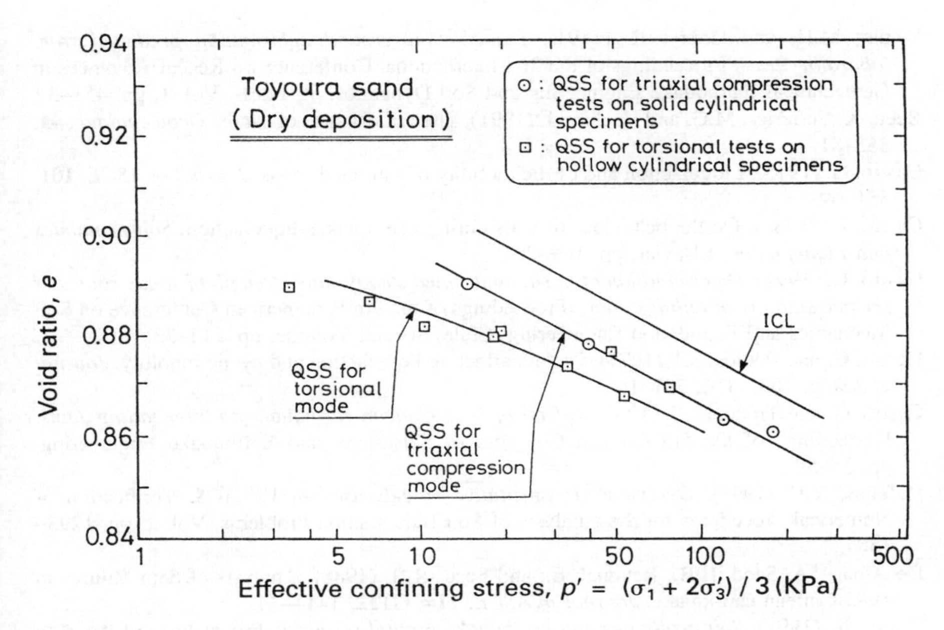

Fig. 11.27 Effects of mode of deformation on the quasi-steady state line.

cylindrical specimens of Toyoura sand prepared by the method of dry deposition. The specimen has a dimension 20 cm in height, and 10 cm in outer and 6 cm in inner diameter. The initial consolidation line (ICL) and the quasi-steady state obtained in the tests are shown in Fig. 11.27. Also shown superimposed on this diagram are the data obtained from the triaxial compression tests on solid cylindrical specimens of Toyoura sand prepared by the method of dry deposition. Thus the two kinds of specimens having an identical fabric were consolidated along the same consolidation line indicated in Fig. 11.27 and then sheared undrained in different modes of deformation. As seen in Fig. 11.27, the QSS for the torsional shear tests is generated at effective confining stresses which are smaller than that in the triaxial compression tests. Therefore, the value of r_c is larger for the torsional test data, leading to a smaller value of normalized residual strength as compared to that obtained from the triaxial compression tests. At the present state of the art, there is no consensus as to which mode of deformation does most precisely represent in situ conditions, but the estimate of residual strength by means of the triaxial compression tests may perhaps lead to somewhat conservative design.

References

Alarcon-Guzman, A., Leonards, G.A., and Chameau, J.L. (1988). Undrained monotonic and cyclic strength of sands. *Journal of ASCE*, **114**, GT10, 1089–1109.

Baziar, M.H. and Dobry, R. (1991). *Liquefaction ground deformation predicted from laboratory tests*. Proceedings of the 2nd International Conference on Recent Advances in Geotechnical Earthquake Engineering and Soil Dynamics, St. Louis, Vol. 1, pp. 451–8.

Been, K., Jefferies, M.G. and Hachey, J. (1991). The critical state of sands. *Geotechnique*, **41**, 365–81.

Castro, G. (1975). Liquefaction and cyclic mobility of saturated sands. *Journal of ASCE*, **101**, 551–69.

Castro, G. (1987). On the behaviour of soils during earthquakes-liquefaction. *Soil Dynamics and Liquefaction*, Elsevier, pp. 169–204.

Castro, G. (1991). *Determination of in situ undrained steady state strength of sandy soils and seismic stability of tailings dams*. Proceedings of the 9th Panamerican Conference on Soil Mechanics and Foundation Engineering, Chile, Special Volume, pp. 111–33.

Castro, G. and Poulos, S.J. (1977). Factors affecting liquefaction and cyclic mobility. *Journal of ASCE*, **103**, GT6, 501–16.

Castro, G. and Troncoso, J. (1989). *Effects of 1985 Chilean earthquake on three tailing dams*. Proceedings of the 5th Chilean Congress of Seismicity and Earthquake Engineering, pp. 35–59.

Dafalias, Y.F. (1993). *Overview of constitutive models used in VELACS*. Verification of Numerical Procedures for the Analysis of Soil Liquefaction Problems, Vol. 2, pp. 1293–1304.

De Alba, P.A. Seed, H.B., Retamal. E., and Seed, R.B. (1988). Analysis of dam failures in 1985 Chilean earthquake. *Journal of ASCE*, **114**, GT12, 1414–34.

Dobry, R. (1991). *Soil properties and earthquake ground response*. Proceedings of the 10th European Conference on Soil Mechanics and Foundation Engineering, Florence, Italy, Vol. 4.

Georgiannou, V.N., Hight, D.W., and Burland, J.B. (1991). Behaviour of clayey sands under undrained cyclic triaxial loading. *Geotechnique*, 41, 383–93.

Hanzawa, H. (1980). Undrained strength and stability analysis for a quick sand. *Soils and Foundations*, **20**, 17–29.

Ishihara, K. (1993). Liquefaction and flow failure during earthquakes. The 33rd Rankine Lecture, *Geotechnique*, 43, pp. 349–415.

Ishihara, K., Tatsuoka, F., and Yasuda, S. (1975). Undrained deformation and liquefaction of sand under cyclic stresses. *Soils and Foundations*, 15, 29–44.

Ishihara, K., Verdugo, and R. Acacio, A.A. (1991). *Characterization of cyclic behavior of sand and post-seismic stability analyses*. Proceedings of the 9th Asian Regional Conference on Soil Mechanics and Foundation Engineering, Vol. 2, Bangkok, Thailand, pp. 45–70.

Konrad, J.M. (1990a), Minimum undrained strength of two sands. *Journal of ASCE*, 116, GT6, 932–47.

Konrad, J.M. (1990b). Minimum undrained strength versus steady-state strength of sands. *Journal of ASCE*, **116**, GT6, 948–63.

Marcuson, III, W.F., Hynes, M.E., and Franklin, A.G. (1990). Evaluation and use of residual strength in seismic safety analysis of embankments. *Earthquake Spectra*, 6, pp. 529–72.

Mohamad, R. and Dobry, R. (1986). Undrained monotonic and cyclic triaxial strength of sand. *Journal of ASCE*, 112, GT10, 941–58.

Poorooshasb, H.B. (1989). Description of flow of sand using state parameters. *Computers and Geotechnics*, **8**, 195–218.

Poulos, S.J., Castro, G., and France J.W. (1985a). Liquefaction evaluation procedure. *Journal of ASCE*, GT6, 111, 772–92.

Poulos, S.J., Robinsky, E.I., and Keller, T.O. (1985b). Liquefaction resistance of thickened tailings. *Journal of ASCE*, **111**, GT12, 1380–94.

Schofield, A.N. and Wroth, C.P. (1968). *Critical state soil mechanics*. McGraw-Hill, London.
Vaid, Y.P., Chung, E.K.F., and Keurbis, R.H. (1990). Stress path and steady state. *Canadian Geotechnical Journal*, 27, 1–27.
Vaid, Y.P. and Thomas, J. (1994). *Post liquefaction behaviour of sand*. Proceedings of the 13th International Conference on Soil Mechanics and Foundation Engineering, New Delhi, Vol. 1, 1305–10.
Verdugo, R.L. (1992). *Characterization of sandy soil behaviour under large deformation*. Ph.D. dissertation to the University of Tokyo.

12

EVALUATION OF LIQUEFACTION RESISTANCE BY IN SITU SOUNDINGS

For accurate evaluation of the cyclic strength of sand, recovery of high-quality undisturbed samples and testing in the laboratory would be the most reliable procedure. However, the sampling of sand samples from deposits below the ground water table is a costly operation and cannot be recommended for every occasion of liquefaction investigation except for an important construction project. Therefore a simpler and more economically feasible procedure to assess the cyclic resistance of sand needs to be established. One of the methods to accomplish this end is to take advantage of the penetration resistance of the Standard Penetration Test (SPT) or Cone Penetration Test (CPT) which have found worldwide use in investigating in situ characteristics of soil deposits. There are basically two different approaches as described below for correlating penetration resistance with the cyclic strength of soils in the field.

12.1 Correlation based on field performances

12.1.1 *SPT correlation*

A method of determining the liquefaction-inducing cyclic stress ratio as a function of the N value in the SPT was proposed by Seed *et al.* (1983) on the basis of a vast amount of field performance data of sand deposits during recent earthquakes. In this method, the magnitude of the cyclic stress ratio believed to have developed at a given site during an earthquake is determined, using the relation of eqn (2.5), based on a recorded or properly assessed value of the maximum horizontal acceleration. The cyclic stress ratio thus obtained is plotted versus the SPT N value of soil deposits at the same site. If there was any evidence of liquefaction observed on the surface, the data points are marked with solid circles. If there was no sign then the data points are inscribed with open circles. Then a line is drawn along the lowest rim of the zone enclosing all open-marked data points. The points with open circles above this line are considered to indicate the case where liquefaction has occurred probably in deep deposits but with no sign manifested on the ground surface. The line is regarded as a boundary differentiating between conditions in which liquefaction can and cannot occur, and thus considered as the desired correlation between the cyclic shear strength and the N value in the SPT. The most recent version of this kind of correlation obtained by Seed *et al.* (1985) based on

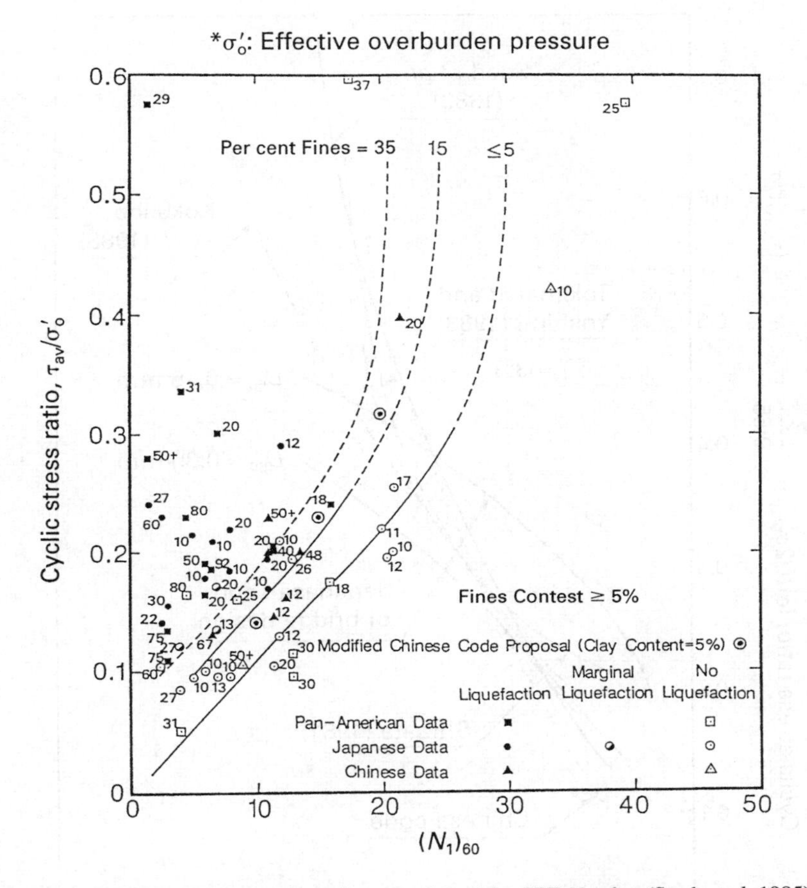

Fig. 12.1 Correlation between cyclic strength and the SPT N value (Seed *et al.* 1985).

experiences of earthquakes with a magnitude around 7.5 is shown in Fig. 12.1 in which the cyclic stress ratio in the ordinate $\tau_{av}/\sigma_v{}'$ is determined as

$$\frac{\tau_{a.v}}{\sigma_v{}'} = 0.65\,\frac{\tau_{max}}{\sigma_v{}'} = 0.65\,\frac{a_{max}}{g}.r_d.\frac{\sigma_v}{\sigma_v{}'} \tag{12.1}$$

where σ_v and $\sigma_v{}'$ denote, respectively, the total and effective overburden pressure, a_{max} is the maximum acceleration, τ_{max} is the maximum shear stress and τ_{av} is the averaged amplitude of cyclic stress. The quantity r_d is called the stress reduction factor which takes a value decreasing with depth from a value of unity on the ground surface. The relation between the middle and the right-hand side of eqn (12.1) is derived from eqn (2.5), but details are described in Section 13.1. The relation between the left-hand side and the middle of eqn (12.1) was derived empirically. The blow count number obtained at the site in question is plotted in the abscissa of

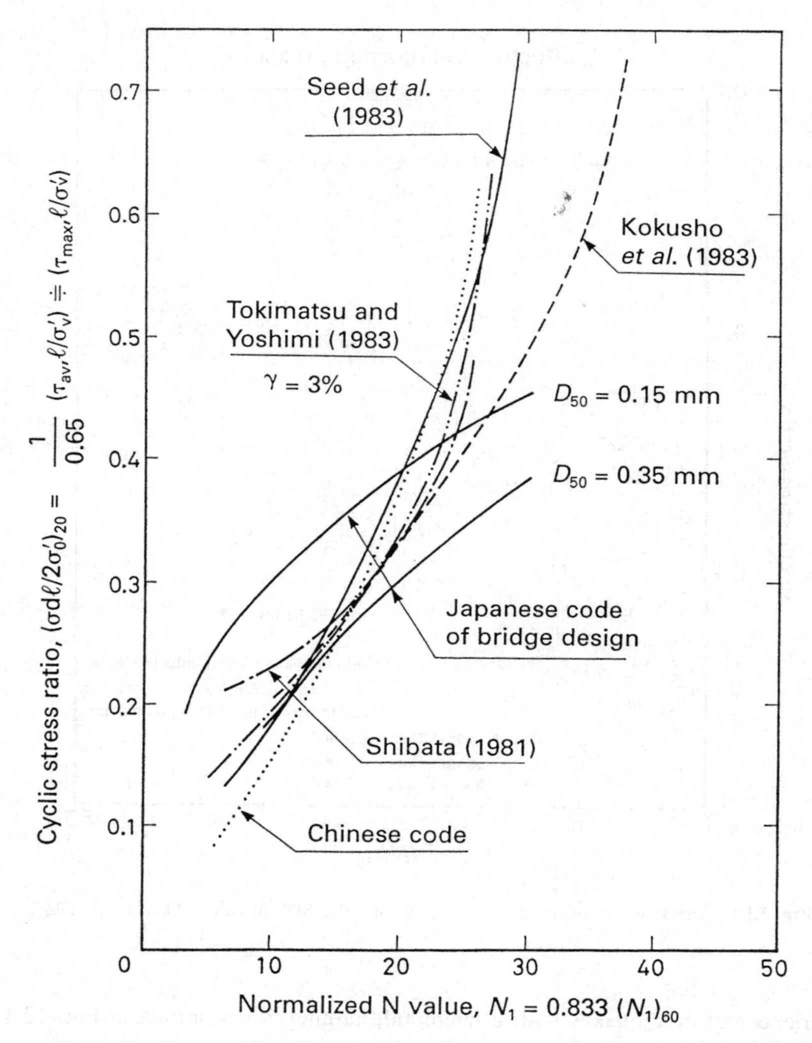

Fig. 12.2 Summary chart for evaluation of the cyclic strength of sands based on the normalized SPT *N* value.

Fig. 12.1. The value of $(N_1)_{60}$ in this figure indicates the corrected blow count which would have been obtained under an effective overburden pressure of 1 kgf/cm^2 or 98 kPa when the driving energy in the drill rods is 60% of the theoretical free-fall energy of the SPT hammer. The SPT practice in the United States is considered to employ this level of energy on the average.

Similar efforts have been made to establish field performance correlations by Shibata (1981) and Tokimatsu and Yoshimi (1983), based on the available body of data obtained mainly at Japanese sites. The results of these studies are shown in Fig. 12.2. According to the recent study by Seed *et al.* (1985), the energy transmission in the Japanese practice of the SPT is considered to be 1.2 times greater on average than that in the US practice. Therefore the relation $N_1 = 0.833\,(N_1)_{60}$ may be used to convert the N value of the SPT between these two countries. With this conversion taken into account, the curve in Fig. 12.1 is rewritten in Fig. 12.2.

On the basis of earthquake experiences in recent years, correlations between the liquefaction resistance of sand deposits and the SPT N value have been developed in China and presented in the form of a code requirement, expressed in terms of a critical penetration resistance N_{cri} as follows,

$$N_{cri} = \overline{N}\,[1 + 0.125\,(Z - 3) - 0.05\,(H - 2)] \qquad (12.2)$$

where Z is the depth to a sand layer under consideration and H is the depth to water table, in meters. The value of $\overline{N}$ indicates the reference N value which is specified in Table 12.1 as a function of the earthquake shaking intensity

The relationship between the liquefaction-inducing acceleration $a_{max,\,\ell}$ and the N value indicated in three steps in Table 12.1 can be expressed as,

$$a_{max,\,\ell} = \frac{5}{6}\,\overline{N}^2 + \frac{70}{6}\,\overline{N}. \qquad (12.3)$$

In eqn (12.3), the $\overline{N}$ value is regarded as a blow count value under a particular overburden pressure corresponding to $Z = 3$ m and $H = 2$ m, that is, $\sigma_v = 57$ kPa and $\sigma_v{}' = 47$ kPa assuming a typical unit weight of soils to be $\gamma_t = 19$ kN m^{-3}. Thus putting these values into eqn (12.1), one obtains

$$\frac{a_{max,\,\ell}}{g} = 0.864\,\frac{\tau_{max,\,\ell}}{\sigma_v{}'}. \qquad (12.4)$$

The reference $\overline{N}$ value at a particular overburden pressure of $\sigma_v{}' = 0.47$ kgf/m^2 can be converted to the N_1 value at $\sigma_v{}' = 1$ kgf/m^2 by using the relation of eqn (12.9),

$$N_1 = \frac{1.7}{\sigma_v{}' + 0.7}\,\overline{N} = 1.45\,\overline{N} \qquad (12.5)$$

Therefore introducing eqns (12.4) and (12.5) into eqn (12.3), one obtains

Table 12.1 Reference blow count value in Chinese code

Chinese intensity	Acceleration (gal)	$\overline{N}$
7	100	6
8	200	10
9	400	16

$$\left(\frac{\sigma_{d\ell}}{2\sigma_0{}'}\right)_{20} = \frac{\tau_{max,\,\ell}}{\sigma_v{}'} = \frac{1}{1000}\,(9.5\,N_1 + 0.466\,N_1{}^2). \tag{12.6}$$

This relation is also shown in Fig. 12.2.

12.1.2 *CPT correlation*

More recently, several attempts have been made to establish field performance correlation with the Dutch cone penetration resistance CPT. Based on compilation of

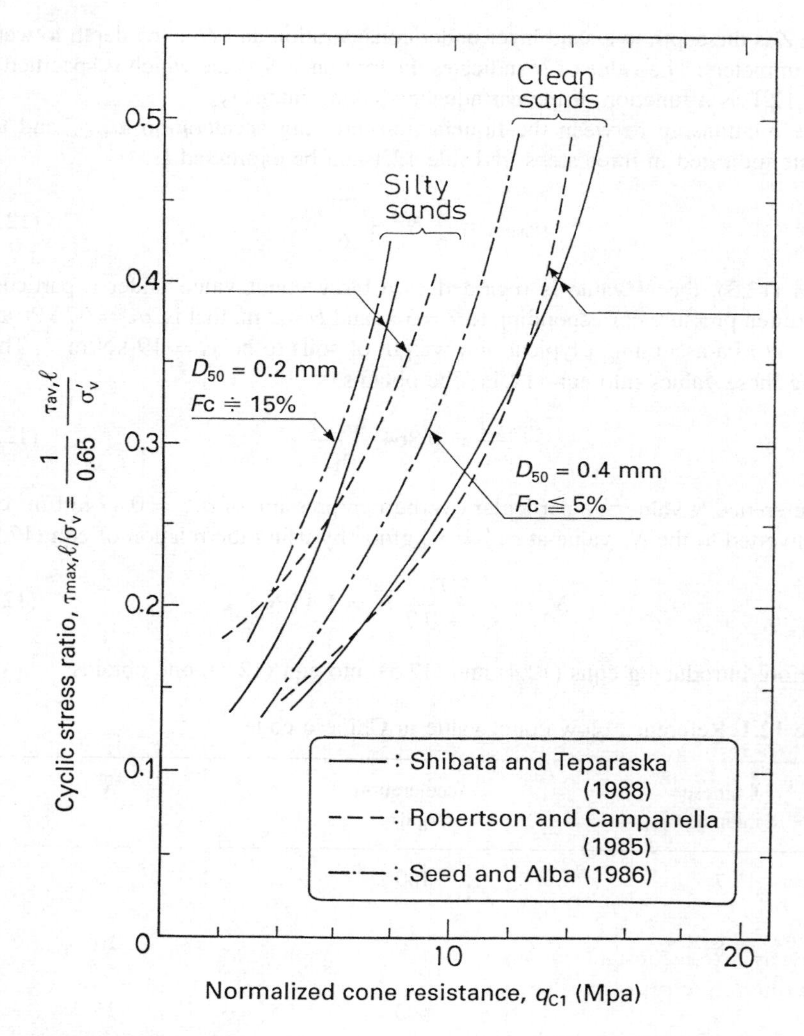

Fig. 12.3 Summary chart for evaluation of the cyclic strength of sands based on the normalized CPT q_c value.

field performance data, Robertson and Campanella (1985) proposed a correlation shown in Fig. 12.3, where the value of q_c is expressed in the abscissa in a form q_{c1} normalized to a confining pressure of 1 kgf/cm^2. Similar correlation was also established by Seed and Alba (1986) as demonstrated in Fig. 12.3. These correlations are however limited in coverage to predominantly sandy soils. Recently, a comprehensive study was made by Shibata and Teparaska (1988) to establish correlations between the cyclic strength and cone resistance including the influence of fines content. The outcome of data arrangements based on field performances is summarized in a form of an empirical correlation which is displayed in Fig. 12.3.

12.2 Correlation based on laboratory tests

12.2.1 *The relation between relative density and penetration resistance*

The correlation between the laboratory-determined cyclic strength and the N value of the SPT was established initially by way of the relatively density D_r. An early attempt in this context was made by Gibbs and Holtz (1957) who performed a series of penetration tests in a pressure chamber in the laboratory. For the results of this tests, Meyerhof (1957) proposed a formula:

$$D_r = 21\sqrt{\frac{N}{\sigma_v{}' + 0.7}} \tag{12.7}$$

where $\sigma_v{}'$ is in terms of kgf/cm^2. If the SPT N value under an effective overburden pressure of $\sigma_v{}' = 1.0$ kgf/cm^2 is denoted by N_1, the relative density under this condition is given by

$$D_r = 16\sqrt{N_1}. \tag{12.8}$$

By equating eqns (12.7) and (12.8), one obtains the following formula which is widely used in Japan to convert a measured N value into an N_1 value.

$$N_1 = C_N \cdot N, \quad C_N = \frac{1.7}{\sigma_v{}' + 0.7}. \tag{12.9}$$

Later, as a result of extensive survey over many sets of existing in situ data on the N value of the SPT, the correlation was expressed in a general form as follows by Skempton (1986):

$$N = (a + b\,\sigma_v{}')\left(\frac{D_r}{100}\right)^2 \tag{12.10}$$

where a and b are constants which depend mainly on the grain size of soils. This can be rewritten in terms of the N_1 value as

$$\frac{N_1}{(D_r/100)^2} = a + b. \tag{12.11}$$

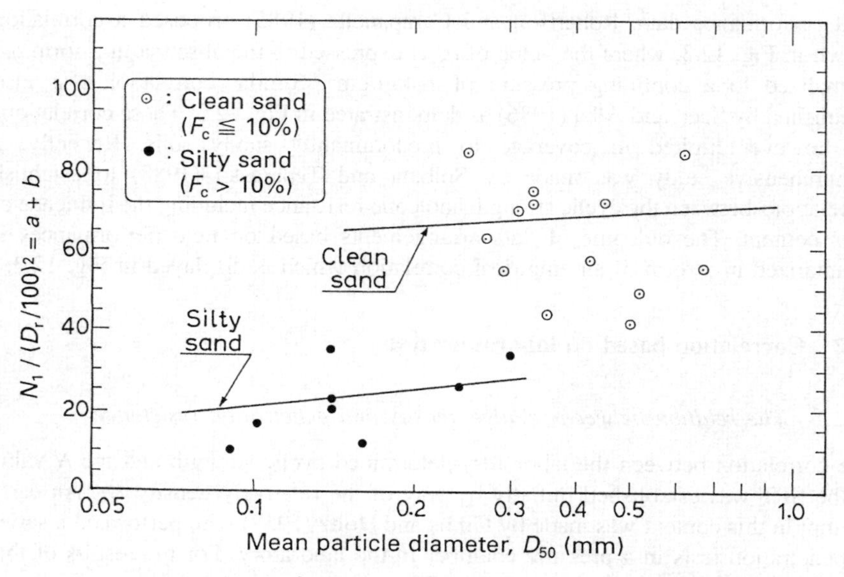

Fig. 12.4 $N_1/(D_r/100)^2 = a + b$ plotted against the mean particle diameter.

The value of $a + b$ obtained from studies by Skempton (1986) are plotted in Fig. 12.4 against the mean diameter D_{50}. Note that the laboratory-obtained data are excluded. In situ data collected recently in Japan are also shown. The silty sands, which contain more than 10% fines, show significantly smaller values of $a + b$ and therefore separate curves are drawn through the two groups of data points. It is to be noted that the formula of eqn (12.7) by Meyerhof (1957) corresponds to the case of $a = 16$ and $b = 23$ which is a gross average encompassing both clean sands and silty sands.

Similar attempts have been made by Jamiolkowski *et al.* (1985) to establish an empirical correlation in the following general form between the relative density and the q_c value in the CPT:

$$D_r = -A + B \log (q_c/\sqrt{\sigma_v{}'}) \tag{12.12}$$

where q_c and $\sigma_v{}'$ are in terms of kgf/cm^2.

While somewhat greater values are suggested by Jamiolkowski *et al.* (1985), Tatsuoka *et al.* (1990) recommended the use of $A = 85$ and $B = 76$ based on their test results using a calibration chamber. If these constants are used, the relation of eqn (12.12) is modified as

$$D_r = -85 + 76 \log q_{c1}$$
$$q_{c1} = C_N q_c, \quad C_N = \frac{1}{\sqrt{\sigma_v{}'}}. \tag{12.13}$$

12.2.2 *Cyclic strength estimated from penetration tests*

A number of cyclic triaxial tests have been performed in recent years on undisturbed samples of sandy soils. In the majority of cases, the samples were recovered from alluvial deposits by means of sophisticated types of tube samplers and the SPTs were carried out at various depths at nearby sampling sites. On the basis of the compilation of a vast amount of such data, an empirical relation was derived, as follows, for typical clean sands:

$$\left(\frac{\sigma_{dl}}{2\sigma_0{}'}\right)_{20} = 0.0042\, D_r. \tag{12.14}$$

Introducing eqn (12.8) into eqn (12.14) a correlation between the cyclic strength and the N_1 value was obtained for clean sands without the effect of fines being considered. Allowing for the effects of particle size, Tatsuoka *et al.* (1980) proposed later a more general correlation as:

$$\begin{aligned} \left(\frac{\sigma_{dl}}{2\sigma_0'}\right)_{20} &= 0.0676\sqrt{N_1} + 0.225\log_{10}\left(\frac{0.35}{D_{50}}\right) \\ &\quad \text{for } 0.04\text{ mm} \le D_{50} \le 0.6\text{ mm}, \\ \left(\frac{\sigma_{dl}}{2\sigma_0'}\right)_{20} &= 0.0676\sqrt{N_1} - 0.05 \\ &\quad \text{for } 0.6\text{ mm} < D_{50} \le 1.5\text{ mm}, \end{aligned} \tag{12.15}$$

where D_{50} indicates the average particle diameter in millimeters. This formula was incorporated into the Japanese Design Code of Bridges. The cyclic strength obtained from the above formula is plotted in Fig. 12.2 for typical particle sizes of 0.15 mm and 0.35 mm. As indicated by eqn (12.14), the relation of eqn (12.15) is based on a linear correlation between the laboratory-determined cyclic strength and the relative density D_r. Therefore, eqn (12.15) is to be regarded as applicable only in the range of relative density less than 70% where the cyclic strength is linearly related with the relative density as indicated in Fig. 10.10. In terms of the SPT blow count, eqn (12.15) should be considered to hold true for N_1 values less than about 20.

A similar correlation was established by Kokusho *et al.* (1983*a,b*) for clean sands as a result of a comprehensive series of laboratory tests. The proposed correlation is also demonstrated in Fig. 12.2.

Looking a the chart in Fig. 12.2, one can note that the two kinds of correlation as epitomized above, one from field performance data and the other from the laboratory tests, may be deemed as representing a relationship of identical nature between the N_1 value and the resistance of sandy soils to liquefaction. Viewed in this manner, it may be mentioned that all the curves in Fig. 12.2 show approximately equal trend in correlating the cyclic strength with the SPT N_1 value. In fact, all the curves yield almost the same correlation in the intermediate range for the N_1 value of between 15 and 20. However for somewhat stiffer soils with larger N_1 values in excess of about 25, the formula by the Japanese Code of Bridge Design gives a lower value of cyclic strength than is actually the case and the use of other curves is recommended. For loose deposits with N_1 values smaller than 15, there is some scatter in the predicted value of cyclic strength.

12.3 The effects of fines on cyclic strength

In most of the correlations epitomized above, the effects of the presence of fines are allowed for in such a way that the penetration resistance becomes smaller with increasing amount of fines if soils were to possess equal cyclic strength. In fact, with equal magnitude of penetration resistance, soils are observed to have increasing cyclic strength with increasing fines content as schematically illustrated in Fig. 12.5. This characteristic tendency is incorporated into the above mentioned correlations in terms of a parameter associated with grading, such as the fines content, F_c, or the average diameter, D_{50}.

Looking at several curves shown in Fig. 12.1, one can recognize that each curve is disposed in such a way that it is shifted approximately in parallel to the left with increasing fines content. Therefore it becomes possible to determine the amount of this shift ΔN_1 as illustrated in Fig. 12.5 as a function of fines content. The value of ΔN_1 is interpreted as a decrease of the N_1 value for silty sands so as to have the same cyclic strength as for clean sand. Thus suppose the cyclic strength for clean sands is given by $f(N_1, 0)$, then the increment ΔN_1 is determined as

$$f(N_1 - \Delta N_1, F_C) = f(N_1, 0). \qquad (12.16)$$

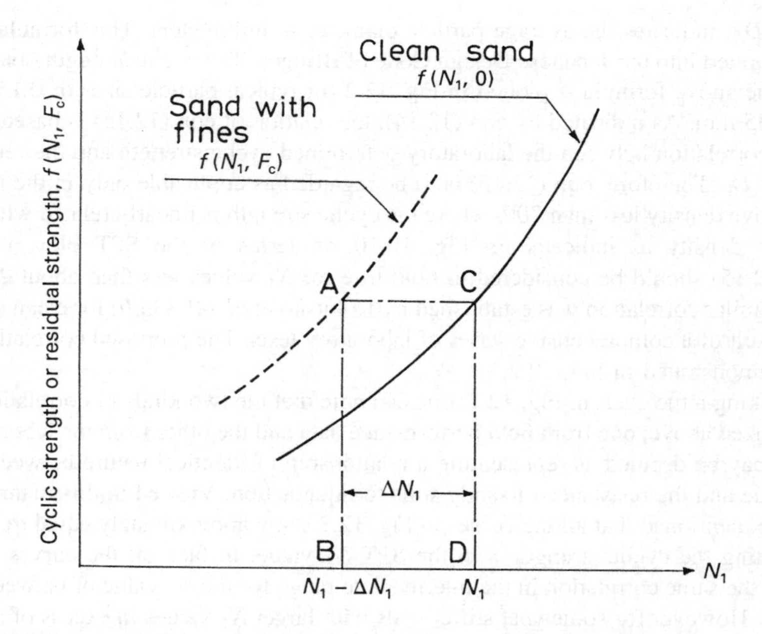

Fig. 12.5 Definition of an increment of N_1 value, allowing for the effects of fines.

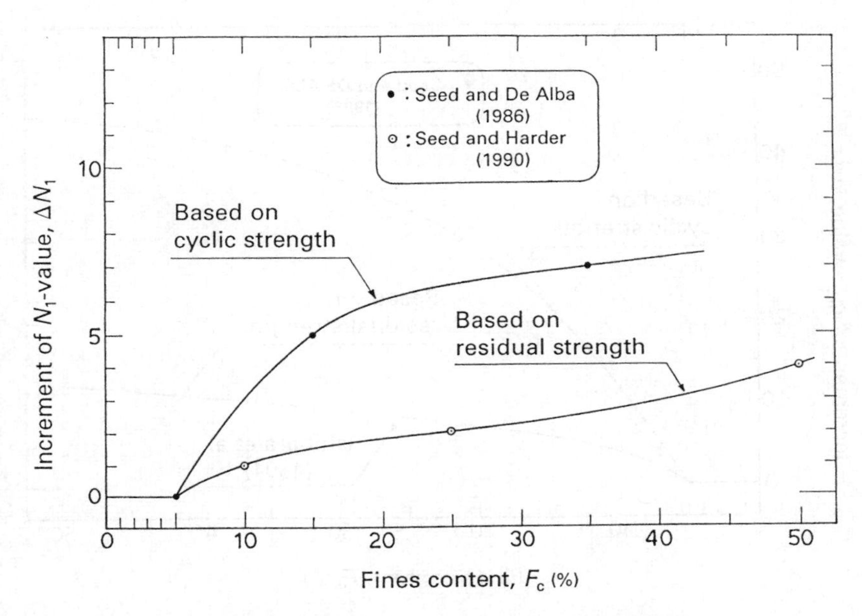

Fig. 12.6 Increment ΔN_1 value as a function of fines content.

Introducing this requirement into the empirical correlation such as eqn (12.14), one can obtain an explicit expression for ΔN_1 but it is also possible to read off ΔN_1 directly from a family of curves such as that shown in Fig. 12.1 compiled for different values of fines content. The increment ΔN_1 thus obtained from the compiled data by Seed and De Alba (1986) is plotted in Fig. 12.6 by solid circles. By connecting these points, a curve is drawn in Fig. 12.6 which can be used for practical purposes. The same argument can also be developed to obtain the increment ΔN_1 associated with the residual strength. In this case, the value of ΔN_1 means the increment of the N_1 value for fines-containing sand required to have the same residual strength as for clean sand. Needless to say the increment related to the cyclic strength is different from the ΔN_1 associated with the residual strength. In passing, a curve of ΔN_1 for the residual strength was obtained from the values suggested by Seed and Harder (1990) and presented in Fig. 12.6. The use of the latter curve will be discussed later. The curves for the increment of Δq_{c1} similarly obtained are also shown in Fig. 12.7. These curves can be used to estimate the cyclic strength or residual strength based on in situ data from the CPT.

The method of correction as described above is based on the assumption that the effects of fines can be taken into account in terms of the grading parameters such as fines content and mean diameter. However, as pointed out above, the grading of soils is not necessarily an essential factor influencing cyclic strength, and the nature of the fines themselves as represented by the plasticity index is physically a more meaningful parameter governing the strength mobilized in cyclic loading. If this effect is to be incorporated into the cyclic strength versus penetration resistance

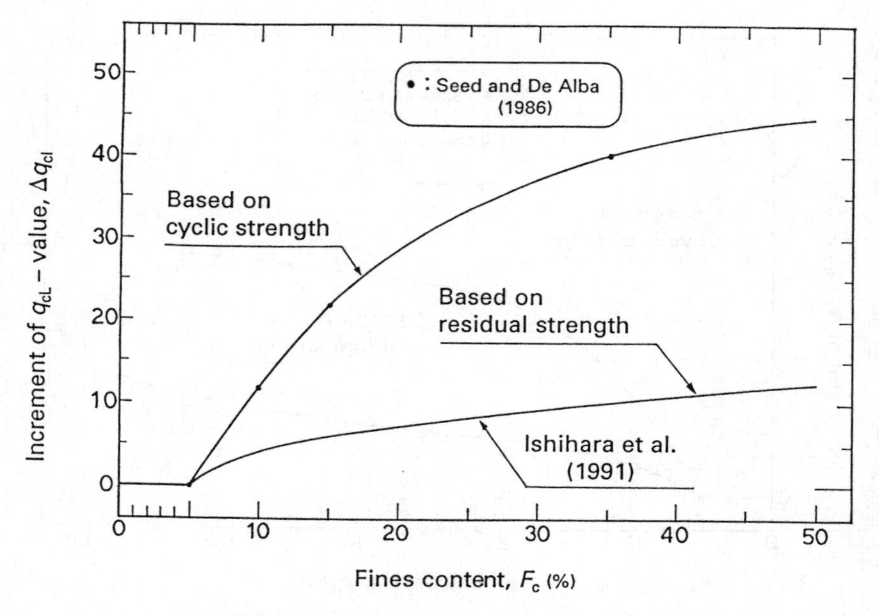

Fig. 12.7 Increment Δq_{c1} as a function of fines content.

relationship, it becomes necessary to know how the penetration resistance is influenced by the plasticity of fines. However, there is no shred of test data in this respect.

Under the circumstances as above, the only conceivable way at present would be to evaluate first the cyclic strength of in situ soil deposits through the procedures described above, where the effects of fines are allowed for in terms of the grading indices, and then to modify it in accordance with the relationship such as that shown in Fig. 10.15. In utilizing this relationship, it would be more expedient to normalize the cyclic strength at any plasticity to the cyclic strength for a low plasticity index below 10. The curve modified in this way is shown in Fig. 12.8. With the background information as above, the procedures to determine the cyclic strength of the soil in a given deposit are summarized as follows:

1. By means of the SPT or the CPT, the penetration resistance, N_1 value or q_{c1} value, is obtained, together with the fines content F_c or mean diameter D_{50} for the soils in question, for various depths at a given site. If necessary, the plasticity index of the fines fraction needs to be determined.

2. If the material is identified as clean sands with fines content less than 5%, the cyclic strength is determined from the charts shown in Fig. 12.2 or Fig. 12.3. If more than 5% fines is shown to exist in the soil under consideration, the measured N_1 value or q_{c1} value should be increased based on the chart in Figs 12.6 or 12.7. Then,

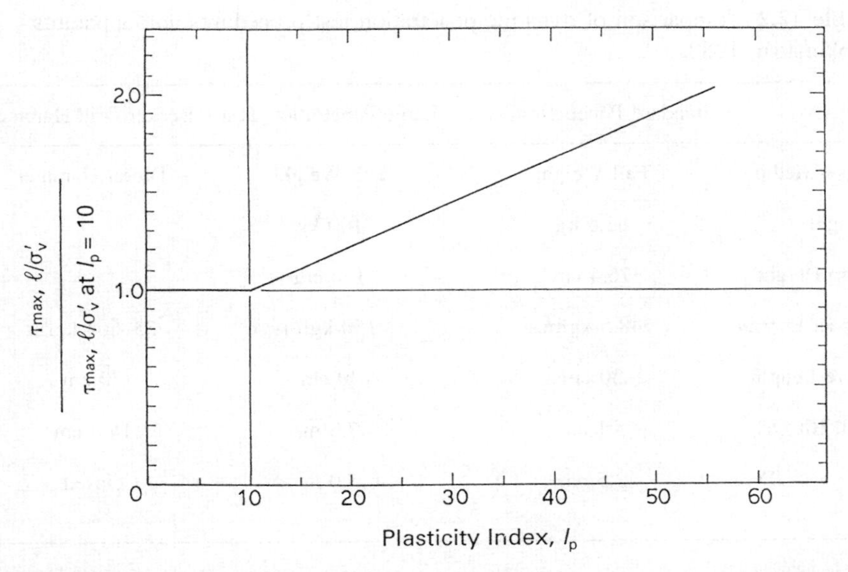

Fig. 12.8 Chart for modification of cyclic strength allowing for the effects of plasticity index.

using the chart in Fig. 12.2 or in Fig. 12.3, the cyclic strength is determined. If the plasticity index of the fines is found to be greater than $I_p = 10$, further correction needs to be made for the cyclic strength by using the chart in Fig. 12.8. Experience has shown that most of the sandy soils existing in alluvial deposits or man-made fills possess a plasticity index less than 15. Therefore the correction in this regard may not appreciably change the value of the cyclic strength.

12.4 Correlation for gravelly soils

One of the problems associated with penetration tests in gravelly soils is that penetration resistance is quite sensitive to particle grain size. Another problem is that the size of the common test probes and the standard driving energy is not sufficient to match the strong penetration resistance of gravelly soils. This means that the standard test method may not produce reliable results unaffected by the local presence of large particles. To overcome these difficulties, different types of specific dynamic penetration tests are now in use in the world. Some of these tests include the large diameter penetration test, and the Becker drill test. The specification of these new penetration tests together with that of the SPT are summarized in Table 12.2.

Among these promising test methods is the Becker drill test (BDT) using the Becker Hammer which was originally developed in Canada. This test method has been extensively studied and standardized by Harder and Seed (1986) who established a correlation with SPT blow counts. It has also been applied to a case

Table 12.2 Comparison of dynamic penetration test procedures and apparatus (Tokimatsu, 1988)

	Standard Penetration Test	Large Penetration Test	Becker Drill Hammer
Drive Method	Fall Weight	Fall Weight	Diesel Hammer
Weight	63.5 kg	100 kg	
Drop Height	76.4 cm	150 cm	
Impact Energy	48.5 kgfm	150 kgfm	35–664 kgfm
Drive Length	30 cm	30 cm	30 cm
Drill Bit OD	5.1 cm	7.3 m	16.8 cm
ID	3.5 cm	5.0 m	closed

study of the liquefaction of gravelly soils in the Borah Peak earthquake as described by Andrus and Youd (1987).

Another test method, the Large Diameter Penetration Test (LPT), was initially proposed by Kaito *et al.* (1971) and has been extensively employed recently in Japan

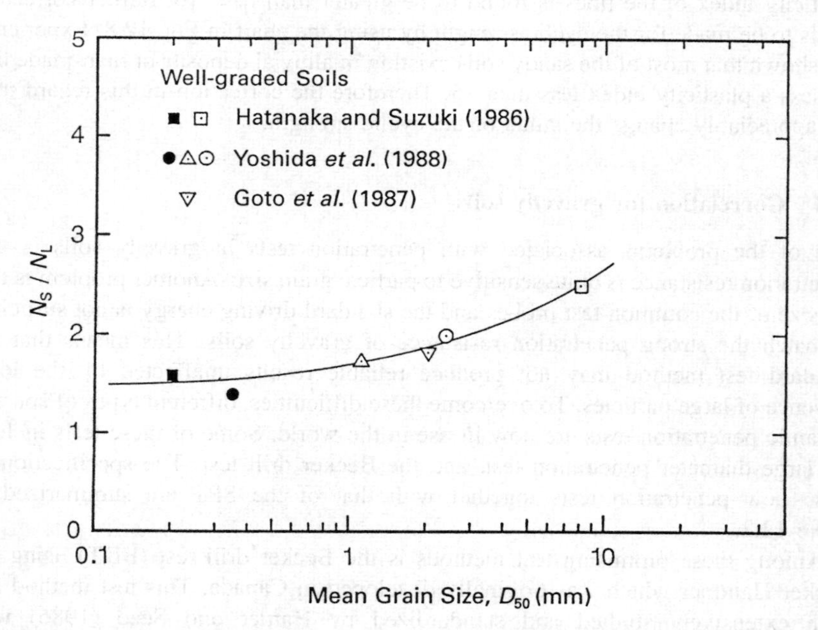

Fig. 12.9 Variation of N_s/N_L ratio with mean grain size for well-graded soils (Tokimatsu, 1988).

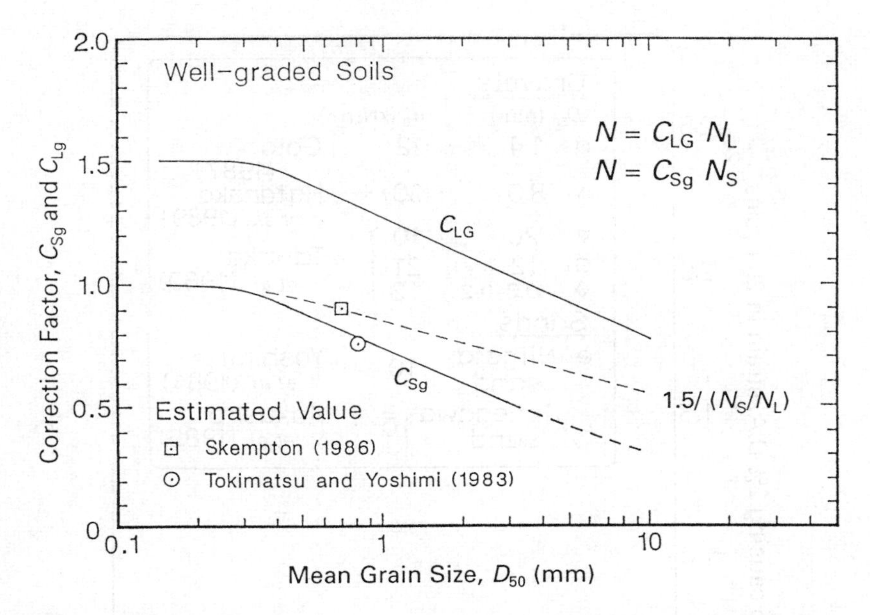

Fig. 12.10 Correlation factor for penetration resistance of gravelly soils (Tokimatsu, 1988).

in conjunction with the evaluation of liquefaction potential of gravelly soils.

Yoshida *et al.* (1988) have recently made systematic laboratory and field tests using SPT and LPT tests on gravelly soils with different grain size distributions. Their goal was to establish correlations between SPT blow counts for gravelly soils, the N_s value, and LPT blow counts, the N_L value, with soil parameters such as relative density and shear wave velocity. By summarizing these data and others, Tokimatsu (1988) has interpreted that the ratio N_s/N_L varies appreciably with the mean grain size of gravelly soil (Fig. 12.9). This is probably due to the increased penetration resistance in the SPT over that in the LPT caused by the presence of particles of large size. Based on this interpretation, correlation factors C_{Sg} and C_{Lg} for the blow count values of N_s and N_L were suggested by Tokimatsu (1988) as shown in Fig. 12.10. These values may be used to convert measured blow counts, N_s or N_L to the sand-related SPT blow count N value to give liquefaction resistance free from unfavorable grain size effects.

Based on undrained cyclic triaxial test data for specimens recovered by the in situ freezing and coring method, combined with the N_s value or N_L value determination at the same site, efforts are currently being made to develop the correlation between the penetration resistance and the in situ undrained cyclic strength for gravelly soils in a manner similar to the existing liquefaction potential evaluation charts established for sandy soils. Figure 12.11 gives a typical correlation of this type between the undrained cyclic strength and the N_1 value obtained for different gravelly soils from natural deposits. Also shown is the data for clean sands by

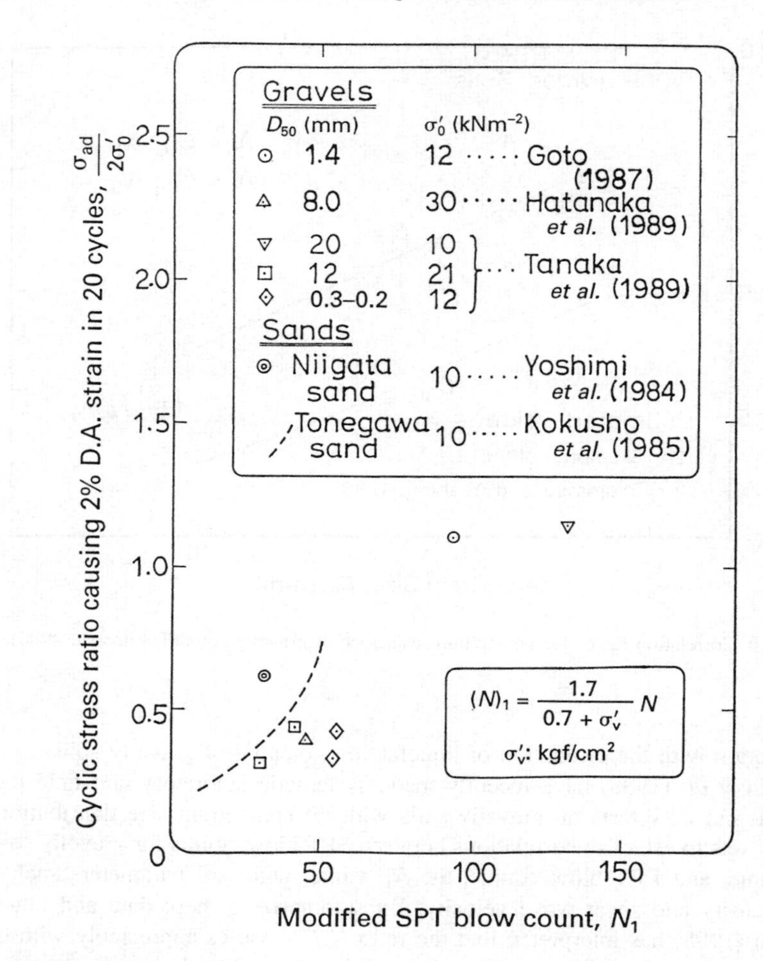

Fig. 12.11 Cyclic strength versus SPT N_1 value for gravelly soils based on undisturbed samples.

Yoshimi *et al.* (1984) based on in situ freezing and sampling and by Kokusho *et al.* (1985) based on tests on reconstituted samples. Despite the limited number of data and their scatter, it may be seen that the strength of gravelly soil is apparently smaller than that for sandy soil if one compares the two kinds of materials with identical N_1 values.

Figure 12.12 shows similar correlations between the undrained cyclic strength and the normalized blow counts $(N_L)_1$ for the large penetration test (LPT) under the effective overburden pressure of 1 kgf/cm^2. For the case where LPT data was not available, conversion of the normal SPT N values was carried out based on empirical relations. Unlike the SPT based correlation, the LPT based correlation appears to be

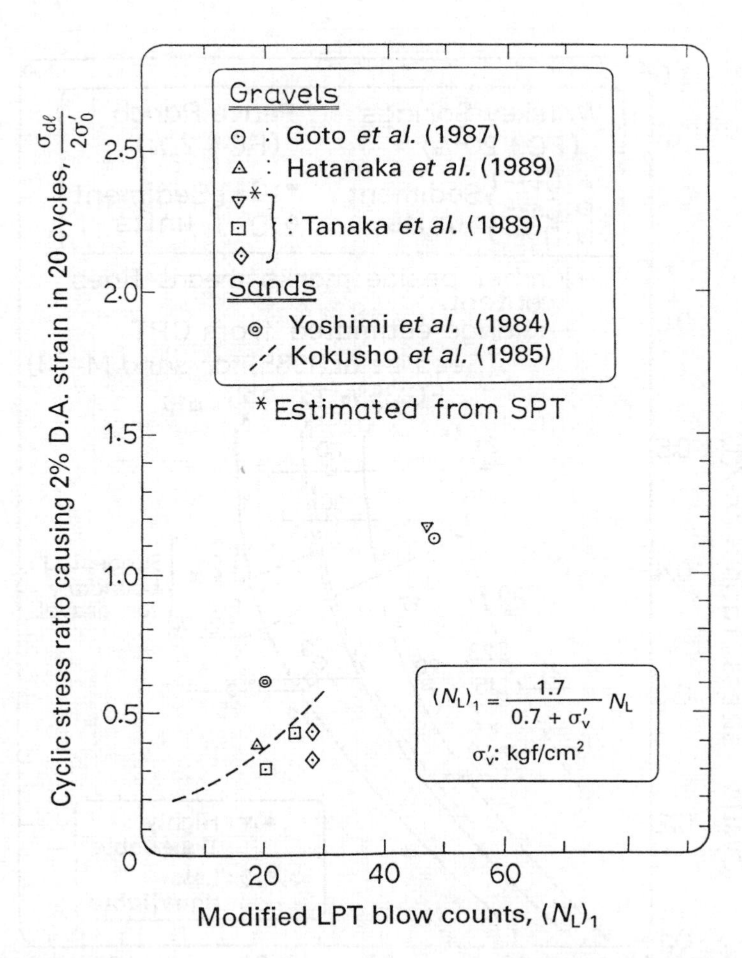

Fig. 12.12 Cyclic strength versus LPT $(N_L)_1$ value for gravelly soils based on undisturbed samples.

more consistent between gravels and sands and also amongst different gravels. This may indicate that the LPT blow count, the N_L value, is less sensitive to increasing particle grain size than the SPT N value and thus better suited as a measure for the penetration resistance of gravelly soils.

Further, Tokimatsu (1988) showed how it is possible to convert the N value for gravelly soils to an N value free from particle grain size effects using the relationship shown in Fig. 12.10. If a similar procedure is followed here for the gravel data in Fig. 12.11, the difference between the results for sands and gravels would appreciably decrease, indicating that essentially the same correlation may be applicable both for sands and gravels.

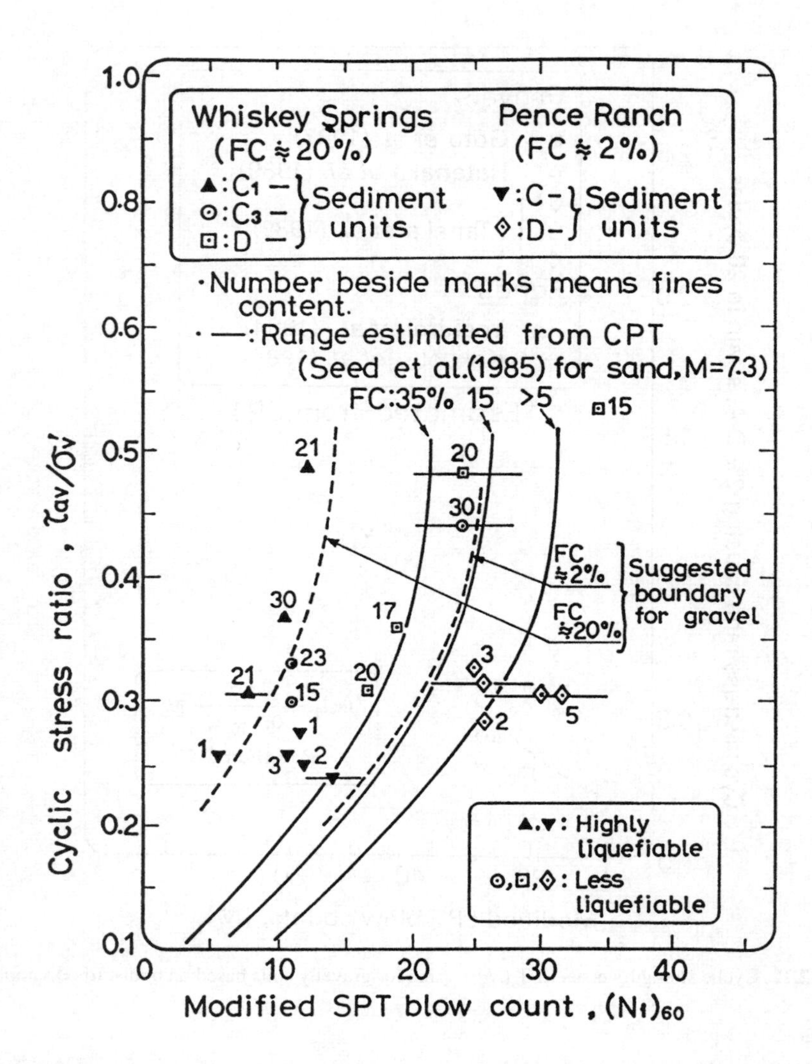

Fig. 12.13 Cyclic stress ratio versus SPT $(N_1)_{60}$ value for gravelly soils based on liquefaction case study (Andrus and Youd, 1989).

In the United States, a similar correlations has been derived recently for the data from gravel soil liquefaction reported for the 1984 Borah Peak earthquake (Andrus and Youd, 1989). This correlation is shown in Fig. 12.13 which plots the cyclic stress ratio τ_{av}/σ_v', defined in the same manner as before, plotted against the modified penetration resistance $(N_1)_{60}$. In this example, detailed field studies at the Whiskey Springs Site and at the Pence Ranch Site attributed high liquefaction susceptibility to what was designated as Sediment C and Sediment C_1 at the two sites, respectively. However, low liquefaction potential was attributed to Sediment D at both sites. In accord with these differentiations, one may tentatively assume a boundary curve between liquefiable and nonliquefiable soil as being the dotted curve drawn in Fig. 12.13 considering the effect of fines content which was about 20% at Whiskey Springs and about 2% at Pence Ranch. If this curve, which corresponds to an earthquake of $M = 7.3$, is compared with the curves proposed by Seed *et al.* (1985) for silty sands for $M = 7.5$, one may assume a higher liquefaction strength for gravel than for sand.

Judging from these results from undisturbed frozen sampling and testing and liquefaction case studies, it may be mentioned that gravelly soils exhibit almost equal or larger undrained cyclic strength than sands.

References

Andrus, R.D. and Youd, T.L. (1987). *Subsurface investigation of a liquefaction induced lateral spread at Thousand Springs Valley, Idaho*. Miscellaneous Paper GL-87-8, U.S. Army Engineer Waterways Experiments Station.

Andrus, R.D. and Youd, T.L. (1989). *Penetration tests on liquefiable gravels*. Proceedings of the 12th International Conference on Soil Mechanics and Foundation Engineering, Vol. 1, pp. 679–82.

Gibbs, H.J. and Holtz, W.G. (1957). Research on determining the density of sand by spoon penetration test. *Proc. 4th International Conference on Soil Mechanics and Foundation Engineering*, Vol. 1, pp. 35–9.

Goto, G., Shamoto, Y., and Tamaoki, K. (1987). *Dynamic properties of undisturbed gravel samples obtained by the in situ freezing method*. Proceedings of the 8th Asian Regional Conference for ISSMFE, Kyoto, Vol. 1, pp. 233–6.

Harder, Jr., L.F. and Seed, H.B. (1986). *Determination of penetration resistance for coarse-grained soils using Becker Hammer Drill*. Report, Earthquake Engineering Research Center, Report no. EERC-86/06.

Hatanaka, M. and Suzuki, Y. (1986). Dynamic properties of undisturbed Tokyo gravel obtained by freezing. *Proc. 7th Japan Earthquake Engineering Symposium*, Tokyo, pp. 649–54 (in Japanese).

Hatanaka, M., Suzuki, Y., Kawasaki, T. and Endo, M. (1988). Cyclic undrained shear properties of high quality undisturbed Tokyo gravel. *Soils and Foundations*, **28**, No. 4, 57–68.

Ishihara, K., Vendugo, R., and Acacio, A.A. (1991). Characterization of cyclic behaviour of sand and post-seismic stability analyses. *Proc. 9th Asian Regional Conference on Soil Mechanics and Foundation Engineering*, Vol. 2, pp. 45–67.

Jamiolkowski, M., Ladd, C.C., Germaine, J.T. and Lancellotta, R. (1985). New developments in field and laboratory testing of soils. *Proc. 9th International Conference on Soil Mechanics and Foundation Engineering*, San Francisco, Vol. 1, pp. 57–153.

Kaito, T., Sakaguchi, S., Nishigaki, Y., Miki, K., and Yukami, H. (1971). Large penetration test. *Tsuchi-to-Kiso*, 629, 15–21 (in Japanese).

Kokusho, T., Yoshida, Y., Nishi, K., and Esashi, Y. (1983*a*). *Evaluation of seismic stability of dense sand layer (part 1) – dynamic strength characteristics of dense sand*. Report 383025. Electric Power Central Research Institute, Japan (in Japanese).

Kokusho, T., Yoshida, Y., and Esashi, Y. (1983*b*). *Evaluation of seismic stability of dense sand layer (part 2) – evaluation method by standard penetration test*. Report 383026. Electric Power Central Research Institute, Japan (in Japanese).

Kokusho, T., Yoshida, Y., and Nagasaki, K. (1985). *Liquefaction strength evaluation of defense sand layer*. Proceedings of the 11th International Conference of Soil Mechanics and Foundation Engineering, Stockholm, Vol. 4, pp. 1897–900.

Meyerhof, G.G. (1957). *Discussion*, Proceedings of the 4th International Conference on Soil Mechanics and Foundation Engineering, Vol. 3, p. 110.

Robertson, P.K. and Campanella, R.G. (1985). Liquefaction potential of sands using the CPT. ASCE, Vol. III, GT3, pp. 384–403.

Seed, H.B. and De Alba, P. (1986). *Use of SPT and CPT tests for evaluating the liquefaction resistance of sands*. Proc. In-Situ Test, ASCE, pp. 281–302.

Seed, R.B., and Harder, L.F., (1990). SPT-based analysis of cyclic pore pressure generation and undrained residual strength. *Proc. of Memorial Symposium of H. B. Seed*, Vol. 2, pp. 351–76.

Seed, H.B., Idriss, I.M., and Avango, I. (1983). Evaluation of liquefaction potential using field performance data. *Journal of Geotechnical Engineering*, ASCE, **109**, GT3 458–82.

Seed, H.B., Tokimatsu, K., Harder, L.F., and Chung, R.M. (1985). Influence of SPT procedures in soil liquefaction evaluations. *Journal of Geotechnical Engineering*, ASCE, Vol. III, No. 12, 1425–45.

Shibata, T. (1981). Relations between *N*-value and liquefaction potential of sand deposits, *Proc. 16th Annual Convention of Japanese Society of Soil Mechanics and Foundation Engineering*, pp. 621–4 (in Japanese).

Shibata, T. and Teparaska, W. (1988). Evaluation of liquefaction potential of soils using cone penetration tests. *Soils and Foundations*, **28**, 49–60.

Skempton, A.W (1986). Standard penetration test procedures and the effects in sands of overburden pressure, relative density, particle size, aging and over consolidation. *Geotechnique*, **36**, (3), 425–47.

Tanaka, Y., Kokusho, T., Yoshida, Y., and Kudo, K. (1989). *Dynamic strength evaluation of gravelly soils*. Special Volume on Influence of Local Soils on Seismic Response, 12 ICSMFE, Rio de Janeiro.

Tatsuoka, F., Iwasaki, T., Tokida, K. Yasuda, S., Hirose, M., Imai, T., and Kon-no, M. (1980). Standard penetration tests and soil liquefaction potential evaluation. *Soils and Foundations*, **20**, (4), 95–111.

Tatsuoka, F., Zhou, S., Sato, T, and Shibuya, S. (1990). Evaluation method of liquefaction potential and its application. Report on Seismic Hazards on the Ground in Urban Areas. Ministry of Education of Japan (in Japanese).

Tokimatsu, K. (1988). *Penetration tests for dynamic problems*. Proceedings of the First International Symposium on Penetration Testing ISOPT, pp. 117–36.

Tokimatsu, K. and Yoshimi, Y. (1983). Empirical correlation of soil liquefaction based on SPT *N*-value and finer content. *Soils and Foundations*, **23** (4) 56–74.

Yoshida, Y., Kokusho, T., and Ikemi, M. (1988). *Empirical formula of SPT blow counts for gravelly soils*. Proceedings of the First International Symposium on Penetration Testing, ISOPT-1, pp. 381–7.

Yoshimi, Y., Tokimatsu, K., and Kaneko, O. (1984). Undrained cyclic shear strength of a dense Niigata sand. *Soils and Foundations*, **24**, 131–45

13

ANALYSIS OF LIQUEFACTION

The cyclic shear stress induced at any point within a level ground during an earthquake due to upward propagation of shear waves can be assessed by means of a simple procedure proposed by Seed and Idriss (1971). If a soil column to a depth z is assumed to move horizontally as shown in Fig. 2.4 and if the peak horizontal acceleration on the ground surface is $a_{\max}$, the maximum shear stress $\tau_{\max}$ acting at the bottom of the soil column is given by,

$$\tau_{\max} = \frac{a_{\max}}{g} \cdot r_{\mathrm{d}} \cdot \gamma_{\mathrm{t}} \cdot z \tag{13.1}$$

$$r_{\mathrm{d}} = 1 - 0.015\, z$$

where γ_{t} is the unit weight of the soil and g is the acceleration due to gravity. The coefficient r_{d} is a stress reduction coefficient, allowing for the deformability of the soil column, which takes a value less than unity. Seed and Idriss (1971) expressed the value of r_{d} in a graphical form, but Iwasaki *et al.* (1978) subsequently recommended the use of an empirical formula as indicated in eqn (13.1) where z is in meters. By dividing both sides of eqn (13.1) by the effective vertical stress σ'_{v} eqn (13.1) is modified to read,

$$\frac{\tau_{\max}}{\sigma'_{\mathrm{v}}} = \frac{a_{\max}}{g} \cdot r_{\mathrm{d}} \cdot \frac{\sigma_{\mathrm{v}}}{\sigma'_{\mathrm{v}}} \tag{13.2}$$

where $\sigma_{\mathrm{v}} = \gamma_{\mathrm{t}}\, z$ denotes the total vertical stress. The above equation has been used widely to assess the magnitude of shear stress induced in a soil element during an earthquake. One of the advantages of using eqn (13.1) is that the vast amount of information on horizontal acceleration recorded on the ground surface can be used directly to assess the shear stress induced by seismic shaking in the horizontal plane within the ground.

The analysis of liquefaction can be made by simply comparing the seismically induced shear stress against the similarly expressed shear stress required to cause initial liquefaction or whatever level of shear strain amplitude deemed intolerable in design consideration. Usually, the occurrence of 5% D.A. axial strain is adopted to define the cyclic strength in consonance with 100% pore water pressure build-up as mentioned above. The externally applied cyclic stress ratio can be evaluated by eqn (13.2) and the corresponding strength may be obtained from any of the procedures mentioned in the foregoing section. If the applied stress ratio is expressed in terms of the equivalent stress ratio $\tau_{\mathrm{av}}/\sigma'_{\mathrm{v}}$, then the corresponding

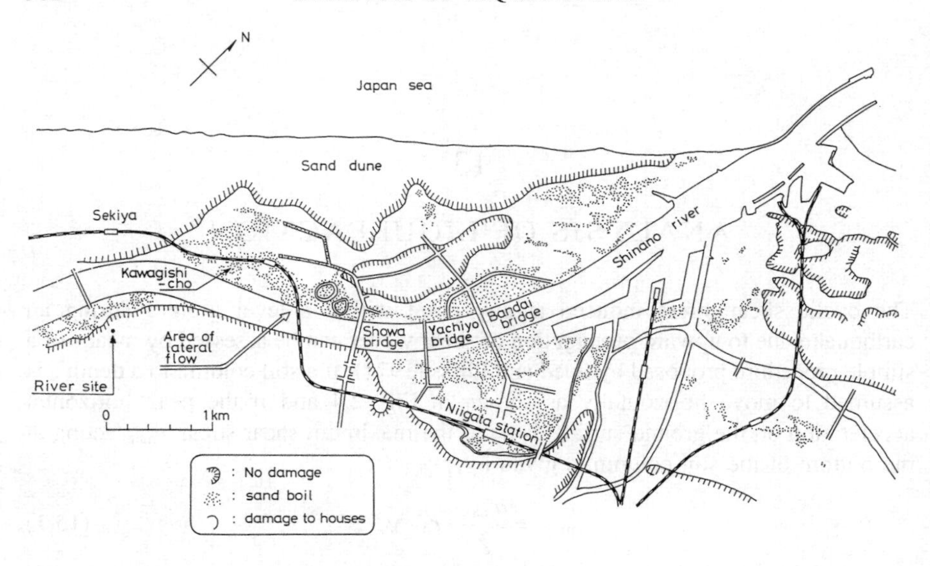

Fig. 13.1 Location of bridges in the city area of Niigata.

stress ratio for strength expression needs to be in a form of equivalent stress ratio $\tau_{\mathrm{av},\ell}/\sigma'_{\mathrm{v}}$ as given by eqn (12.1). Thus, the liquefaction potential of a sand deposit is evaluated in terms of factor of safety F_ℓ which is defined as

$$F_\ell = \frac{\tau_{\mathrm{max},\ell}/\sigma'_{\mathrm{v}}}{\tau_{\mathrm{max}}/\sigma'_{\mathrm{v}}} = \frac{\tau_{\mathrm{av},\ell}/\sigma'_{\mathrm{v}}}{\tau_{\mathrm{av}}/\sigma'_{\mathrm{v}}}. \tag{13.3}$$

Note that whichever expression is used, either the maximum or equivalent stress ratio, the factor of safety is identical. If the factor of safety is equal or less than unity, liquefaction is said to take place. Otherwise, liquefaction does not occur.

The applicability of the above method of liquefaction evaluation is explained below by referring to some case studies. On June 16, 1964, a violent earthquake rocked the Niigata prefecture in Japan, inflicting considerable damage in the city area of Niigata. In the area along the Shinano and Agano rivers where sand deposits were widespread, the damage was primarily associated with the liquefaction of loose sand deposits. Buildings not embedded deeply on firm strata sank or tilted towards the centre of gravity. Underground installations such as septic tanks, sewage conduits, and manholes floated up a meter or two above the ground surface. The area affected by the hazardous liquefaction was mainly located along the Shinano river as indicated in Fig. 13.1. A series of detailed investigations was undertaken at several sites to clarify the cause of the liquefaction. They included the standard penetration test, Dutch cone test, sampling of undisturbed specimens from sand deposits below the ground water table, and testing undisturbed samples in laboratory cyclic triaxial tests to determine the cyclic strength of the intact samples. At the Kawagishicho site

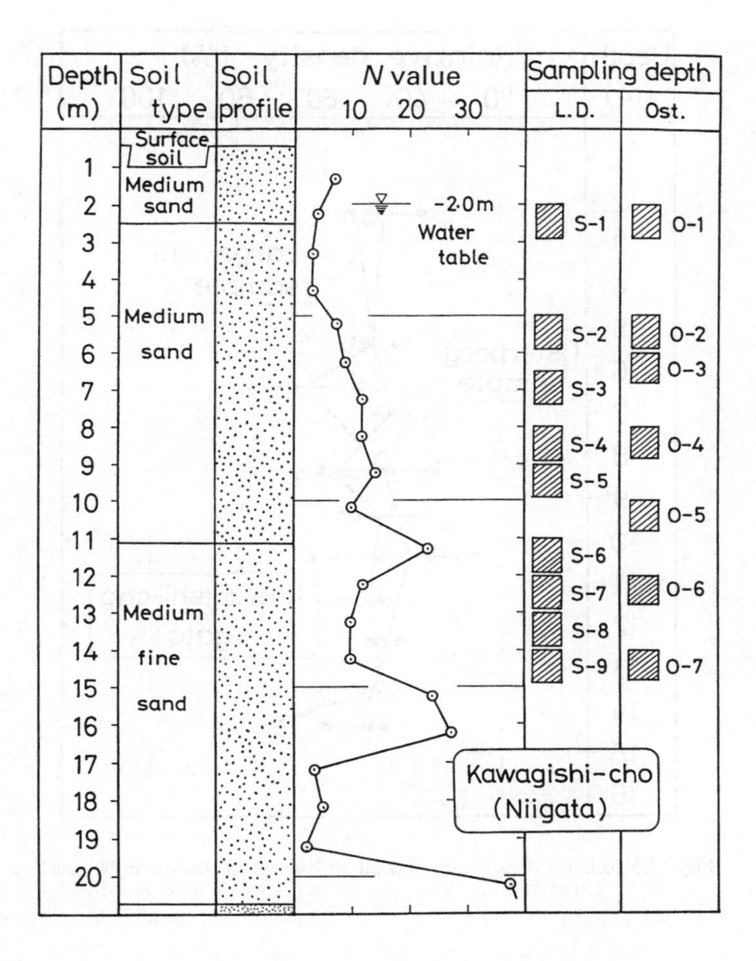

Fig. 13.2 Soil profile and depth of sampling at Kawagishi-cho, Niigata.

where the buildings sank and overturned, in situ exploration disclosed the soil profile as shown in Fig. 13.2 where it is seen that loose sand deposits with an N value less than 15 exist down to a depth of about 10 m. The cyclic triaxial tests on samples recovered by the large diameter sampler and Osterburg samples showed the distribution of relative density and cyclic strength versus depth as shown in Figs 13.3 and 13.4, respectively. It may be seen that the relative density is in the range of 40 to 60% with the cyclic strength taking values in the order of 0.15 to 0.20. On the basis of the cyclic strength data, analysis of liquefaction was performed using the simple procedure described above. During the 1964 earthquake, acceleration records were obtained at the basement of a 4 storey reinforced concrete apartment

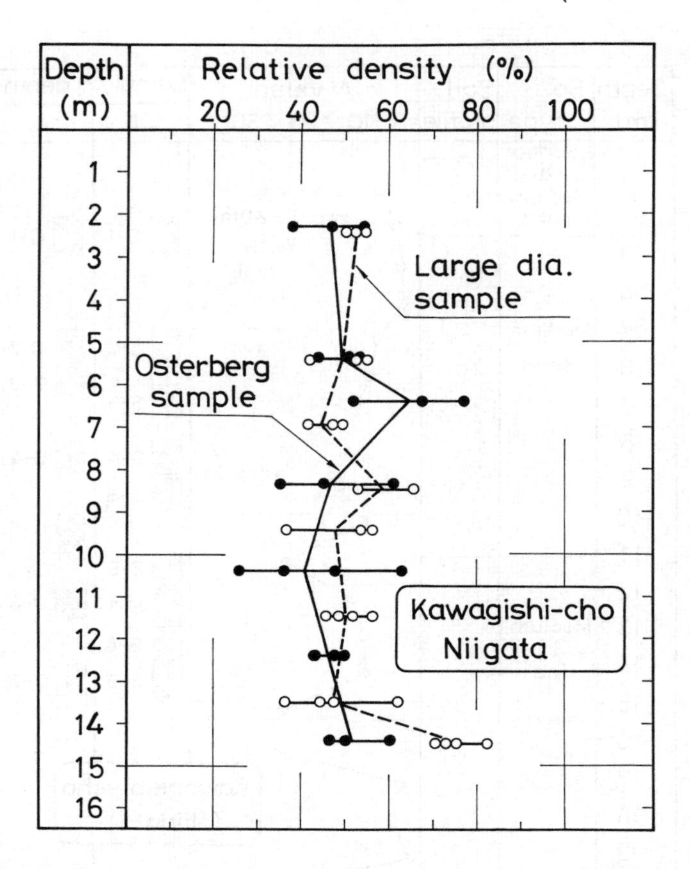

Fig. 13.3 Relative density versus depth at Kawagishi-cho site in Niigata.

building. This building, no. 2 unit, was located about 50 metres away from the site where the soil investigation were performed. The time histories of the recovered accelerations are shown in Fig. 13.5 where it is noted that the peak acceleration was 159 gal and 155 gal in EW and NS directions, respectively.

The results of liquefaction analysis based on these accelerations are presented in Fig. 13.6 in terms of the factor of safety plotted versus the depth (Ishihara and Koga, 1981). The analysis, shown in Fig. 13.6(a), based on the cyclic strength of undisturbed specimens secured by the large diameter sampler indicates that liquefaction had developed in the sand sediment down to a depth of 13 m. Liquefaction of similar extent through the depth is also noted in Fig. 13.6(b) as a result of analysis based on the cyclic strength of the specimens recovered by the Osterberg sampler. These results are consistent with what was observed on the ground surface at the time of the 1964 earthquake. The factor of safety obtained in

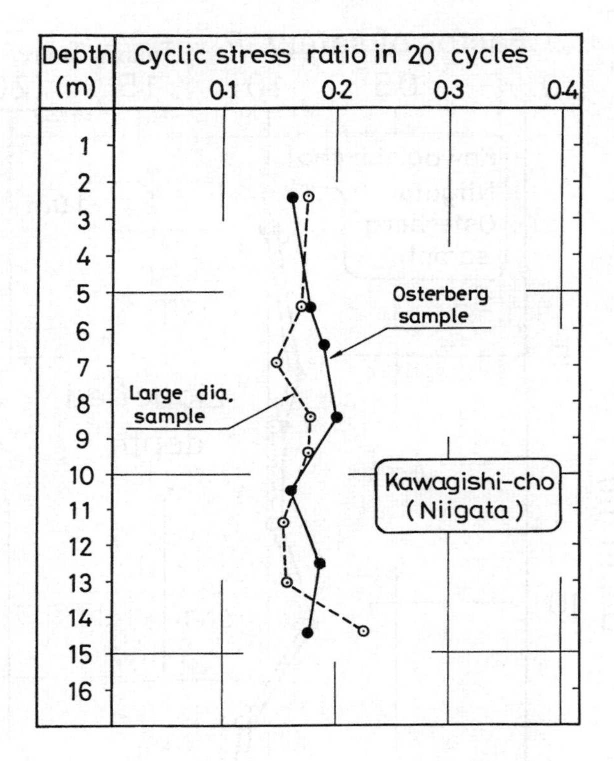

Fig. 13.4 Cyclic strength of sand samples from Niigata obtained by large diameter samples and Osterberg samples.

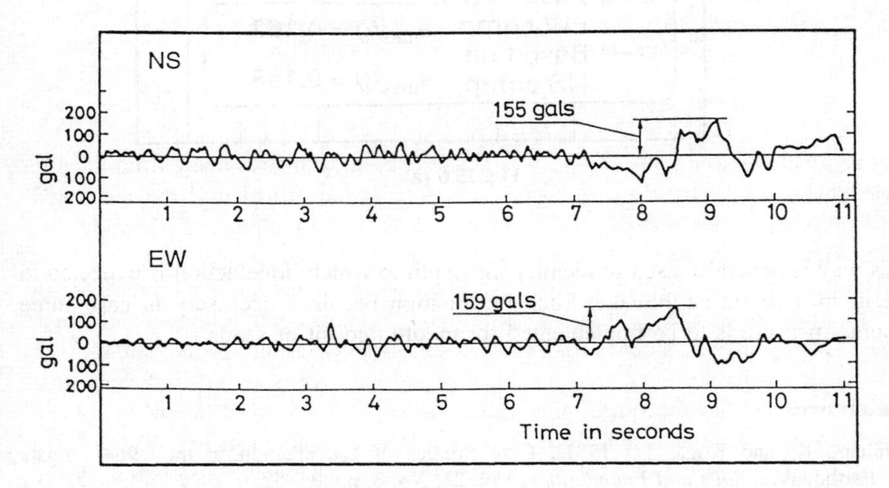

Fig. 13.5 Time histories of horizontal accelerations obtained at Kawagishi-cho at the time of the 1964 earthquake.

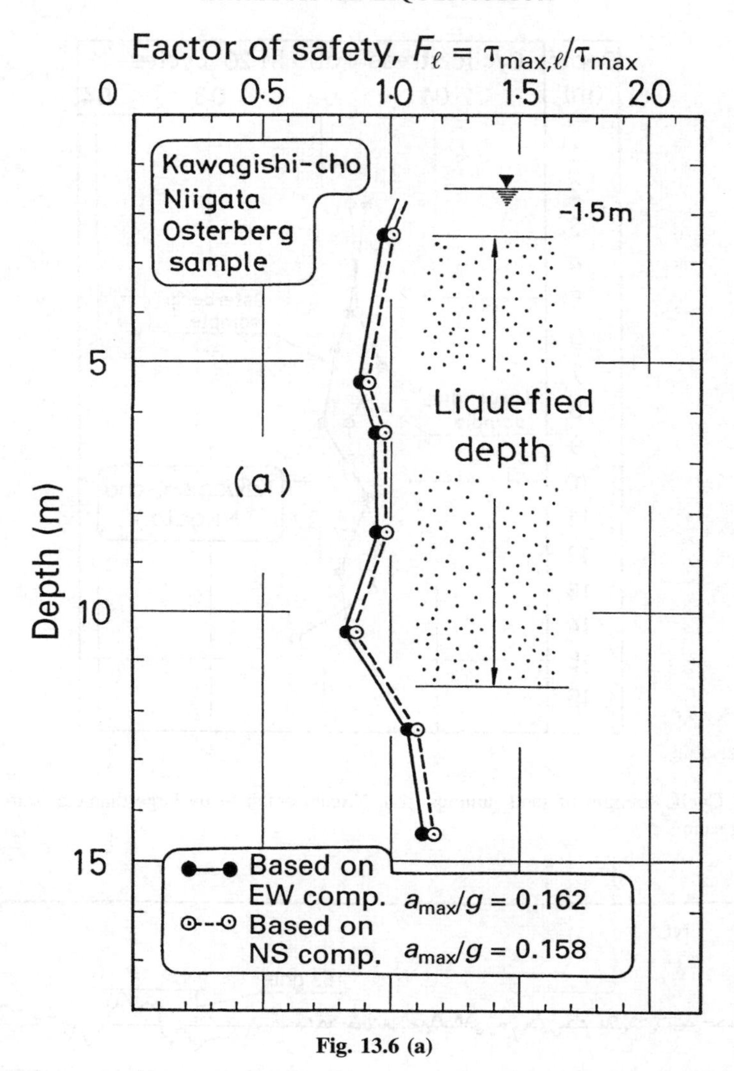

Fig. 13.6 (a)

this way is generally used to identify the depth to which liquefaction is expected to occur in a future earthquake. Such information becomes necessary in case some countermeasure is to be implemented for in situ deposit of sands.

References

Ishihara, K. and Koga, Y. (1981), Case Studies of Liquefaction in the 1964 Niigata Earthquakes, *Soils and Foundations*, Vol. 21, No. 3, pp. 35–52.

Iwasaki, T., Tatsuoka, F., Tokida, K. and Yasuda, S. (1978), *A Practical Method for Assessing Soil Liquefaction Potential Based on Case Studies at Various Sites in Japan,*

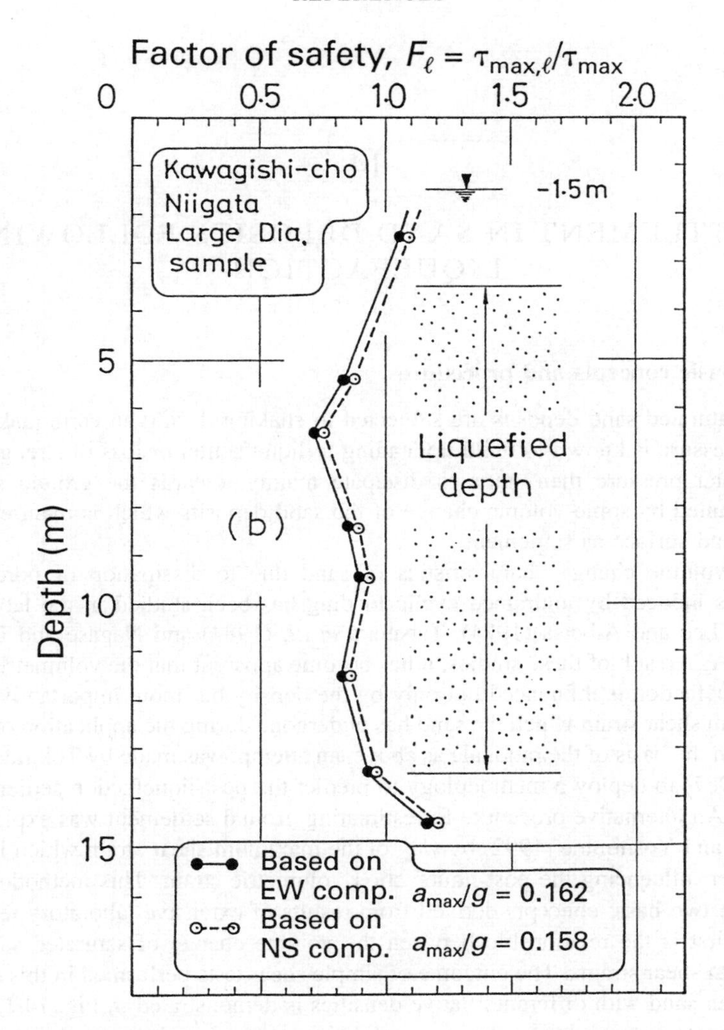

Fig. 13.6 (b) Estimated factor of safety for liquefaction in Niigata.

Proc. 2nd International Conference on Microzonation for Safer Construction-Research and Application, Vol. II, pp. 885–96.

Seed, H.B. and Idriss, I.M. (1971), Simplified Procedures for Evaluating Soil Liquefaction Potential, *Journal of Soil Mechanics and Foundation Engineering*, ASCE, Vol. 97, SM9, pp. 1249–73.

14

SETTLEMENT IN SAND DEPOSITS FOLLOWING LIQUEFACTION

14.1 Basic concepts and procedures

When saturated sand deposits are subjected to shaking during an earthquake, pore water pressure is known to build up leading to liquefaction or loss of strength. The pore water pressure then starts to dissipate mainly towards the ground surface, accompanied by some volume change of the sand deposits which is manifested on the ground surface as settlement.

The volume change characteristics of sand due to dissipation of pore water pressures induced by undrained cyclic loading has been studied in the laboratory tests by Lee and Albeisa (1974), Tatsuoka *et al.* (1984) and Nagase and Ishihara (1988). As a result of these studies, it has become apparent that the volumetric strain after liquefaction is influenced not only by the density but more importantly by the maximum shear strain which the sand has undergone during the application of cyclic loads. On the basis of the rationale as above, an attempt was made by Tokimatsu and Seed (1987) to deploy a methodology to predict the post-liquefaction settlement of ground. An alternative procedure for estimating ground settlement was explored by Ishihara and Yoshimine (1992) by way of the maximum shear strain which is a key parameter influencing the post-liquefaction volumetric strain. This methodology is based on two basic concepts derived from results of extensive laboratory tests.

The first is the relationship between the volume change of saturated sand and maximum shear strains. The outcome of simple shear tests performed in this context on a clean sand with different relative densities is demonstrated in Fig. 14.1, where the volumetric strain during the reconsolidation ε_v is plotted versus the maximum shear strain γ_{max} experienced by the sample during the undrained irregular loading. It was noted that the developed pore water pressure becomes equal to the initial vertical stress when the amplitude of irregular loads was large enough to produce the maximum shear strain of about 3% as accordingly indicated in Fig. 14.1. This is consistent with the results of many other tests indicating that cyclic softening or initial liquefaction with 100% pore water pressure build-up occurs accompanied by maximum shear strain of the order of 2 to 3%. It is of importance to notice in Fig. 14.1 that, even when the maximum shear strain increases beyond 2 to 3% which is the value required to cause initial liquefaction, the volumetric strain during reconsolidation tends to increase significantly.

In order to estimate the liquefaction-induced settlement of a sand deposit using the correlation shown in Fig. 14.1, it is necessary to know the magnitude of the

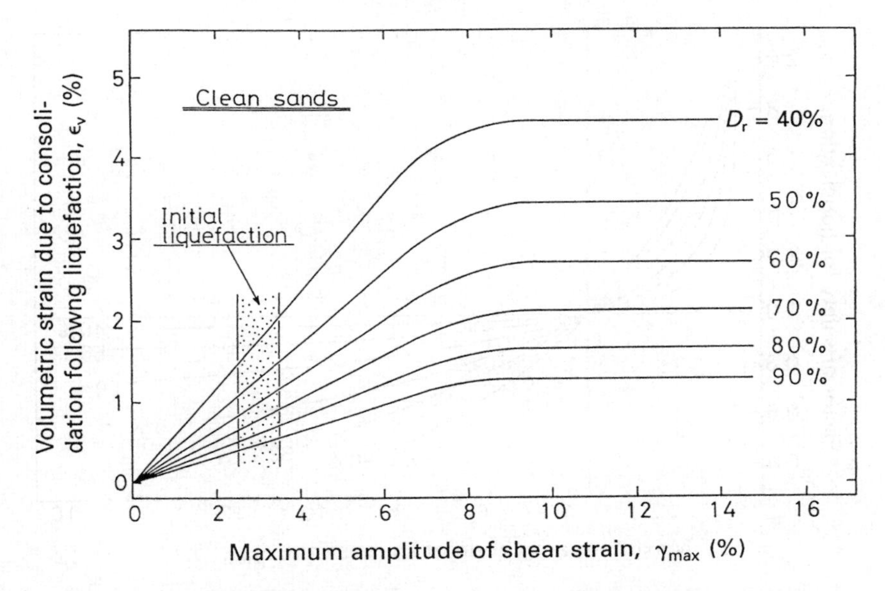

Fig. 14.1 Post-liquefaction volumetric strain plotted against maximum shear strain.

maximum shear strain which the sand will undergo during the application of shaking in a future earthquake. This can be achieved based on the second concept as described in the following. As previously mentioned, the cyclic strength has been customarily defined as as the cyclic stress ratio required for a double-amplitude axial strain of 5% to be developed in the sample in triaxial tests. In the case of very loose sand, the sand starts to greatly deform as soon as such a state of softening is encountered, and therefore whatever amount of double-amplitude strain may be used, almost the same value is obtained for the cyclic strength. However, in the case of medium dense to dense sand, this does not hold true. In fact, an increasing cyclic stress ratio is required with increasing magnitude of shear strain specified to define cyclic softening. In other words, a larger magnitude of cyclic resistance can be actually mobilized over the nominally determined cyclic strength if more than 5% D.A. axial shear strain is allowed to take place in the triaxial test samples. Consequently, when the factor of safety for liquefaction is defined as indicated by eqn (13.3) for the 5% D.A. axial shear strain, cases often occur where the computed factor of safety becomes less than unity. Defined in the above fashion, the factor of safety of unity implies a state of cyclic softening producing a 5% D.A. axial strain, and a factor of safety less than unity means that the soil has been softened to a state in which more than 5% D.A. axial strain is produced. Thus the factor of safety F_ℓ is considered to be a function of the double-amplitude axial strain, and conversely speaking, if the factor of safety is known for a sand deposit at a given site, the D.A. axial strain developing in the sand during liquefaction can be made known. Half of

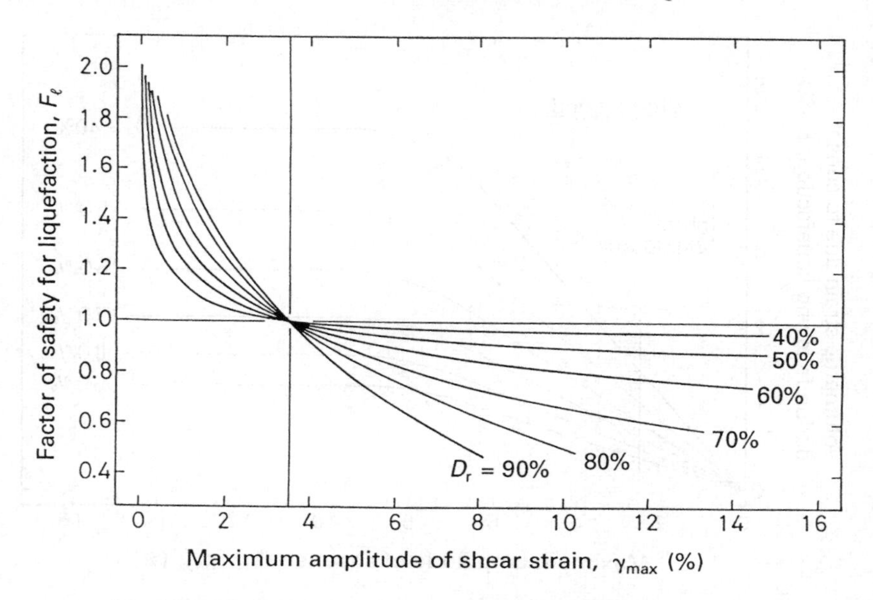

Fig. 14.2 Relation between factor of safety and maximum shear strain.

this shear strain, i.e. the single-amplitude axial strain, is regarded as the maximum shear strain which the sand has undergone in the course of liquefaction during earthquakes.

The relationship in the above context between the factor of safety and the shear strain amplitude can be established on the basis of available test data which have been obtained in laboratory tests. A number of simple shear test data on a clean sand obtained by Nagase (1985) was processed in this context as represented by a family of curves in Fig. 14.2 where the factor of safety is shown versus the maximum shear strain γ_{max} for the sand with different relative densities. It may be seen in Fig. 14.2 that, at a given value of the factor of safety less than unity, the larger the relative density, the smaller the maximum shear strain. The family of curves in Fig. 14.2 can be used to assess the maximum amplitude of shear strain for a known value of factor of safety. If the value of maximum shear strain is known in this manner, the post-liquefaction volumetric strain can be determined through the use of the already established curves shown in Fig. 14.1.

At this stage, suppose the factor of safety is known as a result of liquefaction analysis as described in Chapter 13. It will be possible to circumvent the determination of the maximum shear strain and to directly estimate the amount of post-liquefaction volumetric strain. For this purpose, combinations of the factor of safety F_ℓ and the volumetric strain ε_v giving equal magnitude of maximum shear strain were read off from each family of curves shown in Figs 14.1 and 14.2. The combinations of F_ℓ and ε_v thus obtained are plotted to establish a family of curves as

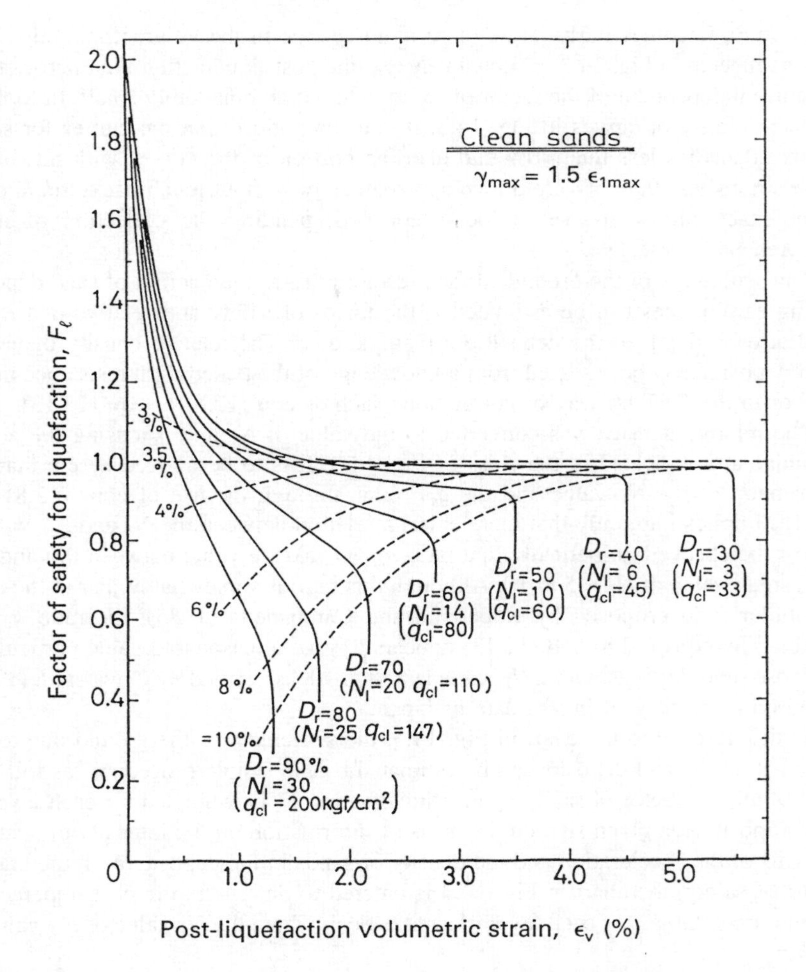

Fig. 14.3 Chart for determination of post-liquefaction volumetric strain as a function of factor of safety.

demonstrated in Fig. 14.3 in which the maximum shear strain $\gamma_{\max}$ is taken as a tracking parameter. If these curves are to be used for practical purposes, the axial strain in the triaxial mode should be converted to shear strain in the simple shear mode through the use of the relation, $\gamma_{\max} = 1.5\,\varepsilon_{1\,\max}$. Note that in the case of constant-amplitude axial strain in the triaxial test, $\varepsilon_{1\,\max}$ is taken to be equal to ε_1. Thus the maximum shear strain $\gamma_{\max}$ indicated in Fig. 14.3 is the one converted accordingly.

As can be seen in Fig. 14.1, there exists an upper limit in the reconsolidation volumetric strain for each given density and therefore even when the maximum shear strain becomes fairly large, there is no change in the volumetric strain. This fact is reflected in Fig. 14.3 in such a way that the post-liquefaction volumetric strain becomes independent of the factor of safety when it is remarkably small. In looking over the family of curves in Fig. 14.3, it is to be noticed that the curves for small relative densities less than 40% and also the portion of the curves with maximum shear strains less than 5% are drawn approximately without much background data. More exact curves are yet to be established, pending the collection of more comprehensive test data.

The settlement of the ground surface resulting from liquefaction of sand deposits during earthquakes can be estimated if the factor of safety and relative density of sand at each depth of the deposit are made known. The relative density of in situ sand deposits may be assessed from a knowledge of the penetration resistance in the SPT or in the CPT by way of correlations such as eqn (12.8) or eqn (12.13).

The relative density was converted to the value of N_1 and q_{c1} using the above formulae and indicated in the chart of Fig. 14.3. It is to be noticed herein that the conversion of the N_1 value and the q_{c1} value through the use of eqns (12.8) and (12.13) implies implicitly that there exists a relation between the N_1 and q_{c1} values. As can be checked numerically, the ratio q_{c1}/N_1 takes a value between 0.6 and 0.8 for a small value of $N_1 = 5$ to 10. Although this ratio is somewhat higher in the light of similar ratio proposed by Robertson and Campanella (1985), the ratio q_{c1}/N_1 obtained by eqns (12.8) and (12.13) appears to give a reasonable value particularly for loose sand deposits with a N_1 value less than 10, as verified by Ohya *et al.* (1985) based on a majority of in situ data in Japan.

With reference to the chart in Fig. 14.3, the settlement of the ground due to the liquefaction of sand deposits can be estimated by the simple procedures as follows. First of all, the factor of safety against liquefaction F_ℓ is evaluated for each layer of sand deposits at a given site, on the basis of information on the intensity of shaking in terms of the acceleration and the density of sands in the deposit. With the known factor of safety, the chart in Fig. 14.3 is entered to determine the post-liquefaction volumetric strain ε_v for each layer of sand deposit where the N_1 value or q_{c1} value is known.

With the volumetric strains established for each layer, the amount of settlement on the ground surface can be obtained by adding the vertical displacements produced in each layer of the deposit.

14.2 Evaluation of settlement

Let the case of the Niigata liquefaction be considered as an example of application of the above procedure. As seen in Fig. 13.6, the factor of safety of the loose sand deposit at Kawagishi site is about 0.9 on the average through the depth of 3 to 12 m. The relative density determined from undisturbed samples is shown in Fig. 13.3 where an average value of approximately 50% may be read off through the depth being considered. Entering into the chart of Fig. 14.3 with these values, the post-liquefaction volumetric strain is estimated to be about 2.5 to 3.5% whereby producing the maximum shear strain of about 5 to 7%. Thus the surface settlement is

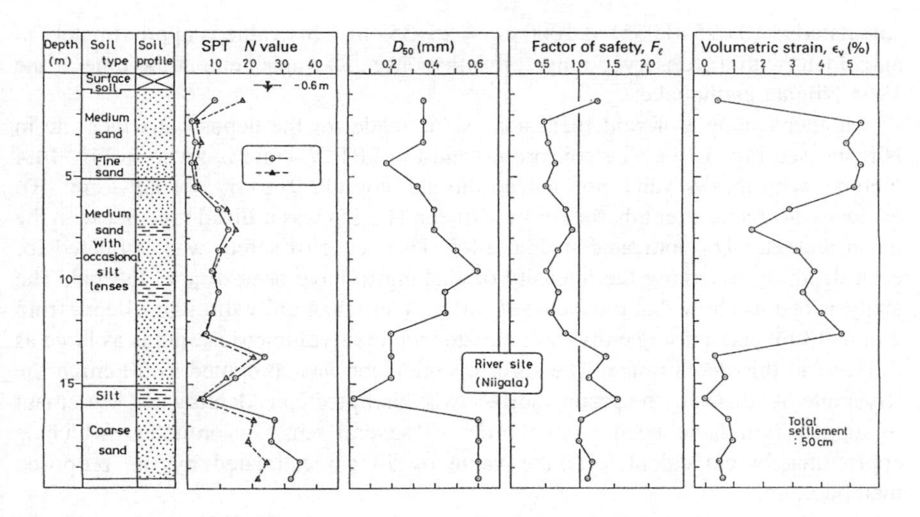

Fig. 14.4 Settlements analysis of a sandy deposit at Niigata in the 1964 earthquake.

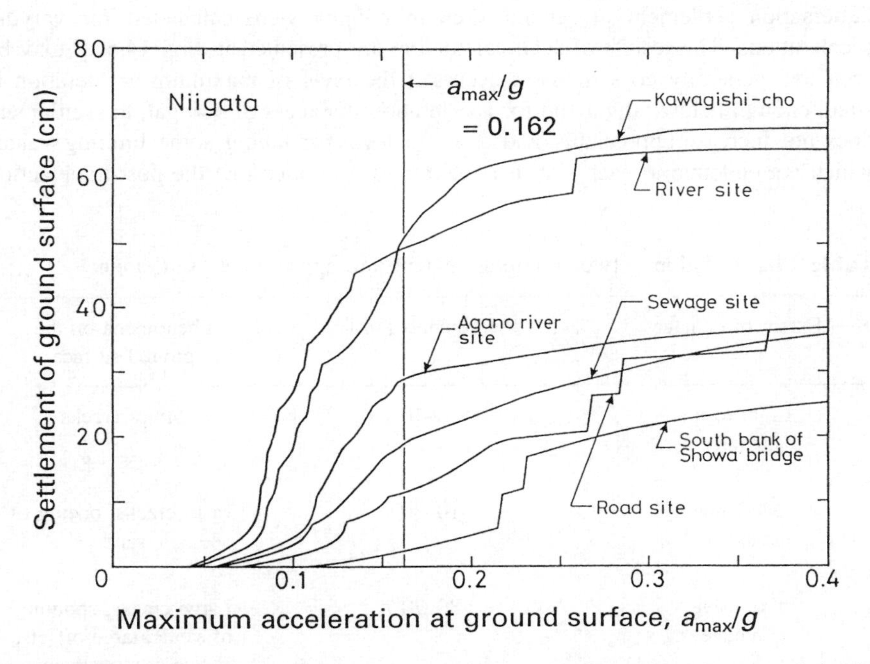

Fig. 14.5 Increasing settlements with increase in ground acceleration estimated at several sites in Niigata.

calculated as (0.025–0.035) × 1000 cm = 25–35 cm. This value is approximately in agreement with the observed range of settlement at Kawagishicho at the time of the 1964 Niigata earthquake.

Another example of settlement analysis is made for the deposit at River site in Niigata (see Fig. 13.1). The soil profile and the SPT N value is given in Fig. 14.4 together with the N_1 value normalized through eqn (12.9) to $\sigma'_v = 1$ kgf / cm^2. To evaluate the cyclic strength, the formula of eqn (12.15) was utilized together with the mean diameter D_{50} indicated in Fig. 14.4. The factor of safety was estimated for each depth by assuming the intensity of shaking to have been $a_{max} = 158$ gal. The analysis results show that the factor of safety is less than unity through a depth from 2 m to 13 m, and consequently the estimated value of volumetric strain is as large as 2 to 5% in this depth range. The surface settlement was computed as 50 cm at the River site. At this site, the ground surface was disrupted considerably and settlement is supposed to have been on the order of several tens of centimeter which is approximately coincident with the value of 50 cm estimated by the proposed methodology.

It is apparent that the factor of safety defined by eqn (13.3) decreases with the increasing level of maximum acceleration and, consequently, the settlement induced by liquefaction tends to increase as the intensity of shaking becomes stronger. However, the feature of settlement increase with increasing level of acceleration is different for each site being considered, depending upon the soil profile and the nature of soils composing the deposits. In order to look into this aspect, the post-liquefaction settlement at several sites in Niigata were calculated for varying accelerations. The results of such calculation are presented in Fig. 14.5. It may be seen that generally no settlement occurs if the level of maximum acceleration is small enough below 50 gal and for acceleration in excess of 300 gal, the settlement does not increase appreciably and tends to level off taking some limiting values which are endemic to each site. It is of interest to notice that the post-liquefaction

Table 14.1 Relation between damage extent and approximate settlement

Extent of damage	Settlements (cm)	Phenomena on the ground surface
Light to no damage	0–10	Minor cracks
Moderate damage	10–30	Small cracks, oozing of sand
Extensive damage	30–70	Large cracks, spouting of sands, large offsets, lateral movement

settlement in loose sand deposits tends to increase sharply with an increase in acceleration in the range of 100 and 200 gals.

Interpreted overall in the light of the observed performances of the ground at many other sites during past earthquakes, it may be mentioned that settlement on the order of 10 cm or less correspond to the area where there is little or no destruction, and in moderately damaged areas settlement is roughly between 10 to 20 cm, and if it becomes greater than 30 cm, considerable destruction always occurs on the ground surface, such as sand spurting, fissures, and large offsets. The qualitative correspondence between damage extent and settlement as above may be summarized as shown in Table 14.1.

References

Ishihara, K. and Yoshimine, M. (1992). Evaluation of settlements in sand deposits following liquefaction during earthquakes. *Soils and Foundations*, **32**, 173–88.

Lee, K.L. and Albeisa, A. (1974). Earthquake induced settlements in saturated sands. *Journal of Geotechnical Engineering*, ASCE, **100**, GT4, 387–406.

Nagase, H. and Ishihara, K. (1988). Liquefaction-induced compaction and settlement of sand during earthquake. *Soils and Foundations*, **28**, 66–76.

Nagase, H. (1985). *Behavior of sand in multi-directional irregular loading*. PhD dissertation, Department of Civil Engineering, University of Tokyo (in Japanese).

Ohya, S., Iwasaki, T., and Wakamatsu, M. (1985). *Comparative study of various penetration tests on ground that underwent liquefaction during the 1983 Nihonkai–Chubu and 1964 Niigata earthquake*. Proceedings of the US–Japan Joint Workshop on In situ Testing Methods for Evaluation of Soil Liquefaction Susceptibility, San Francisco, pp. 56–88.

Robertson, P.K. and Campanella, R.G. (1985). Liquefaction potential of sands using the CPT. *Journal of Geotechnical Engineering*, ASCE, 111, GT3, 384–403.

Tatsuoka, F., Sasaki, T., and Yamada, S. (1984). *Settlements in saturated sand induced by cyclic undrained simple shear*. Proceedings of the 8th World Conference on Earthquake Engineering, San Francisco, Vol. 3, pp. 95–102.

Tokimatsu, K. and Seed, H.B. (1987). Evaluation of settlements in sands due to earthquake shaking. *Journal of Geotechnical Engineering*, ASCE, 113, GT8, 861–78.

15

FLOW AND NON-FLOW CONDITIONS AND RESIDUAL STRENGTH

Flow-type failure has been known to occur in loose deposits of sandy soils when they are put into a steady state of deformation following the onset of liquefaction due to seismic shaking. For the flow-type failure to occur, it is necessary therefore, that the sand be loose enough to exhibit contractive behaviour during shear stress application. Under the usual range of overburden pressures around 100 kPa, there appears to be an upper limit in relative density above which there is practically no possibility for sand to become contractive and hence to be put in a steady state with a reduction in shear resistance. Conversely, if the sand exists with a relative density below this limit, the likelihood would be high for developing steady-state deformation and hence a flow-type failure. Such a limit may be properly expressed in term of any of the density parameters such as relative density. If these parameters are correlated with the penetration resistance of any sounding tests in the field such as the SPT or CPT, the limiting condition may be expressed alternatively in terms of a threshold value of the penetration resistance. In compliance with the limiting conditions or threshold values as above, there would also be an upper limit in the normalized residual strength. This aspect of the problem will be addressed in the following.

15.1 Flow conditions in SPT and CPT

In Section 11.8 the fabric formed during deposition was shown to have an important influence on the response of sandy soils. Suppose there are three possible sets of depositional state, as illustrated in Fig. 15.1, for an element of sand subjected to the same overburden pressure p_c'. Let it be assumed that the soil deposit has been formed under either of the conditions as represented by the method of the moist placement (M.P.), dry deposition (D.D.), or water sedimentation (W.S.). For each of the depositional conditions, the sand could exist at different states which are however specified by varying the value of the initial state ratio $r_c = p_c'/p_s'$. For example, under the depositional condition of moist placement, the sand could be put in an initial state of consolidation with large values of r_c, but it can also take a state with $r_c = 2.0$ if the sand is compacted initially to some extent. This state is assumed to have a void ratio of e_m as accordingly indicated in Fig. 15.1. Under the condition of water sedimentation, the initial state with $r_c = 2.0$ could be produced anyway with the void ratio e_W. Thus for the initial state with $r_c = 2.0$ undergoing the same overburden pressure p_c', there are three states with different void ratios resulting

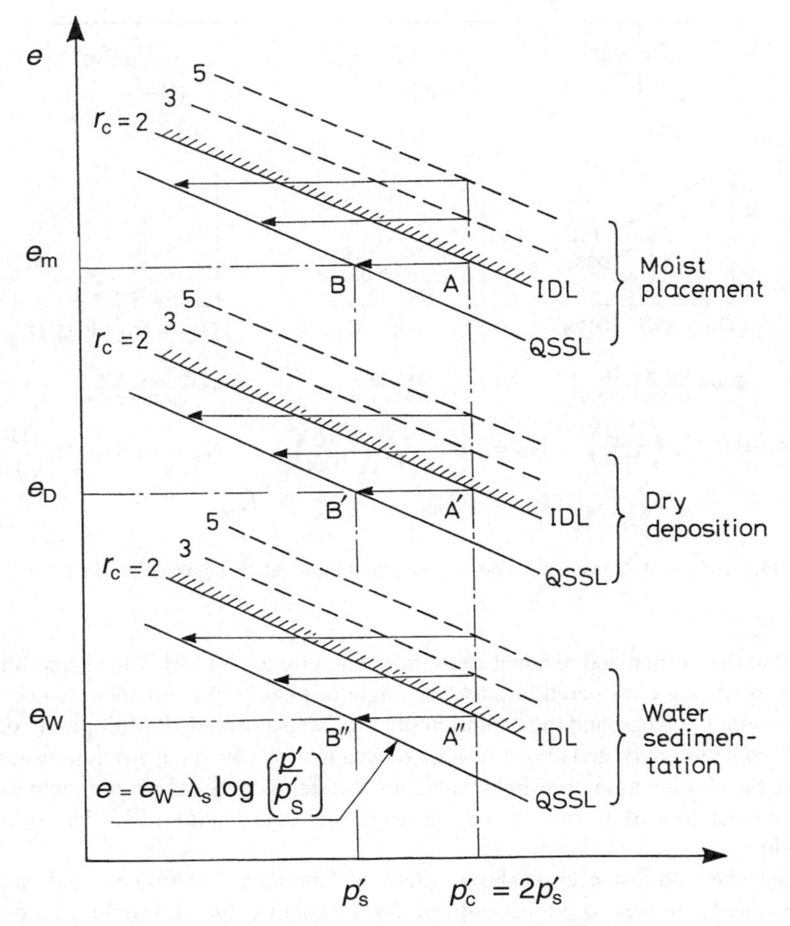

Fig. 15.1 QSSL and IDL for samples with different fabrics.

from different conditions of deposition, as indicated in Fig. 15.2. As a result of numerous laboratory tests, the void ratio e_M produced by the method of moist placement has been known to be the highest among those that would be obtained under the three depositional environments. This situation may be described by the inequality $e_M > e_D > e_W$. If it is expressed in terms of the relative density, one obtains another inequality $D_{rW} > D_{rD} > D_{rM}$ where D_{rD} denotes, for example, the relative density obtained from the void ratio e_D corresponding to the dry-deposited sand deposit. If each of the relative densities D_{rW}, D_{rD}, and D_{rM} is introduced into the formula of eqn (12.10), one obtains three different N values of the SPT as illustrated in Fig. 15.2.

The above reasoning implies that there are three possible values of void ratio and hence SPT blow count which is conceivably the case in a given deposit in the field.

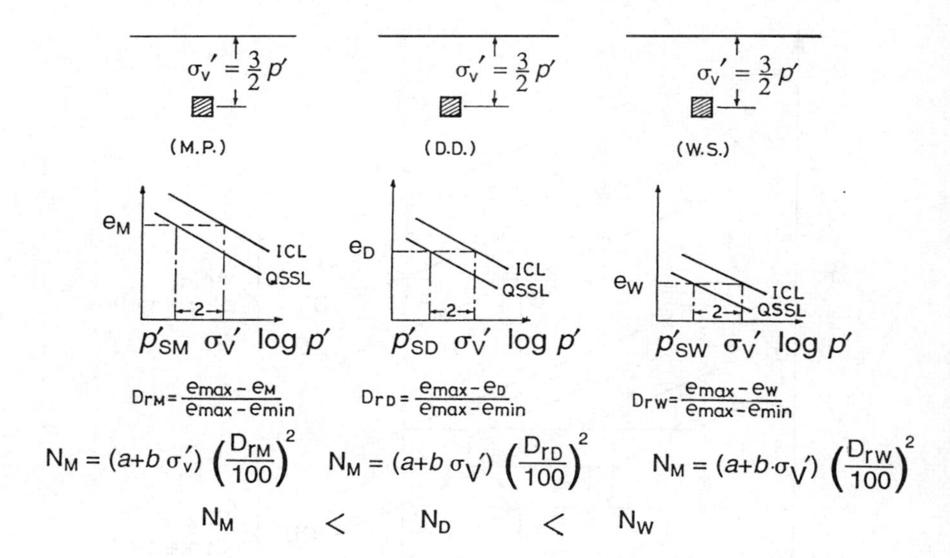

Fig. 15.2 Effects of depositional modes on the void ratio, relative density and SPT *N* value.

Note that the normalized residual strength defined by eqn (11.4) is the same for the three cases being considered because the angle of phase transformation ϕ_s does not change with the fabric and the condition of $r_c = 2.0$ is postulated. Multiplicity of the SPT N value as envisaged above raises a question as to which penetration resistance should be chosen as a threshold value to differentiate between contractive and dilative behaviour of a sand without knowing the exact feature of its deposition in the field.

Under the circumstances as above where the laboratory-determined void ratio on reconstituted samples is to be utilized for estimating the threshold penetration resistance, it would be most legitimate to choose the initial state or QSSL occurring at the lowest void ratio. In the illustrative diagram in Fig. 15.1, this can be achieved by choosing the IDL obtained from samples prepared by the method of water sedimentation, for example. This choice will indeed lead to the most conservative estimate of in situ conditions of sandy soils regarding the occurrence or non-occurrence of flow-type failure due to a shaking by an earthquake.

When performing tests, it is difficult to prepare several specimens so as to have an identical void ratio even under supposedly identical conditions for sample preparation. Therefore initial consolidation lines (ICL) tend to deviate from each other even if specimens are prepared in the same manner. However the QSSL is determined more consistently, given the condition for sample preparation. Thus it is more expedient to choose the QSSL and assume the ICL from the postulated r_c-value. In what follows, this line of reasoning will be adopted.

Looking back at the test results shown in Fig. 11.25, one can notice that a curved line is most fitting to represent the QSS in the e–log p' diagram. However, the QSS

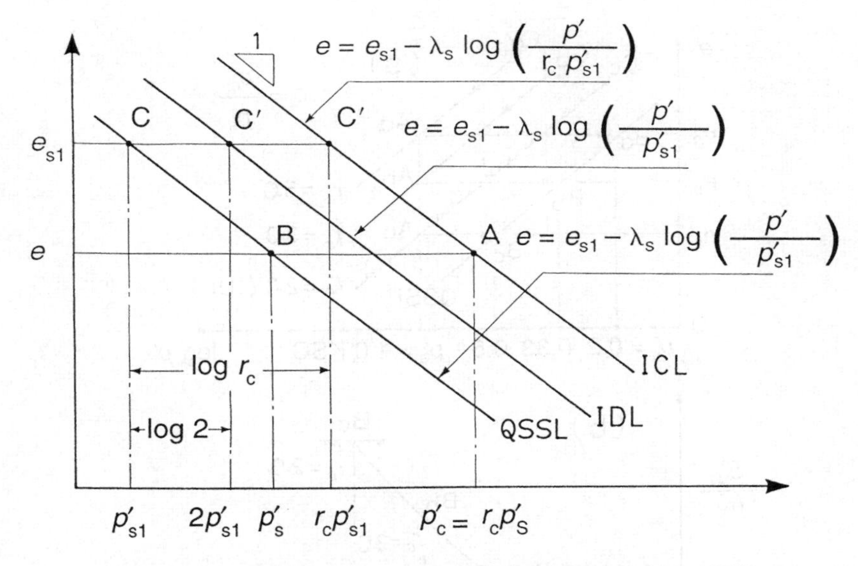

Fig. 15.3 Location and expressions of ICL and IDL relative to QSSL.

may also be represented approximately by a straight line as

$$e = e_{s1} - \lambda_s \log_{10}\left(\frac{p'}{p'_{s1}}\right) \tag{15.1}$$

where e_{s1} and p'_{s1} are an arbitrarily chosen set of void ratio and effective confining stress on the QSSL to fix its location in the e–log p' diagram and λ_s denotes the slope of the line, as illustrated in Fig. 15.3. If a set of values of e_{s1} and p'_{s1} is chosen for a QSSL with the lowest void ratio, eqn (15.1) gives an expression for the QSSL which is associated with the most conservative estimate of the penetration resistance for in situ sounding tests. In what follows, all discussions will be made with reference to the QSSL as defined above.

As illustrated in Figs 11.6 and 15.1, samples can be prepared at various initial void ratios producing several initial consolidation curves. The location of these initial consolidation lines (ICL) relative to the QSSL is represented by the initial state ratio r_c, as illustrated in Fig. 15.3, on the grounds that the ICL is parallel to the QSSL. Thus with reference to eqn (15.1), an ICL with a given value of r_c is expressed as

$$e = e_{s1} - \lambda_s \log_{10}\left(\frac{p'}{r_c p_{s1}'}\right). \tag{15.2}$$

Introducing this relation into the definition of relative density, the ICL can be expressed in terms of the relative density as

$$\frac{D_r}{100} = \frac{1}{e_{max} - e_{min}}\left[e_{max} - e_{s1} + \lambda_s \log_{10}\left(\frac{p'}{r_c\, p_{s1}'}\right)\right]. \tag{15.3}$$

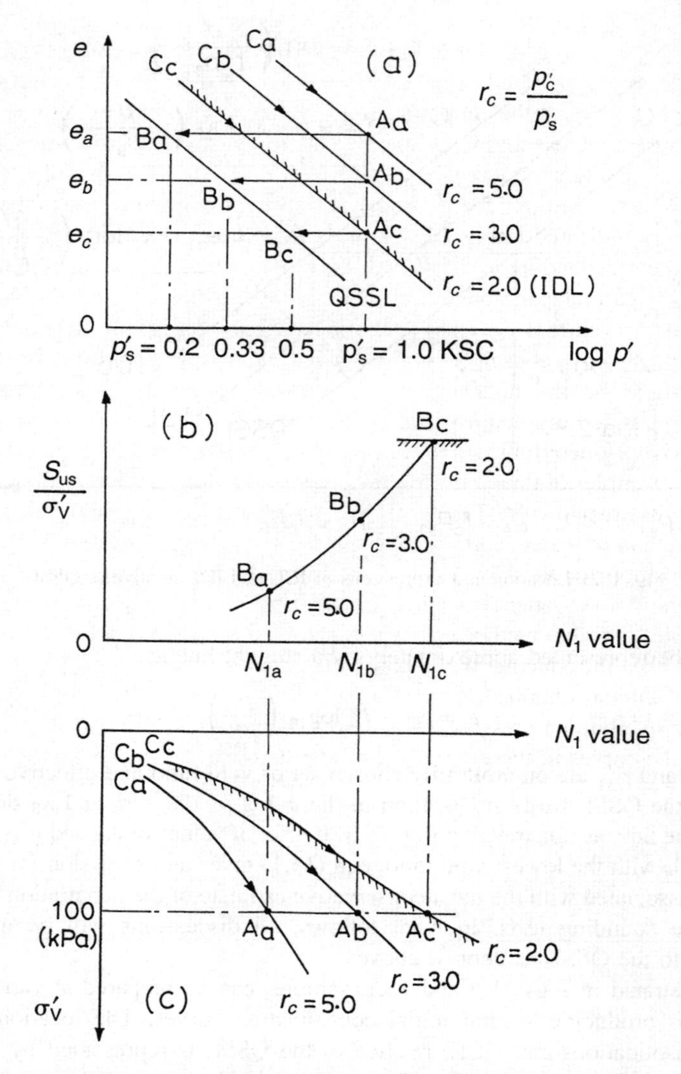

Fig. 15.4 Illustration for establishing the relation between the normalized residual strength and N_1 value of the SPT.

Introducing eqn (15.3) further into eqn (12.10), the ICL can now be expressed in terms of the N value of the SPT as

$$N = (a + b\,\sigma'_{\mathrm{v}}) \left[\frac{e_{\max} - e_{\mathrm{s1}} + \lambda_{\mathrm{s}} \log\left(\dfrac{2}{3p'_{\mathrm{s1}}} \dfrac{\sigma'_{\mathrm{v}}}{r_{\mathrm{c}}} \right)}{e_{\max} - e_{\min}} \right]^2 \tag{15.4}$$

where p' is replaced by $p'_c = \sigma'_0$ and further σ'_0 is rewritten in terms of σ'_v through the relation $\sigma'_0 = 2\,\sigma'_v/3$. The implication of eqn (15.4) is illustrated in Fig. 15.4(a) and (c). If a loosely prepared sample, say, with $r_c = 5.0$ is consolidation from point C_a to A_a in Fig. 15.4(a), the corresponding N value increases from point C_a to A_a in Fig. 15.4(c) with increasing initial effective confining stress. If a sample is prepared to exhibit marginally contractive behaviour with $r_c = 2.0$, the consolidation process from point C_c to A_c in Fig. 15.4(a) produces a corresponding increase in the N value, as accordingly indicated in Fig. 15.4(c). It is important to remember that the curve for $r_c = 2.0$ corresponding to the IDL in the e–log p' diagram yields the maximum possible N value under the condition that the sand exhibits contractive behaviour. Therefore if the value of $r_c = 2.0$ is introduced into eqn (15.4), together with the values of e_{s1} and p'_{s1} chosen for the QSSL at the lowest void ratio, the resulting N value would be the highest among those attainable by any state of initial consolidation under any depositional environments.

The laboratory-determined constants to quantify the QSS of clean sands were obtained for samples of three typical sands prepared by the method of dry deposition and the results are shown in Table 15.1. The constants for silty sands were obtained from the method of water sedimentation for Tia Juana silty sand, and by the method of dry deposition for the silty sands from Chiba Japan and Dagupan, Philippines. The constants for the latter two soils were adjusted so that they correspond to those from samples by the method of water sedimentation, and they are shown in Table 15.2. By introducing these constants into the formula of eqn (15.4), the threshold N value is obtained separately for clean sands and silty sands as a function of effective overburden pressure σ'_v and presented in Figs 15.5 and 15.6. It is to be noted that the curves in these figures indicate the highest N value with $r_c = 2.0$ for

Table 15.1 Constants for clean sands by the method of dry deposition

Constants	Toyoura	Chiba	Kiyosu
Fines content F_c : %	0	3	0
Grain size D_{50} : mm	0.170	0.170	0.310
e_{max}	0.977	1.271	1.206
e_{min}	0.597	0.839	0.745
e^*_{s1}	0.876	1.095	1.015
λ_S	0.022	0.085	0.050
a	33	33	33
b	37	37	37
M	1.24	1.37	1.20
ϕ_S	30°	30°	30°
N_1 value	4	7	9

*Void ratio at the QSS for $P'_{s1} = 1$ kgf/cm^2.

Table 15.2 Constants for silty sands by the method of water sedimentation

Constants	Tia Juana	Chiba	Dagupan
Fines content F_c : %	12	18	15
Grain size D_{50} : mm	0.160	0.150	0.180
e_{max}	1.099	1.307	1.454
e_{min}	0.620	0.685	0.600
e^*_{s1}	0.820	0.940	0.835
λ_S	0.075	0.090	0.180
a	10	8	8
b	15	12	12
M	1.22	1.37	1.25
ϕ_S	30.5°	34°	31°
N_1 value	6.5	6	8

* Void ratio at the QSS for $P'_{s1} = 1$ kgf/cm^2.

each of the clean sands and silty sands used in the tests. Thus each curve has the same meaning as the curve C_c–A_c shown in Fig. 15.4(c). The values of a and b for the curves in Figs 15.5 and 15.6 were estimated from the chart of Fig. 12.4 based on the mean diameter D_{50} of individual sands. Because of the marked difference in relative density of individual sands at which the QSS occurs, the curves in Fig. 15.5 separating conditions of flow and non-flow for clean sands are somewhat diverse.

To check the validity of the laboratory-based curves, case studies were performed by collecting in situ N values from sites where flow-type failures are presumed to have taken place during actual earthquakes in the past. The data points shown in Figs 15.5 and 15.6 indicate field performance data thus obtained. It may be seen that the laboratory-based threshold N value is more or less in the same range as the field N value obtained directly from actual sites of apparent flow failure. With both of these two sources of data taken together, it may be considered appropriate to choose a curve giving the maximum N value as a bounding curve for clean sands. The curves of similar nature but for silty sands containing more than 10% fines are demonstrated in Fig. 15.6 where the curve for Dagupan soil may be taken as a representative bounding curve.

It is of interest to observe in Figs 15.5 and 15.6 that the bounding curve chosen for silty sands is located only slightly to the left giving nearly the same threshold curve as that adopted for clean sands. Generally speaking, the values of a and b are significantly smaller for dirty sands as compared to clean sands as displayed in Fig. 12.4, whereas the relative density at which QSS occurs is notably larger for dirty sands as against clean sands. These differences tend to act as mutually compensating factors to make the threshold N values practically identical for clean

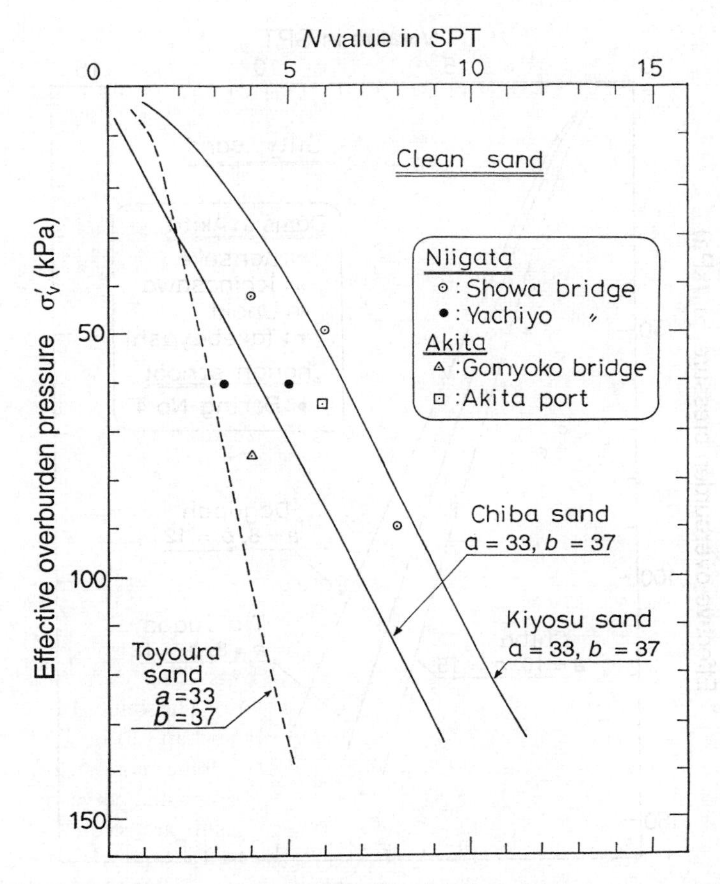

Fig. 15.5 Laboratory-determined threshold conditions for three clean sands converted to the SPT *N* value compared with in situ data from sites of liquefaction-induced flow failure.

sands and dirty sands. In view of these considerations, the two boundary curves giving the maximum possible *N* value for any sandy soils were taken from Figs 15.5 and 15.6 and displayed in Fig. 15.7 as a narrow zone which may be deemed as a boundary differentiating between conditions in terms of *N* value in which flow-type failure can or cannot occur. This zoned boundary may be used for clean sands as well as for silty sands but with fines no more than 30%.

The boundary curve in the above vein was suggested by Sladen and Hewitt (1989) in terms of q_c value in CPT on the basis of in situ investigations of man-made islands in the Canadian Beaufort Sea which have or have not experienced flow slide. The outcome of this study is shown in Fig. 15.8. In the absence of actual data on sites of earthquake-induced flow failure investigated by CPT, it might be difficult to explore precise comparison, but if the bounding curve in Fig. 15.7 is converted to

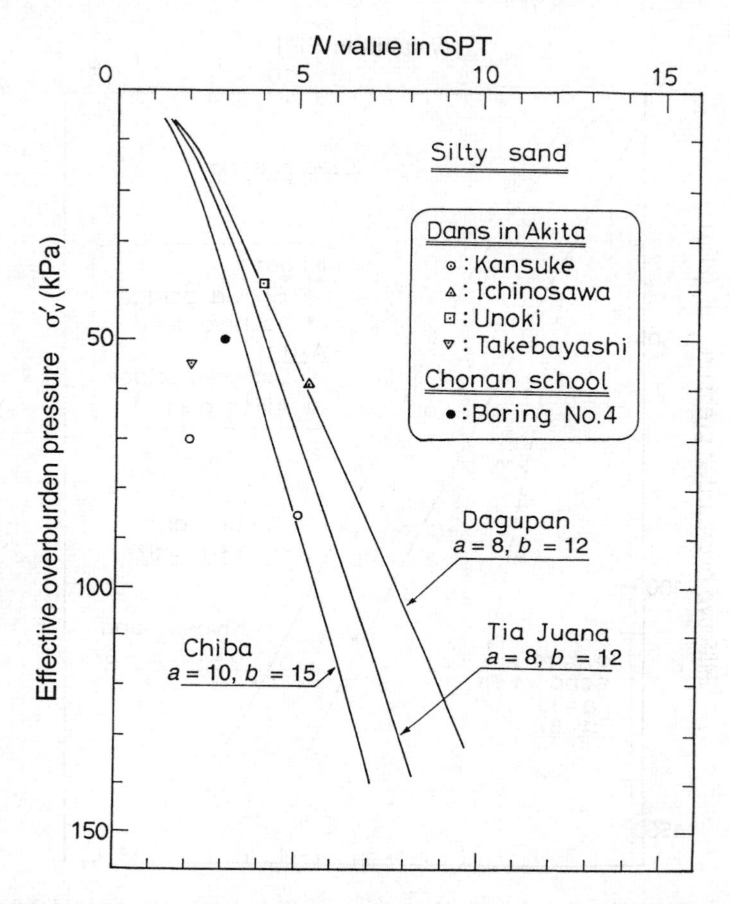

Fig. 15.6 Laboratory-determined threshold conditions for there silty sands converted to the SPT *N* value compared with in situ data from sites of liquefaction-induced flow failure.

the one in terms of q_c value, it would provide an interesting comparison. The conversion from *N* value to q_c value may be made by using the linear relationship proposed by Robertson and Campanella (1985), but compilation of data by Ohya *et al.* (1985) from deposits of liquefied sites in Japan indicated that the ratio q_c/N tends to increase with decreasing *N* value and takes a value by and large between 0.5 and 0.7 for the *N* value around 10. By drawing on these results, the *N* value corresponding to a median line in the zoned boundary in Fig. 15.7 was converted to the boundary in term of q_c value as indicated in Fig. 15.8. The bounding line suggested by Sladen and Hewitt (1989) is taken from Fig. 15.8 with a modification of $\sigma'_v = 1.5\,\sigma'_0$ and shown together in Fig. 15.9. Also superimposed in this figure is the boundary zone which was obtained by Robertson *et al.* (1992) as a summary of several proposed curves in the same context. It may be seen in Fig. 15.9

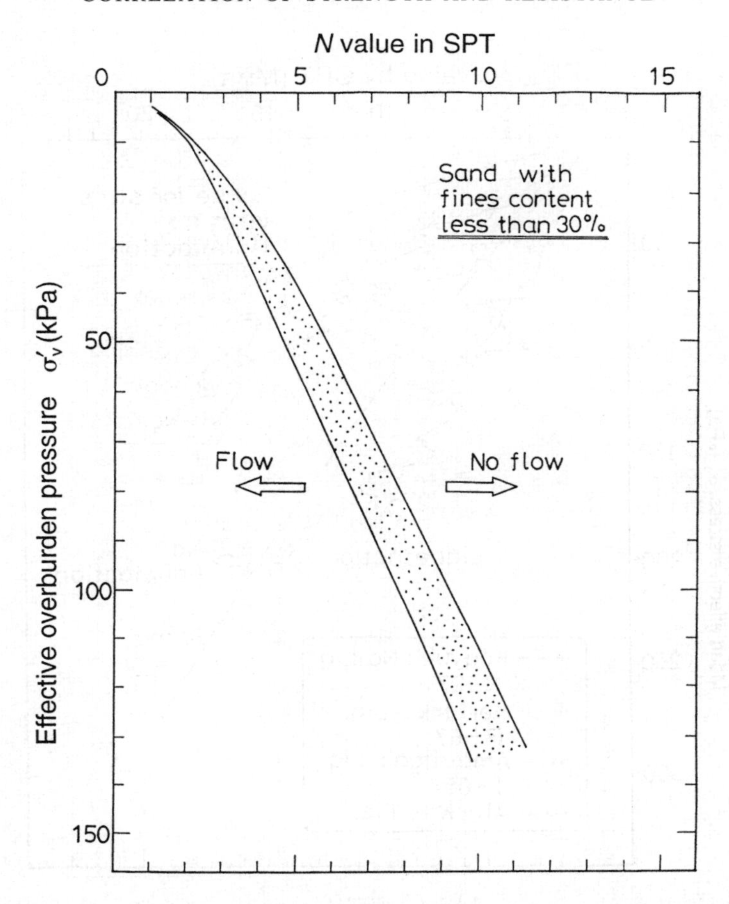

Fig. 15.7 Proposed boundary in SPT *N* value differentiating conditions of flow and non-flow.

that the boundary obtained from *N* value through the conversion as above is roughly in coincidence with the zone suggested by Robertson *et al.* (1992), and it is also located in proximity to the boundary proposed by Sladen and Hewitt (1989).

15.2 Correlation of residual strength and penetration resistance

In view of the difficulty in determining the residual or steady-state strength of once-liquefied soils through the accurate measurement of void ratio of sands in the field deposit, an attempt was made by Seed (1987) to establish a relationship between the SPT *N* value and residual strength which is assessed from back-analysis of many cases of liquefaction-induced failures. Back-analyses of failure cases were also made by Ishihara *et al.* (1990) and integrated into a chart in which the back-calculated residual strength is plotted versus the q_c value of the CPT. Thus it would be of value

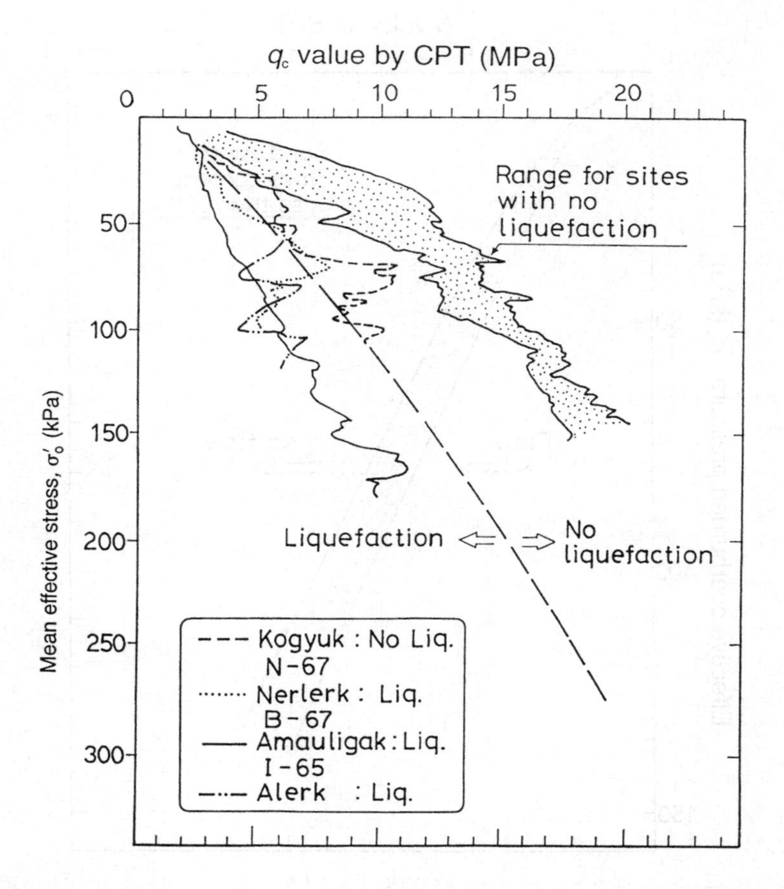

Fig. 15.8 Boundary in CPT q_c value between flow and non-flow obtained from sites in the Canadian Beaufort Sea (Sladen and Hewitt, 1989).

to examine the physical implication of these relationships and to provide correct interpretation in the context of the steady-state concept as discussed above. It is to be recalled, first of all, that the residual strength as defined in eqn (11.2) is a function of the effective confining stress p'_s at QSS. Therefore the value of the residual strength can be interpreted as representing the effective confining stress p'_s at which the QSS is attained. On the other hand, it has been customary to associate the penetration resistance of the SPT or CPT with the relative density. Since the relative density is alternatively expressed by the void ratio, the correlation between the residual strength and the penetration resistance is considered in turn to be essentially the same as the relationship between the effective confining stress and void ratio at which the quasi-steady state deformation takes place in the sand. In establishing such a correlation, the QSS with the lowest void ratio will be taken up for consideration for the same reason as argued above. Since the N value for such a

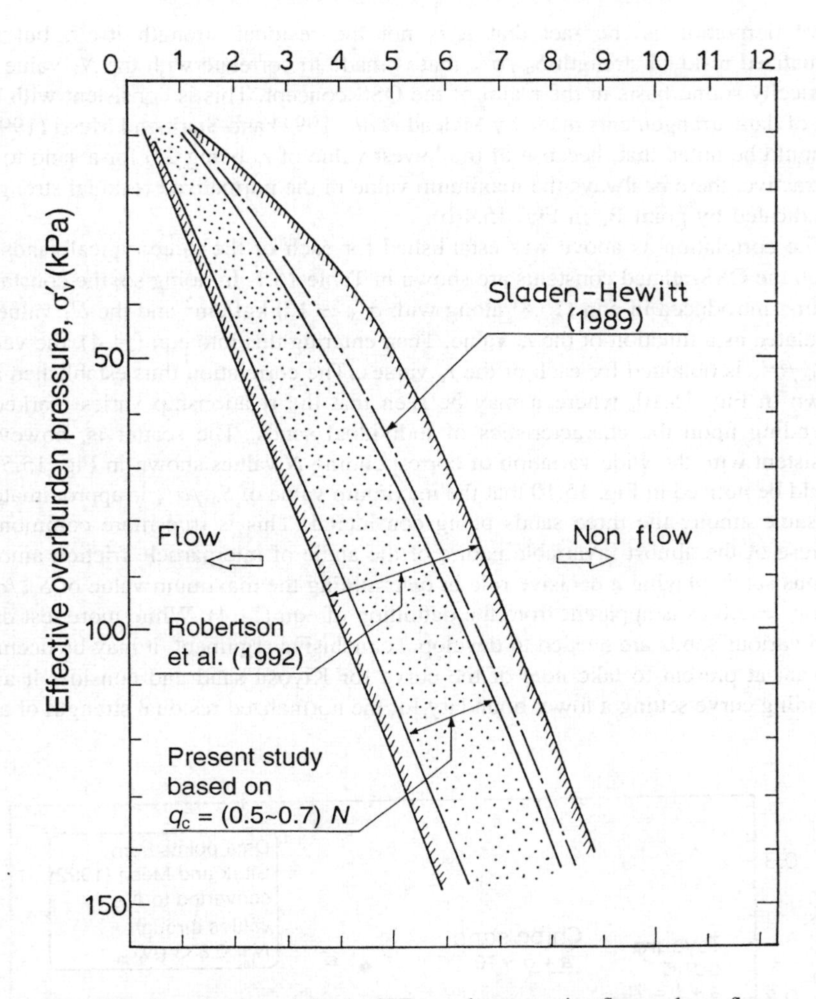

Fig. 15.9 Compiled boundaries in CPT q_c value separating flow and non-flow.

condition has been thoroughly discussed in the preceding section, what is needed additionally at this stage is the determination of the residual strength S_{us}. This can be achieved directly by the procedure illustrated in Fig. 15.4. The N_1 value corresponding to the penetration resistance at $\sigma'_v = 1$ kg/cm^2 (= 98 kPa) can be determined by locating points such as A_a, A_b, A_c in Fig. 15.4(c). Since the value of the initial state ratio r_c is already prescribed, the normalized residual strength S_{us}/σ'_o can be obtained readily by means of the relation of eqn (11.3). Then the residual strength normalized to σ'_v is obtained by using the relation $\sigma'_o = 2\sigma'_v/3$. The plot of S_{us}/σ'_v thus obtained for each of the r_c value is shown in Fig. 15.4(b) as a function of the N_1 value of SPT. This is the desired relationship. Noteworthy in the above derivation is the fact that it is not the residual strength itself, but the normalized residual strength S_{us}/σ'_v that is made to correlate with the N_1 value on physically sound basis in the realm of the QSS concept. This is consistent with the idea of data arrangements made by Mclead *et al.* (1991) and Stark and Mesri (1992). It should be noted that, because of the lowest value of r_c being 2.0 for a sand to be contractive, there is always the maximum value in the normalized residual strength, as indicated by point B_c in Fig. 15.4(b).

The correlation as above was established for each of the three typical sands of which the QSS-related constants are shown in Table 15.1. In doing so, the constants are first introduced in eqn (15.4) along with $\sigma'_v = 1.0$ kgf/cm^2 and the N_1 value is calculated as a function of the r_c value. Then entering this into eqn (11.4), the value of S_{us}/σ'_v is obtained for each of the r_c values. The correlation thus established are shown in Fig. 15.10, where it may be seen that the relationship varies markedly depending upon the characteristics of individual sands. The scatter is, however, consistent with the wide variation of corresponding N values shown in Fig. 15.5. It should be noticed in Fig. 15.10 that the maximum value of S_{us}/σ'_v is approximately the same among the three sands being considered. This is true more commonly, because of the almost invariable nature of the angle of interparticle friction among various sands playing a decisive role in determining the maximum value of S_{us}/σ'_v with $r_c = 2.0$ as is apparent from the definition of eqn (11.4). While more test data from various sands are needed to develop a conclusive argument, it may be deemed relevant at present to take note of the curve for Kiyosu sand and consider it as a bounding curve setting a lower boundary for the normalized residual strength of any

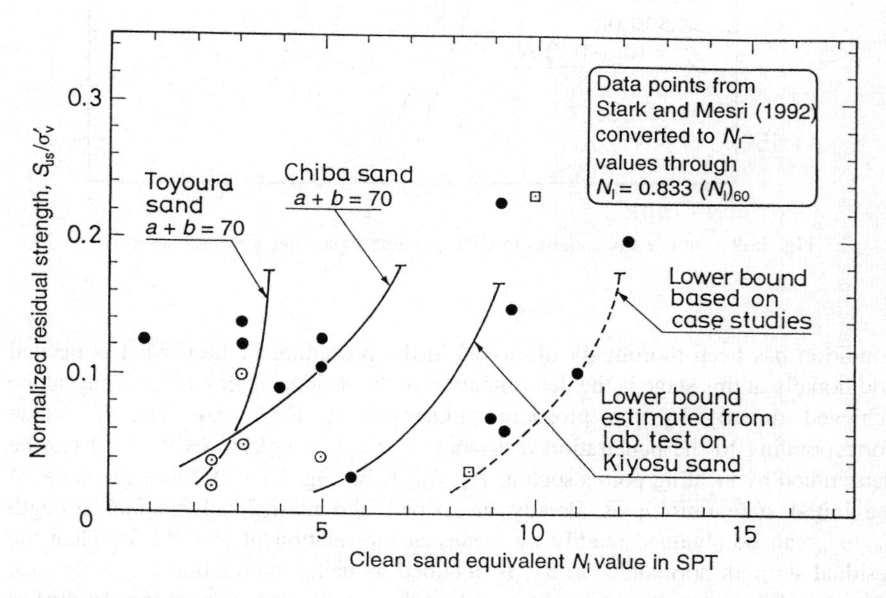

Fig. 15.10 Relation of normalized residual strength and SPT N_1 value based on laboratory-determined QSSL compared with data from back-analyses of actual failure cases.

kind of sand at a give N_1 value. Shown superimposed in Fig. 15.10 are the data quoted from the work by Stark and Mesri (1992) who compiled many case-studied data in the form of a S_{us}/σ'_v versus N_1 value correlation. Note that the value of $(N_1)_{60}$ in the abscissa is transformed into the N_1 value through the relation $N_1 = 0.8333\ (N_1)_{60}$. It may be seen in Fig. 15.10 that a majority of the case-studied data points lie within the zone above the boundary line established above, except for the six data obtained from failures case in North America. If all the case-studied data are to be encompassed, an alternative boundary may need to be established as accordingly indicated by a dashed line in Fig. 15.10. However, it appears judicious at present to adopt the former bounding curve based on Kiyosu sand behaviour as being a practically applicable boundary line.

Case studies have also been made by Ishihara *et al.* (1990) by assembling records of earthquake-caused failures in embankments, tailings dams, and river dykes. The outcome of these case studies is shown in Fig. 15.11 in terms of the normalized residual strength plotted versus q_{c1} value of CPT. It is to be noticed that the q_{c1} value originally obtained from silty sand deposits has been converted to a value equivalent to clean sands based on the correction indicated in Fig. 12.7. Also shown superimposed in Fig. 15.11 are the possible boundary curves obtained by converting the Kiyosu sand-based lower bound in Fig. 15.10 through the empirical correlation $q_{c1}/N_1 = 0.4$ and 0.5. It may be seen in Fig. 15.11 that even with the relation of $q_{c1} = 0.4N_1$ will be taken as the lower bound enclosing all the data points pertaining to silty sands which were case-studied in terms of q_{c1} value. It should be remembered that the boundary lines suggested in Figs 15.10 and 15.11 are not based

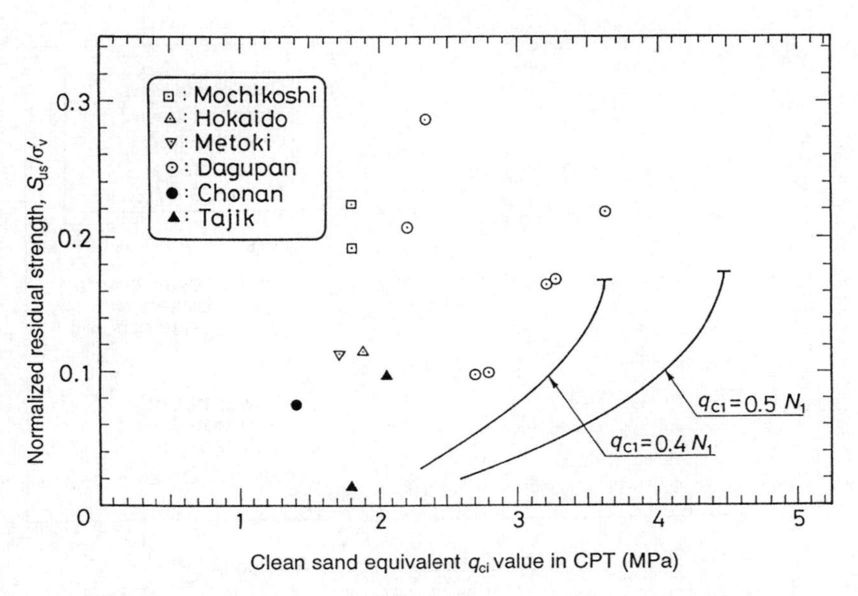

Fig. 15.11 Relation of normalized residual strength and CPT q_{c1} value based on laboratory-determined QSSL compared with data from back-analyses of actual failure cases.

on widely confirmed data and as such they may need to be modified pending collection of more data both from laboratory and the field.

References

Ishihara, K., Yasuda, S., and Yoshida, Y. (1990). Liquefaction-induced flow failure of embankments and residual strength of silty sands. *Soils and Foundations*, **30**, 69–80.

Mclead, H., Chambers, R.W., and Davis, M.P. (1991). *Seismic design of hydraulic fill tailings structures*. Proceedings of the 9th Panamerican Conference on Soil Mechanics and Foundations Engineering, Vina de Mar, Chile, pp. 1063–81.

Ohya, S., Iwasaki, T., and Wakamatsu, M. (1985). *Comparative study of various penetration tests on ground that underwent liquefaction during the 1983 Nihonkai–Chubu and 1964 Niigata earthquake*. Proceedings of the US–Japan Joint Workshop on In situ Testing Methods for Evaluation of Soil Liquefaction Susceptibility, San Francisco, pp. 56–88.

Robertson, P.K. and Campanella, R.G. (1985). Liquefaction potential of sands using the CPT. *Journal of ASCE*, **111**, GT3, pp. 384–403.

Robertson, P.K. Woeller, D.J., and Finn, W.D.L. (1992). Seismic cone penetration test for evaluating liquefaction potential under cyclic loading. *Canadian Geotechnical Journal*, **29**, 686–95.

Seed, H.B. (1987). Design problems in soil liquefaction. *Journal of ASCE*, 113, GT8, 827–45.

Sladen, J.A. and Hewitt, K.J. (1989). Influence of placement method on the in-situ density of hydraulic sand fills. *Canadian Geotechnical Journal*, **26**, 453–66.

Stark, T.D. and Mesri G. (1992). Undrained shear strength of liquefied sands for stability analysis. *Journal of ASCE*, 118, GT.11, 1727–47.

16

ONSET CONDITION FOR LIQUEFACTION AND CONSEQUENT FLOW

16.1 Interpretation of laboratory tests to assess in situ strength

In the preceding chapters, laboratory testing on reconstituted samples has been recognized as a useful tool for making the most conservative estimate of the residual strength of in situ sands. A similar argument may be developed as to the cyclic strength controlling the triggering of liquefaction. The relationship between the cyclic strength and the N_1 value summarized in Fig. 12.2 is reproduced in Fig. 16.1 in the form of a belt zone where most of the proposed correlations are clustered. Also shown superimposed in Fig. 16.1 are the points derived from the results of the coalescent tests conducted in the United States and Japan. In doing so, the cyclic

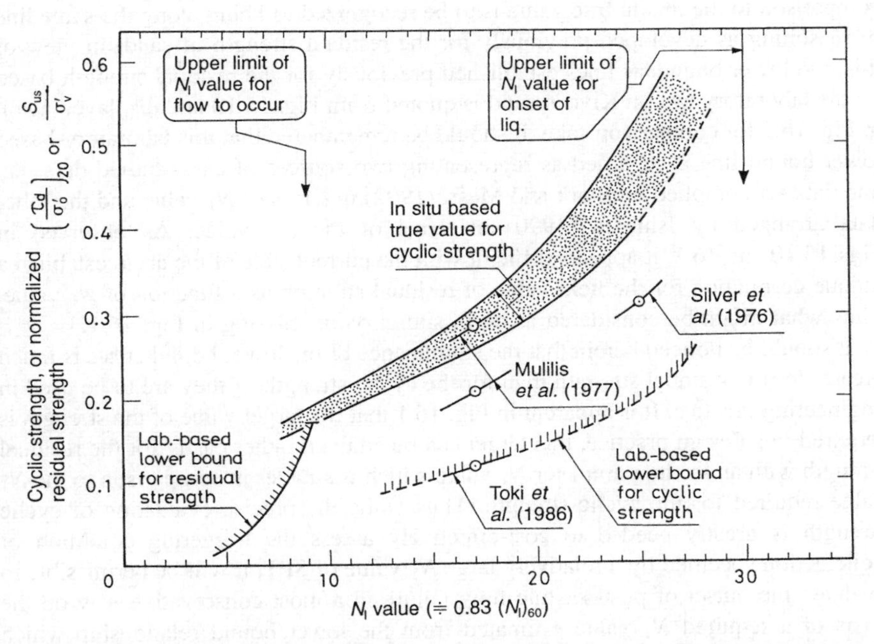

Fig. 16.1 Interpretation of laboratory test-based correlations of cyclic strength and residual strength compared with correlations based on field performance.

stress ratio required to cause 5% D.A. strain in 20 cycles of load application was read off from the test data shown in Figs. 10.12 and 10.13 and plotted in the ordinate of the diagram in Fig. 16.1. The corresponding N_1 value was computed by an empirical formula $N_1 = 70\,(D_r/100)^2$ based on the relative density used in the tests. Since the materials used were apparently clean sands, the value of $a + b = 70$ was chosen from Fig. 12.4 to evaluate the N_1 value by way of eqn (12.11). Note that this choice leads to a relation $D_r = 12\sqrt{N_1}$ which is different from that in eqn (12.8). The test data by Mulilis *et al.* (1977) shown in Fig. 10.11 was also processed in the same manner and displayed in Fig. 16.1. The observation of the test data compiled in Fig. 16.1 indicates that the cyclic strength obtained in the laboratory tests tends to vary significantly because of the difference in the fabric of sand samples reconstituted by different methods of preparation. However, if note is taken of the test data by Toki *et al.* (1986) giving the smallest cyclic strength, the line passing through this point may be taken as the lowest boundary below which the cyclic strength could never occur for any kind of sand with any fabric. The cyclic strength obtained through testing of high-quality undisturbed samples or based on the field performances of in situ sand deposits takes a value which is far larger than this bounding value, as is apparently seen in Fig. 16.1. Thus it may be mentioned conclusively that any effort to determine the cyclic strength for reconstituted samples in the laboratory is to be valued at least as being important in estimating the lowest value of cyclic strength that could possibly occur for a given sand in the field.

The interpretation as above for the laboratory-determined cyclic strength in comparison to the in situ true value is to be recognized as being along the same line of reasoning as developed previously for the residual strength of sand. In view of this, the lower boundary lines established previously for the residual strength based on the laboratory test on Kiyosu sand is quoted from Fig. 15.10 and displayed again in Fig. 16.1 for comparison sake. It should be remembered that this laboratory-based lower bound line is regarded as representing two sources of case-studied data, i.e. one data set complied by Stark and Mesri (1992) in terms of N_1 value and the other data arranged by Ishihara (1990) in terms of the q_{c1} value. As observed in Figs 15.10 and 16.1, it appears difficult with the current state of the art to establish a unique correlation for the field value of residual strength as a function of N_1 value. Thus what might be considered as an in situ curve is missing in Fig. 16.1.

It should be noticed herein that the significance of the lower bound curve is much greater for the residual strength than for the cyclic strength, if they are to be used in engineering practice. It is apparent in Fig. 16.1 that if a target value of the strength is required in a design practice, this target can be attained rather easily for the residual strength with an in situ parameter N_1 value which is smaller in comparison to the N_1 value required for the cyclic strength. Thus while the precise evaluation of cyclic strength is greatly needed to cost-effectively assess the triggering condition of liquefaction specified by a relatively large N_1 value of SPT, it will be permissible to evaluate the outset of post-seismic flow failure in a most conservative way on the basis of a required N_1 value estimated from the lower bound relationship which generally takes a value still small enough to be implemented in the design practice.

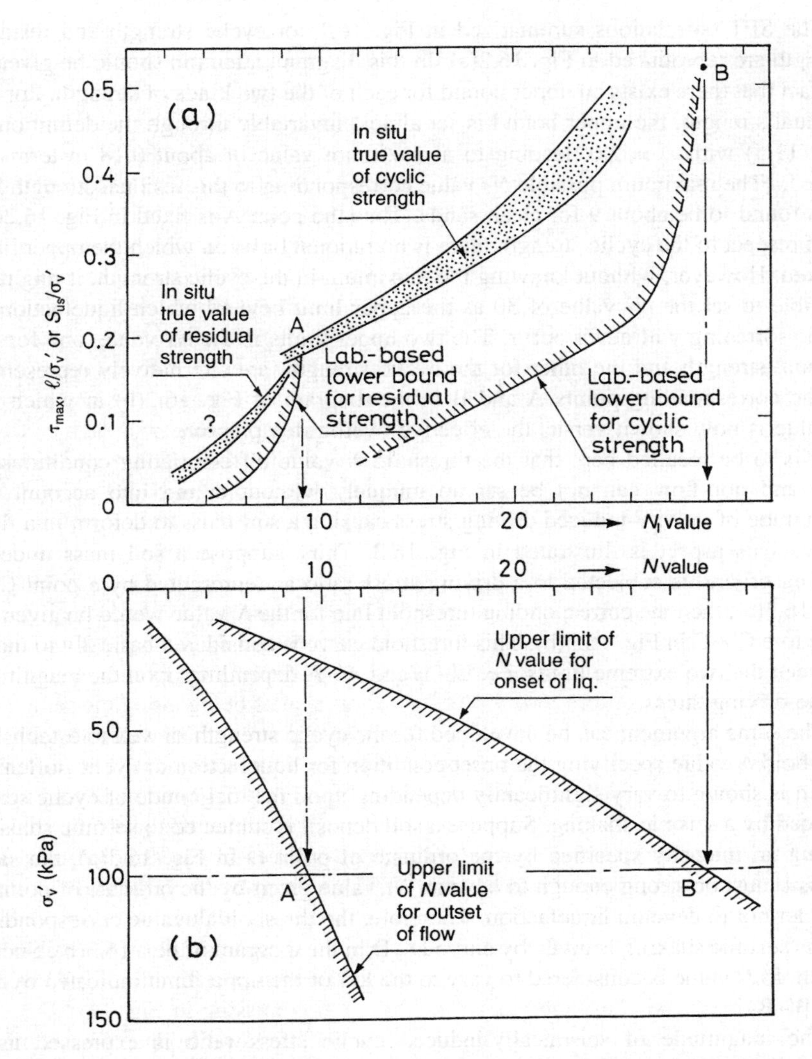

Fig. 16.2 Upper limit boundaries for separating flow and non-flow, and liquefaction and non-liquefaction.

16.2 Onset conditions for liquefaction and consequent flow

In the preceding section, the condition to cause or not to cause flow failure has been established in terms of the penetration resistance of SPT or CPT. At this stage it would be of interest to re-examine the above condition from a broader perspective in the light of another kind of well-known criterion which is related with the onset of liquefaction.

The SPT correlations summarized in Fig. 16.1 for cyclic strength and residual strength are reproduced in Fig. 16.2(a). In this diagram, attention should be given to the fact that there exists an upper bound for each of the two kinds of strength. For the residual strength, the upper bound is set almost invariably through the definition of eqn (11.5) with $r_c = 2.0$, leading to a maximum value of about 0.18 in terms of S_{us}/σ'_v. The maximum possible N_1 value corresponding to this residual strength has been found to be about 9 for clean sands. Thus the point A is fixed in Fig. 16.2(a). With respect to the cyclic strength there is no rational basis on which the upper limit is fixed. However, without knowing the maximum in the cyclic strength, it might be possible to set the N_1 value of 30 as the upper limit beyond which liquefaction or cyclic softening will never occur. The two upper limits in the N_1 value, one for the residual strength and the other for the cyclic strength, are alternatively represented by the corresponding points A and B in the diagram of Fig. 16.2(b) in which the N value is now shown versus the effective overburden pressure σ'_v.

It is to be recalled here that the threshold N value differentiating conditions of flow and non-flow can not be set up uniquely without taking into account the magnitude of gravity-induced driving stress causing a soil mass to deform in a flow mode. This aspect is illustrated in Fig. 16.3. Thus, suppose a soil mass under a sloping ground is subjected to a driving stress ratio as represented by a point C in Fig. 16.3(a), then the corresponding threshold line for the N value would be given by the curve C′–C in Fig. 16.3(b). This threshold curve is considered basically to move between the two extreme limits, i.e. E′–E and A′–A depending upon the magnitude of the driving stress.

The same argument can be developed for the cyclic strength as well to establish a threshold N value specifying the onset condition for liquefaction or cyclic softening which is shown to vary significantly depending upon the magnitude of cyclic stress induced by a seismic shaking. Suppose a soil deposit is subjected to seismic shaking having an intensity specified by the ordinate of point D in Fig. 16.3(a), the sand deposit must be strong enough to have an N_1 value given by the ordinate of point D in order not to develop liquefaction. Therefore, the threshold N value corresponding to this seismic shaking is given by curve D′–D in the diagram of Fig. 16.3(b). Such a threshold N value is considered to vary to the left of the upper limit indicated by the line B′–B.

The magnitude of seismically-induced cyclic stress ratio is expressed as a function of peak acceleration as given by eqn (13.2). Therefore, the N_1 value great enough to mobilize the cyclic strength to be balanced against this externally induced cyclic stress ratio can be determined directly from a chart such as that shown in Fig. 16.3(a), or using the empirical correlation of eqn (12.15). The N value obtained in this way using eqn (12.15) for a peak acceleration of $a_{max} = 0.2g$ (20% of gravity acceleration) is demonstrated in Fig. 16.4 as a plot versus the overburden pressure σ'_v. Superimposed on this diagram is the zoned boundary curve quoted from Fig. 15.7 which is identical to the upper limit boundary line A′–A shown in Fig. 16.3(b). Generally speaking, this flow-related upper limit may be used for practical purposes without considering the driving stress for the purpose of identifing whether or not a given sand deposit has a potential to be exposed to the danger of

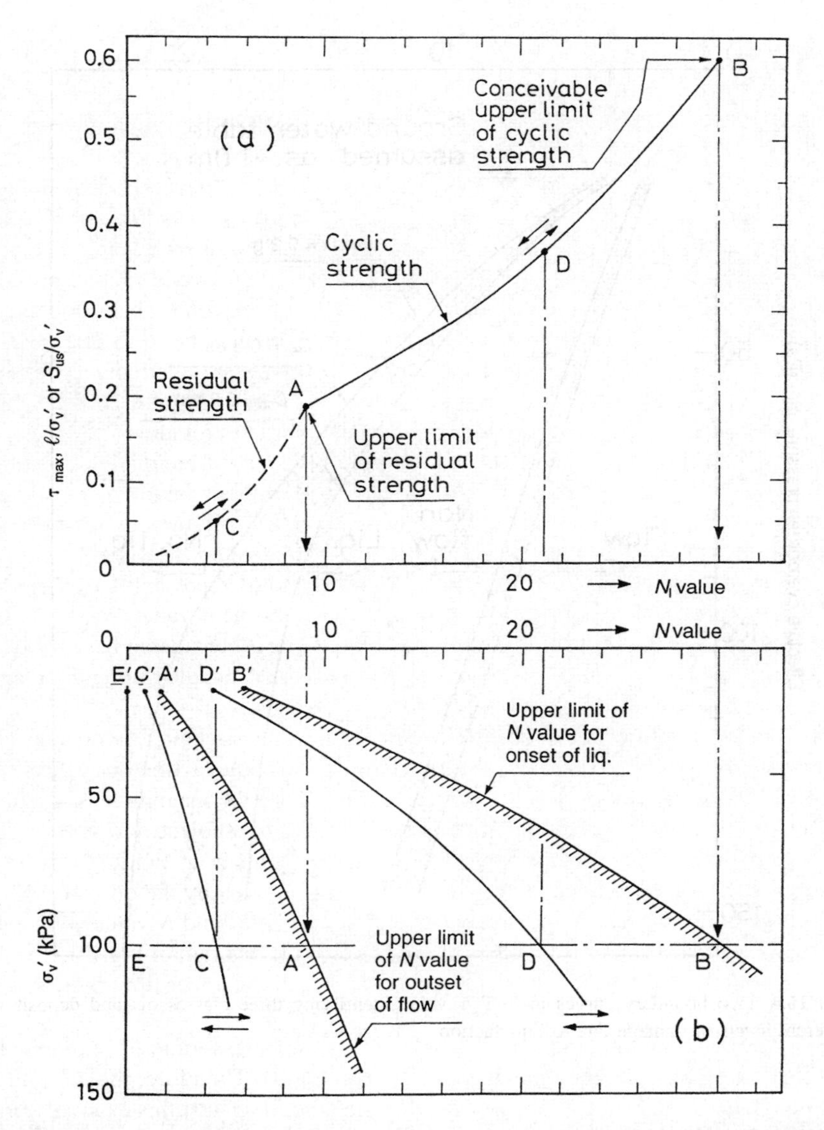

Fig. 16.3 Boundary curves for gravity-induced driving force and intensity of seismic shaking.

flow slide. The boundary N value associated with the triggering of liquefaction needs to be set up in compliance with the magnitude of acceleration specified for design or prediction purpose. It can be seen in Fig. 16.4 that the threshold N value differentiating between onset and non-onset of liquefaction is generally large as

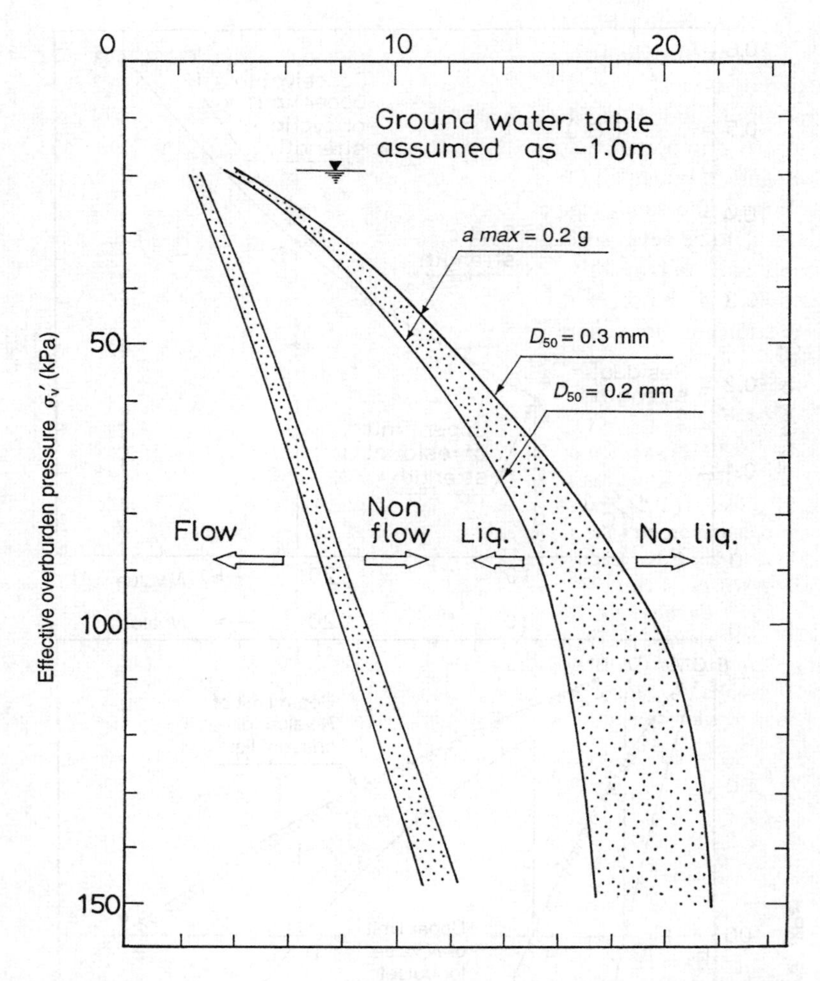

Fig. 16.4 Two boundary curves in SPT *N* value identifying three classes of sand deposit with different levels of damage due to liquefaction.

compared to the bounding values separating conditions to cause or not to cause flow-type failure consequent upon the onset of liquefaction.

The implication of the two boundary curves in Fig. 16.4 may be stated as follows. If a sand deposit is loose with an *N* value below the flow-related threshold, the deposit is identified as not only susceptible to the triggering of liquefaction but also in potential danger of flow slide. If the deposit is moderately loose with an intermediate *N* value between the two threshold lines, the possibility will be high for the deposit to be subjected to the onset of liquefaction, but free from the potential threat of flow slide. Interpreted more semantically, the condition of a sand deposit may be classified into three categories in accordance with different levels of their vulnerability to the liquefaction-induced ground damage, that is catastrophic destruction by flow slide, moderate damage by liquefaction, and little to no damage. The concept of subdividing the damage extent as above will offer a useful rationale of engineering importance in setting up a design policy for implementing soil improvements in potentially liquefiable sand deposits. For an important structure to rest thereupon, the sand deposit must be improved so as to have a density great enough for it to be never susceptible to liquefaction. For the sand deposit underlying less important and readily repairable structures, the target to achieve for improvement may be set with good reasons to the lower threshold N value based on the flow or non-flow condition.

References

Ishirara, K. (1990). *Evaluation of liquefaction potential and consequent deformations in sand fills.* Proceedings of the Seismic Workshop on the port of Los Angeles.

Mulilis, J.P., Seed, H.B., Chan, C.K., Mitchell, J.K., and Arulanandan, K. (1977). Effects of sample preparation on sand liquefaction. *Journal of ASCE*, **103**, GT2 91–108.

Silver, M.L., Chan, C.K., Ladd, R.S., Lee, K.L. Tiedemann, D.A., Townsend, F.C., Valera, J.E., and Wilson, J.H. (1976). Cyclic triaxial strength of standard test sand. *Journal of Geotechnical Engineering, ASCE*, **102**, GT5 511–23.

Stark, T.D. and Mesri, G. (1992). Undrained shear strength of liquefied sands for stability analysis. *Journal of Geotechnical Engineering, ASCE*, **118**, GT11 1727–47.

Toki, S., Tatsuoka, F., Miura, S., Yoshimi, Y., Yasuda, S., and Makihara, Y. (1986). Cyclic undrained triaxial strength of sand by a cooperative test program. *Soils and Foundations*, 26, 117–28.

Appendix

METHODS OF SAMPLE PREPARATION

There are three kinds of procedures widely used for preparing samples of sand for laboratory testing. The basic requirements for any of the methods are firstly to obtain homogeneous samples with uniform distribution of void ratio and secondly to be able to prepare samples having the lowest possible density. The second requirement is needed to cover a wide range of density in the sample reconstituted by an identical method. It has been known that different methods of sample reconstitution create different fabrics, thereby yielding different responses to load application.

A.1 Moist placement method (wet tamping)

Five to six equal preweighed, oven dried portions of sand are mixed with de-aired water at a water content of about 5%. A membrane is stretched taut to the inside face of a split mould which is attached to the base pedestal of the test apparatus. Each portion of the slightly moist sand is strewed with fingers to a predetermined height in five to six lifts, as illustrated in Fig. A1. At each lift stage, tamping is gently applied with a small flat-bottomed tamper. Because of capillary effects between particles, the moist sand can be placed at a very loose structure well in excess of the maximum

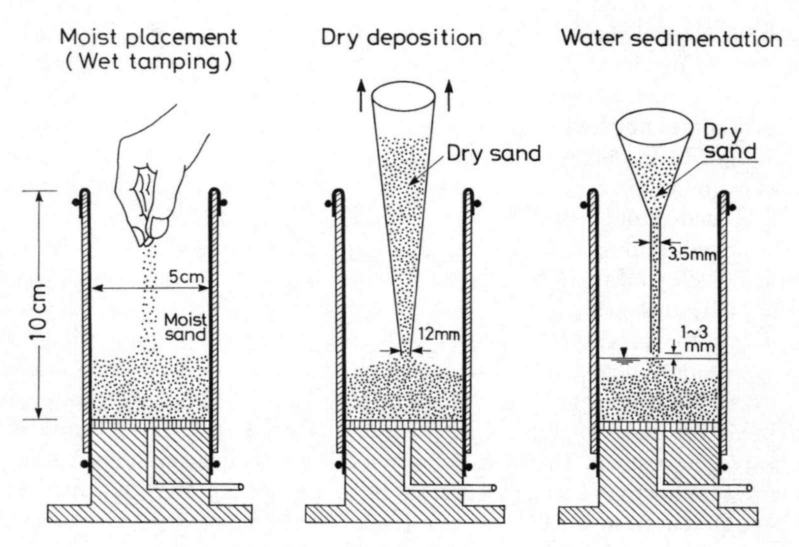

Fig. A1 Methods of sample preparation.

void ratio of dry sand which is determined by the procedures stipulated in the ASTM or JSSMFE standard test method. An optimum amount of energy for preparing the loose initial structure of the sample is a matter to be decided upon for the individual sand used for the test. Generally speaking, if the tamping energy is small, the sample is placed so loose that the volume contraction upon subsequent saturation becomes unduly large and consequent reduction in sample diameter tends to create vertical wrinkles in the membrane. As a rule of thumb, it is recommended that the tamping energy be adjusted so that the volume contraction upon saturation would be about 5% for preparing the loosest specimen by this method. If a denser sample is to be prepared, a larger amount of energy needs to be applied, for example by increasing the amount of tamping during compaction at each stage of the lift. After the sample is enclosed by the membrane with the top cap, a vacuum of 2 to 5 kPa is applied and the mould is dismantled. Carbon dioxide gas is percolated through the sample and de-aired water flushed. During this saturation process, volume reduction takes place due to the collapse of the initial structure of the sample. The void ratio measured after saturation under vacuum pressure is taken as the initial void ratio of the sample. Consolidation is then performed to the desired confining stress. By this procedure, a state of sand with any combination of void ratio and confining stress could be produced by employing varying amounts of compacting energy during tamping, as illustrated in Fig. 25. One of the advantages of this method is the versatility in permitting any sample to be prepared within a wide range of void ratios. Therefore the sample could be very loose and highly contractive or dilative in subsequent loading, depending upon the void ratio at the time of sample preparation.

A.2 Dry deposition method

Oven-dried sand is filled in a cone shaped slender funnel having a nozzle about 12 mm in diameter as illustrated in Fig. A1. This funnel is the same as that used for determining the maximum void ratio of sand based on the JSSMFE method. The sand is spread in the forming mould with zero height of fall at a constant speed until the mould becomes filled with the dry sand. Tapping energy is applied by hitting the side of the mould to obtain a desired density. After the sample is encased in the membrane with the top cap, a vacuum of 20–50 kPa is applied and carbon dioxide gas is percolated through the sample. De-aired water is then flushed making the sample saturated. Full saturation is checked by ensuring attainment of a B value greater than 0.98 by means of back pressure application. Any state of a sample could be produced by this method as illustrated in Fig. 11.6 by adjusting the tapping energy during the process of sample preparation. The sample prepared by this method is generally denser as compared to that prepared by the moist placement method. Even if it is prepared in its loosest state without imparting any tapping, the sample could be only slightly contractive in its behaviour in the subsequent application of shear stress. Therefore this method is not adequate to prepare samples with highly contractive characteristics. It is to be noticed that this method is somewhat different from what is generally called air pluviation in which dry sand is discharged vertically in air from a small nozzle into the mould (Mulilis *et al.* 1977;

Tatsuoka *et al.* 1986). The air pluviation method is known to produce samples which are always dilative and has not been used in the present study.

A.3 Water sedimentation method

Sand is mixed with de-aired water and put in a funnel with a plastic tube attached to the end. The mixture of sand and water is poured, through the plastic tube, in four layers, in the sample forming mould at zero height of fall at a constant speed, so that the surface of water is always held coincident with that of the sand sediment. This procedure is illustrated in Fig. A1. In another method, dry sand is poured through a 1–2 mm diameter nozzle from just above water surface and left to sediment through a height of 2 to 3 cm under water. In either of the above methods, the sand is deposited continuously under water without causing appreciable segregation of the material. If a denser sample is to be prepared, compacting energy is applied by hitting the side of the mould stepwise during the process of sample placement.

In a special method developed by Dobry (1991), a predetermined amount of sand is dumped into water and some length of time, from 30 minutes to 24 hours, is allowed to elapse until the material has completely settled. In this episode of deposition, coarse-grained material sediments at the bottom, grading to fines upwards. On top of this layer, an equal amount of the same soil is poured again and let to sediment under water in the same fashion as before. This stepwide deposition is repeated 4 to 5 times to construct a complete test sample. This apparently discontinuous sample is sometimes considered to best represent the behaviour of in situ deposits of silty sands with stratified structure. If a denser specimen is to be prepared, the side of the sample forming mould is hit with a hammer as many times as desired.

In any of the procedures mentioned above, after the placement in the mould is over, the top cap is mounted and carbon dioxide gas is circulated through the sample. While vacuum is applied, the split mould is dismantled. Any state of a sample having a desired void ratio and confining stress could be created by the method of water pluviation as illustrated in Fig. 11.6. The continuous sample prepared by the first method tends to possess a lower void ratio and it is practically impossible to produce samples with contractive characteristics by this method.

References

Dobry, R (1991). *Soil properties and earthquake ground response*. Proceedings of the 10th European Conference on Soil Mechanics and Foundation Engineering, Florence, Italy, Vol. 4.

Mulilis J.P., Seed, H.B., Chan, C.K., Mitchell, J.K., and Arulanandan, K. (1977). Effects of sample preparation on sand liquefaction. *Journal of ASCE*, **103**, GT2, 91–108.

Tatsuoka, F., Ochi, K., Fujii, S., and Okamoto, M. (1986). Cyclic undrained triaxial and torsional shear strength of sands for different sample preparation methods. *Soils and Foundations*, **26**, 23–41.

INDEX

Figures and Tables are indicated by *italic page numbers.*

ACOT test, *see* anisotropically consolidated oedometer-conditioned torsion test
ACT test, *see* anisotropoically consolidated torsion test
air pluviation [sand sample preparation] method 221, 224
  compared with air deposition method 339–40
  effect on cyclic stress ratio *223, 225*
  see also dry deposition method
alluvial clays, shear moduli *112*
alluvial gravels
  grain size distribution curves *101*
  strain-dependent dynamic properties *139*
alluvial sands
  quasi-steady state studied 262–8
  shear moduli *113, 117*
  shear wave velocity relations *120*
anisotropically consolidated oedometer-conditioned torsion (ACOT) test 208, 209–12
  effective stress paths
    in torsional mode 212, *215*
    in triaxial mode 212, *216*
  lateral-stress changes *210*, 212
  pore water pressure changes *210*, 212
  stress–strain relation 212, *214*
anisotropically consolidated torsion (ACT) test 208, 213–17
  effective stress paths
    in torsional mode *215*
    in triaxial mode *216*
  lateral-strain changes *211*, 216–17
  pore water pressure changes *211*, 217
  stress–strain curve *214*

backbone curves 32, 150, 173
ballast
  grain size distribution curves *101*
  shear moduli
    empirical formulae *100*
    as function of void ratio *102*
Becker Drill Test 293, *294*
bedding error 57
bender elements 54–5
  details *57*
  disadvantages 55
  use in triaxial test apparatus 54, *56*
bentonite clays
  shear moduli
    empirical formulae *89*, 93
    overconsolidation effects *94*
blasting [explosives], loading during *2*, 3
body [earthquake] waves 6
  stresses induced *7*
  see also compressional...; longitudinal...; shear wave
Borah Peak earthquake [USA], liquefaction phenomenon 228, 298–9
Boston blue clays *91*, 93

Canadian Beaufort Sea islands, flow-slide studies 323, *326*
Chiba sand
  characteristic parameters listed *321*
  SPT N curves *323, 328*
Chiba silty sand
  characteristic parameters listed *321*
  SPT N curve *324*
Chinese Code, correlation between liquefaction resistance and SPT N value *291*, 292
clays
  shear moduli
    effects of overconsolidation 93–8
    empirical formulae *89*
    as functions of void ratio and confining stress *91*
  strain-dependent dynamic properties *134, 137*
coarse-grained soils
  shear moduli 98–107
  see also gravelly materials
cohesionless soils
  low-amplitude shear moduli 87–92
    as function of void ratio and confining stress *90*
  see also sands
cohesive soils
  cyclic loading 162–5
  dynamic strength 180–5
    multi-stage cyclic loading test 182–4
    single-stage cyclic loading test 181–2
  low-amplitude shear moduli 92–8
    effect of overconsolidation 93–8
    as function of void ratio and confining stress *91, 92, 94, 95*
  stiffness degradation 152, 165–71
  strain-dependent dynamic properties 133–8
  strength affected by time of loading *187, 188*
  transient loading 154–62
  see also clays; loams; silts

combined static and cyclic loading, effect on strength of soils 187–92
complex modulus 19
complex variables method 18–19
compressional [earthquake] wave 6
  propagation velocity
    compared with that of shear wave 65
    as function of Poisson's ratio *78*
  reflection survey using 60
  stresses induced by propagation 6, *7*
  see also longitudinal wave; P-wave
compression maximum (CM) test 194
  examples of use *197*
  stress–residual-strain relationships in volcanic clays 197, *199*, *200*
Cone Penetration Test (CPT)
  flow conditions studied using 323–5, *326*, *327*
  liquefaction resistance of sands evaluated using 286–7
    effect of fines *292*
  q_c values
    normalized residual strength correlated with *329*
    relative density correlated with 288
confining stress
  low-amplitude shear modulus affected by
    cohesive soils *91*, *92*, *94*, *95*
    sands *90*
  reference strain affected by *131*
  strain-dependent dynamic properties affected by
    cohesive soils *135*
    sands *129*
  stress–residual-strain relationship affected by, irregular loading of volcanic clays 201, *202*, 203, *203*
consolidation [of clays]
  low-amplitude shear moduli affected by 93–8
  strain-dependent dynamic properties affected by 136–8
contractive sand 247, *248*
correction factor, strain-dependent shear modulus 146, *147*
Coulomb failure criterion 91; *see also* Mohr–Coulomb...
CPT, *see* Cone Penetration Test
critical angle of incidence 60, *61*, 63
crosshole [velocity logging] method 66–8
  compared with SASW method *84*
  schematic of procedure *67*
  spacing between boreholes 68
crushed rock
  low-ampitude shear modulus vs void ratio *102*
  strain-dependent dynamic properties *139*
cyclic loading
  deformation characteristics of soils 162–5
  degraded hysteresis behaviour 16
  energy considerations 20–2
  laboratory apparatus
    with precise strain measurements 56–8
    resonant column tests 46–54
    simple shear tests 43–4
    torsional tests 44–5
    triaxial tests 40–3, 181
  multi-stage test 182–4
  non-degraded hysteresis behaviour 16
  resistance of sands 208–44
  single-stage test 181–2
  stress–strain relations 16–39, 153, *154*
cyclic mobility characteristics 236
cyclic-(to)-monotonic loading 180, *181*
  behaviour of undrained sand *249*, *250*
cyclic resistance
  gravelly soils 228–31, *232*
  sands
    in situ deposits 225–7
    reconstituted clean sands 221–5
    triaxial tests used 208–18
  silty sands 227–8
cyclic softening, definition 212, 218–21
cyclic stiffness degradation 152
  evaluation 165–71
  threshold strains 172, *175*
cyclic strength
  correlation with SPT N value *283*
    effect of fines content *283*, 290–1
  estimation from penetration tests *283*, 289–90, 292–3
    effect of fines 290–3
  gravelly soils *230*, *232*, *296*, *297*, *298*, 299
  as indicator of liquefaction potential 220–1
  lower bound *331*, 332, *333*
  sands
    co-operative test program results 222–4, *224*, *225*, 331
    density effects *222*
    effect of K_o conditions *233*, *234*
    effects of confining stress *242*
    effects of initial shear stress *243*
    sample-preparation effects *223*
    undisturbed samples *226*
  upper limit *335*
cyclic strength ratio 187
  plots vs initial shear stress ratio *190*, *191*, *192*
cyclic stresses
  earthquake-induced 6–10
  traffic-induced 10–13
  wave-induced 13–15
cyclic stress ratio, *see* cyclic strength
cyclic torsional tests
  effective stress paths
    in torsional mode *215*
    in triaxial mode *216*
  with lateral confinement 209–12
  stress–strain curves *214*
  without lateral confinement 213–17
    on isotropically consolidated samples 217–18

Dagupan silty sand
 characteristic parameters listed *322*
 SPT *N* curve *324*
damage extent, post-liquefaction, relation to settlement size *314*
damping ratio 32
 confining-stress effects *129*, *136*
 for hyperbolic model 35–6
 for Ramberg–Osgood model 37–8
 relation to shear modulus ratio 36, 38, *132*
 in resonant column test 50–1, 52–3
 strain-dependent
  cohesive soils *134*, *135*, *137*
  gravelly materials *139*, *141*, *145*
  sands *128*, *129*, *143*
dashpot constant 24, 161; *see also* spring-and-dashpot models
degradation index 165–6
 definitions 166
 plot vs number of cycles *167*
degradation parameter 166
 determination 167
 effects of overconsolidation *169*
 effects of plasticity index *171*
 as function of strain amplitude *168*
dilatancy 247
dilative sand 247, *248*
diluvial clays, shear moduli *115*
diluvial sands, shear moduli *117*
direct wave 60–1, 63
 propagation velocity 61–2
 travel time curve *62*
dispersion curve, R-wave characteristics 77–8, *83*
downwhole [velocity logging] method 65–6
 example of data obtained *67*
 schematic of method *66*
 shear moduli determined by 113–19
dry deposition [sand sample preparation] method 251, *338*, 339–40
 compared with air pluviation 339–40
 consolidation characteristics resulting *252*, *264*, *267*, *317*
 effect on residual strength *271*, *272*, *273*, 276
 field conditions simulated 277–8
 quasi-steady state studied using 263, 264, *265*, *267*
 see also air pluviation
dynamic loading
 compared with static loading 2–3
 types of load patterns 180, *181*
dynamic problems
 classification 2
 loading-speed effects 2–3
 load-repetition effects 3
dynamic strength
 multi-stage cyclic loading test 182–4
 single-stage cyclic loading test 181–2

earthquake-induced stresses 6–10
elastic modulus 19
energy considerations
 linear viscoelastic model 20–2
 nonlinear model 30–2
exponential [stress–strain] model 37
extension maximum (EM) test 194
 examples of use *198*
 stress–residual strain relationships in volcanic clays 197, *199*, *200*

factor of safety [in liquefaction analysis] 302
 maximum shear strain plotted against *310*
 Niigata earthquake *306–7*, *313*, 314
 post-liquefaction volumetric strain plotted against *311*
filter tube [in suspension method] 69, *70*
fines-containing sand
 cyclic/liquefaction resistance/strength 227–8, *283*, 290–3
 quasi-steady state 262–8
 residual strength 268–71, *272–3*
 see also sandy silts; silty sands
flow conditions, and penetration tests 316–25, 333, 334–7
flow-type behaviour [of sand] 247, *248*
 with limited deformation 248, 257
flow-type failure 316
freezing-and-coring [sampling] technique
 compared with tube sampling 107, *108*, *116*, *118*, *147*, 226
 gravelly soils 106, 144, 229
 sands 142, 226
Fuji river sand, load irregularity factor as function of relative density *239*
Fujisawa sand, strain-dependent dynamic properties *143*
Fukui earthquake [Japan], liquefaction phenomenon 228

gap sensor 56
 limitations 57
 use in triaxial test *41*, 56
grain size distribution curves
 gravelly materials *101*, *105*, *231*
 sandy soils *101*, *251*
gravelly materials
 cyclic/liquefaction resistance/strength 228–31, *232*, *296*, *297*, *298*, 299
 grain size distribution curves *101*, *105*, *231*
 Poisson's ratio as function of shear modulus *123*
 sampling in undisturbed state 106, 144, 229
 shear moduli
  empirical formulae *100*
  as function of void ratio *102*, *103*
  intact samples 104–8
  reconstituted samples 98–104

gravelly materials *cont.*
strain-dependent dynamic properties 138–42
void ratio ranges *105*, *106*
gravity-induced driving force 334
boundary curves *335*

head wave 64
Hertz contact theory 91
hollow-cylindrical torsional test apparatus 44, *45*
hyperbolic [stress–strain] model 33–7
hysteresis loop 29
hysteretic stress–strain curve
linear model 19–22
nonlinear model 28–9
construction using skeleton curve 30, *31*, 32
decomposition into elastic and energy-dissipating components *29*

ICOT test, *see* isotropically consolidated oedometer-conditioned torsion test
impulse loads 3
inertia force 1
initial consolidation line (ICL) 259, 261
algebraic expressions 319–20
in terms of relative density 319
in terms of SPT *N* value 320
Lagunillas sandy silt *267*, *271*
location relative to IDL and QSSL *260*, *263*, *264*, *267*, *270*, *271*, *319*
Tia Juana silty sand *264*, *270*
Toyoura sand *260*, *263*
see also isotropic consolidation line
initial dividing line (IDL) 259, 260–1
Lagunillas sandy silt *267*
location relative to ICL and QSSL *260*, *263*, *319*
Tia Juana silty sand *264*
Toyoura sand *260*, *263*, *276*
initial state ratio 269
determination 269–70
implications of concept 269, 276, *277*, 278, 319
Lagunillas sandy silt *271*
Tia Juana silty sand *270*, *271*
in situ deposits
evaluation of strain-dependent soil properties 142–8
sampling techniques 106, 142, 229
sands
cyclic resistance 225–7
sampling techniques 142, 226
in situ tests
low-amplitude shear modulus determination 112–19
vibration test *4*, 5
wave propagation methods 60–84

irregular loading conditions
cyclic resistance of sand 233–41
strength of soils 192–9
isotropically consolidated oedometer-conditioned torsion (ICOT) test 217–18
effective stress paths, in torsional mode *215*
pore water pressure changes *213*, 218
stress–strain curve *214*
isotropic consolidation curves
Lagunillas sandy silt *267*, *271*
sample-preparation effects *252*, 275
Tia Juana silty sand *264*, *270*
Toyoura sand *252*, *259*, *260*, *261*, *262*, *263*, *276*, *279*
isotropic consolidation line (ICL) *252*, 259, *260*; *see also* initial consolidation line

Japanese Code of Bridge Design, evaluation of cyclic strength of sands *284*, 289–90
Japanese Society of Soil Mechanics and Foundation Engineering (JSSMFE)
cooperative [cyclic triaxial] test program 223–4, *225*
void ratio determination method 251, 263, 264

kaolinite clays
pore water pressure vs shear strain amplitude *174*
shear moduli
empirical formulae *89*
as functions of void ratio and confining stress *91*
overconsolidation effects *94*
time dependency 107, 109
stress–strain curves compared for slow/static cf. rapid loading 157
Kelvin [viscoelastic] model *23*, 24–5
loading rate effects explained using 160–2
non-viscous type 2–8
Kiyosu sand
characteristic parameters listed *322*
SPT *N* curves *323*, *328*

laboratory tests *4*
apparatus and procedures 40–58
liquefaction mechanism studied using 208–18
low-amplitude shear modulus determinations 87–107
cohesive soil *91*, 92–8
gravelly materials 98–107
sands 87–92
simulation of field conditions 208, *209*, 277–8
strength estimates based on *328*, *329*, 331–2
see also resonant column...; shear...; triaxial tests; wave propagation method

Lagunillas sandy silt 264
consolidation characteristics 267
grain size distribution curve *251*
initial state ratio *271*
normalized residual strength *271*, *273*
quasi-steady state 264, *267*
Large [diameter] Penetration Test (LPT)
comparison with standard test *294*
cyclic strength evaluation using 295–7
details of procedures and apparatus *294*
linear viscoelastic model 17–28
liquefaction
analysis 301–7
definition 212, 218–21
factor of safety 302, 310–11
onset conditions 333–7
settlement after 308–15
liquefaction mechanism, laboratory torsional tests used in study 208–18
liquefaction resistance
effects of confining stress and initial shear stress 241–4
evaluation by in-situ tests 282–99
sands 221–8
effect of K_o conditions 231, 233
in situ deposits 225–8
reconstituted clean sand 221–5
loading rate
effect on soil behaviour 2–3, *17*, 154–62
rapid/transient loading 152, *153*
slow/static loading 152, *153*
loading schemes, classification 152–3
load irregularity factor/coefficient [in irregular loading], as function of relative density *239*, *240*
load repetition, effect on soil behaviour 3, *17*, 162–5
loams, strength affected by time of loading *187*
local deformation transducer (LDT) 57–8
longitudinal wave 6
propagation velocity
compared with that of shear wave 65
as function of Poisson's ratio *78*, 122–3
see also compressional wave; P-wave
loss coefficient
calculation 22, *23*
definitions 19, 22
Kelvin and Maxwell models compared 26
loss modulus 19
low-amplitude shear moduli
in situ determination 112–19
compared with laboratory determination 113–18
penetration test used 119–20
laboratory determination 87–107
cohesive soils *91*, 92–8
compared with in situ determination 113–18
gravelly materials 98–107
sands 87–92
time dependency 107–12

Masing law/rule 17, 30, *31*, 32
Maxwell [viscoelastic] model *23*, 25–7
loss coefficient as function of frequency *26*
mine tailings, *see* tailings
modulus reduction curves *133*
factors affecting 148
Mohr circle representation 203–4
Mohr–Coulomb failure criteria 130, 203
moist placement/tamping [sand sample preparation] method 221, 222, 251, 338–9
consolidation characteristics resulting 251, *252*, 253, 267, 275, *276*, *277*, *317*
effect on cyclic stress ratio *223*, *224*
effect on residual strength 276
field conditions simulated 277
quasi-steady state studied using 257–61, *264*, 267, *267*
steady state studied using 253–6
monotonically increasing cyclic loading 180, *181*
monotonically increasing rapid loading, effect on strength of soils 186, *187*, *188*
monotonic loading 180, *181*
dynamic moduli determined using 58
sand behaviour 247–79
Monterey sand, cyclic stress ratio *223*, *224*
multi-stage cyclic loading test 182–4

Nihonkai-chubu earthquake, acceleration time histories 236, *237*
Niigata earthquake 302
acceleration time histories 233, *235*, *236*, *305*
liquefaction analysis 302–5
estimated factors of safety *306–7*
SPT *N* data *323*
map of liquefied sites *302*
settlement analysis 312–15
non-flow-type sand 247, *248*
nonlinear stress–strain models 28–9
framework for model 28–33
hyperbolic model 33–7
Ramberg–Osgood model 37–9
nuclear energy facilities design 104, 138, 229

Ottawa sand *89*, *90*
overburden pressure, flow/liquefaction boundary affected by *335*, *336*
overconsolidation of clays
degradation parameter affected by *169*, *171*
shear moduli affected by 93–8
strain-dependent dynamic properties affected by *137*

penetration tests
  comparison of test procedures and apparatus *294*
  flow conditions assessed using 316–25
  gravelly soils 293–5
penetration tests *cont.*
  residual strength correlated to penetration resistance 325–30
  sandy soils
    liquefaction resistance evaluation 282–7, 289–90
    effect of fines 290–3
  see also Cone...; Standard Penetration Test
phase transformation angle [for fines-containing sand] 268
  determination 270
  values listed *271*
phase transformation [contractive-to-dilative behaviour] state 253, 257–8, 260
  quasi-steady state defined as particular case 258–9, 260
pile driving, loading during *2*, 3
plasticity index
  clays 93, *94*, *95*, *96*, 133
  fines-containing sand
    cyclic [liquefaction] resistance affected by *228*, 293
    normalized residual strength affected by *274*, 275
  overconsolidation effects influenced by *97*
  shear modulus increase affected by 109, *111*
  stiffness degradation parameter affected by *171*
  strain-dependent dynamic properties affected by 148, *149*, *150*
  strain-dependent stiffness degradation affected by 173, *176*, *177*
  threshold shear strain affected by 172, *175*
  various cohesive soils *96*
Poisson's ratio
  calculation using suspension [velocity-logging] method *76*
  depthwise distribution *76*
  as function of shear modulus 122, *123*
  sands 87
  water-saturated soils 6, 120–3
pore water pressure
  in liquefaction of sands *210*, *211*, *213*, 218, 219–20
    under irregular loading conditions 234, *235*
    volume change resulting 308
  and threshold strain 172, *173*, *174*
proximity transducer 56
P-wave
  downhole survey data, example *67*
  propagation velocity compared with that of S-wave 65
  see also compressional...; longitudinal wave
quartz, shear modulus, empirical formulae *89*
quasi-steady state line (QSSL) 261
  Lagunillas sandy silt *267*, *271*
  location relative to ICL and IDL *263*, *267*, *270*, *271*, *319*
  Tia Juana silty sand *264*, *270*
  Toyoura sand *261*, *263*
quasi-steady state (QSS)
  clean sands 257–62
  silty sands 262–8

Ramberg–Osgood (R–O) [stress–strain] model 33, 37–9
rate-dependent damping 22
rate effect 3
Rayleigh wave (R-wave)
  amplitude vs depth *78*
  dispersion curve 77, *83*
  monitoring of propagation 73–85
  practical technique using 77
  propagation velocity
    as function of Poisson's ratio *78*
    as function of wavelength *80*, *83*
  see also surface wave
reconstituted gravel, *see* gravelly materials
reconstituted sand, *see* sand
reference strain 33, *34*
  effect of confining stress 130, *131*
reflected wave
  propagation paths *61*
  travel time calculation 62
reflection, differentiated from refraction 60, *61*
reflection survey 60–2
  limitations 62
refracted wave
  propagation paths *61*
  travel path *63*
  travel time calculation *62*, 63–4
refraction *60*
  differentiated from reflection 60, *61*
refraction survey 62–5
relative density
  correlation with
    ICL 319
    penetration resistance 287–9
relaxation time 25
repeated-loading test *4*, 5
repetitive-loading effects 3
repetitive-loading test 153
residual strain, *see* stress–residual strain relationship
residual strength 253
  definition 268
  effect of deformation mode 278–9
  effect of [sample] fabric 275–8
  estimating value 272–5
  fines-containing sand 268–71, *272–3*

normalized 268–9
determination 270
effect of plasticity index *274*, 275
effect of sample-preparation method 270, *271*, *272*, *273*, 276
lower bound *328*, *331*, *333*
upper limit *335*
values listed *271*, *272*, *273*
and penetration resistance *320*, 325–30
plot vs initial confining stress *272*, *273*
void ratio effects 272–3
resonant column test 46–54
bottom-excited type *46*, *52*
shear modulus determination using *89*, 93, 107, 113, 114, 133
top-excited type *46*, *55*
retardation time 25
roll-over mechanism 247
R-waves, *see* Rayleigh waves

safety factor, *see* factor of safety
sample-preparation methods [for sand] 221, 222, 224, 251, 338–40
appropriateness to field conditions 277–8
consolidation curves affected by *252*, *264*, *267*, *276*, *277*
cyclic strength affected by *223*
residual strength affected by *271*, *272*, *273*, 275–8
see also dry deposition; moist placement; water sedimentation
sand
consolidation characteristics *252*, 253, *276*, *279*
cyclic resistance 221–8
density effects *222*
sample-preparation effects *223*
flow/non-flow in undrained samples 247–50
in situ deposits
cyclic resistance 225–7
liquefaction-induced settlement 308–15
liquefaction resistance 221–5
effect of K_o conditions 231, 233
low-amplitude shear moduli 87–92
Poisson's ratio as function of shear modulus *123*
quasi-steady state 257–62
sample preparation methods 221, 222, 224, 251, 338–40; *see also* dry deposition; moist placement; water sedimentation
steady state of 253–7
strain-dependent dynamic properties 127–32
see also fines-containing sands; silty sands
sandy gravels
grain size distribution curves *101*
sampling in undisturbed state 106, 144
void ratio ranges *105*
sandy silts
grain size distribution curves *251*
quasi-steady state 264, *267*
shear-stress–residual strain relationship 184–5, *186*
see also Lagunillas sandy silt; silty sands
saturated soils, Poisson's ratio 6, 120–3
secant modulus 30, 32, 35, 153; *see also* shear modulus ratio
seismic shaking, cyclic changes in shear stress 10
settlement analysis
basic concepts and procedures 308–12
evaluation after Niigata earthquake 312–15
SHANSEP procedure 169
shear [earthquake] waves 6
propagation velocity
compared with that of longitudinal waves 65
as function of Poisson's ratio *78*, 122–3
stresses induced by propagation 7–9
shear modulus 30, 32, 35
cohesive soils *91*, 92–8
gravelly materials
intact samples 104–7, *108*
reconstituted samples 98–104
low-amplitude 87–123
sands 97–2
strain-dependent
cohesive soils 133–8
gravelly materials 138–42
sands 127–32
see also low-amplitude...; secant shear modulus
shear modulus ratio
and damping ratio 36, 38, *132*
for hyperbolic model 35
modulus reduction curves *133*
for Ramberg–Osgood model 37, *38*
strain dependency, sands *127*
shear strains, soil deformation characteristics affected by 3–5
shear tests
simple shear test 43–4
torsional shear test 44–5
shock loads 3
silts, Poisson's ratio as function of shear modulus *123*
silty sands
grain size distribution curves *251*
liquefaction resistance 227–8
correlation with CPT q_c values *286*
quasi-steady state 262–8
see also sandy silts; Tia Juana silty sand
simple shear test 43–4
multi-directional test, sand *239*
with two-directional loading *44*
single-stage cyclic loading test 181–2
skeleton curve 29, 153, *155*
hysteresis curve constructed using 30, *31*, 32
slip-down movement 247
sloping ground, laboratory simulation 213–14
Snell's law 60, 63

spectral analysis of surface waves (SASW) 73–85
advantages 84
compared with crosshole method *84*
disadvantages 84–5
general scheme of testing *80*, 81
inversion process 82
forward modelling procedure used 82, *83*
speed effect 3
spring-and-dashpot [viscoelastic] models 22–8; *see also* Kelvin model; Maxwell model
spring constant 24
Standard Penetration Test (SPT)
comparison with other penetration tests *294*
details of procedures and apparatus *294*
example of data *67*
flow conditions studies using *323*, *324*, *325*
liquefaction resistance evaluated using
effect of fines *291*
field tests 282–6
laboratory studies 287–8
N values
cyclic strength correlated with *283*, *331*, *335*
ICL expressed in terms of 320
normalized residual strength correlated with *328*, *331*, *335*
relative density correlated with 287–8
shear modulus determination using 119–20
velocity logging carried out at same time 66
static loading, compared with dynamic loading 2–3
steady state [of sand], 253–7; *see also* quasi-steady state
steady-state line (SSL) 257, *259*
Lagunillas sandy silt *267*, *271*
Tia Juana silty sand *264*, *270*
Toyoura sand *259*, *262*, *263*
steady-state strength 253
stiffness degradation during cyclic loading 152, 162; *see also* degradation index; ...parameter
strain-dependent dynamic properties
cohesive soils 133–8
effect of confining stress *129*
estimation in field 144, 146, 148
gravelly materials 138–42
sands 127–32, *143*
strain-hardening behaviour, sand under monotonic loading 247, *248*
strain ranges 1, 4–5, 16, *17*, 28
laboratory measurement 54, 56, 57
strain-softening behaviour, sand under monotonic loading 247, *248*
stress–residual strain relationship 182
construction of curve
from multi-stage loading test 182, *184*
from single-stage loading test 182, *183*
effect of confining stress 201, *202*, *203*
effect of initial shear stress 184, *185*, 199–201
effect of number of load cycles 184, *185*
example 184–5, *186*
irregular loading of volcanic clays *199*, *200*, *202*, *203*
stress–strain models
nonlinear models 28–9
framework for model 28–33
hyperbolic model 33–7
Ramberg–Osgood model 37–9
stress–strain relations
for cyclic torsional testing of sands *214*
during cyclic loading 16–39
linear model 17–28
nonlinear model 28–39
for hyperbolic model 34–5
for Kelvin model 24
for Maxwell model 25
for Ramberg–Osgood model 37
surface waves, spectral analysis 73–85
suspension [velocity logging] method 68–72
advantages 72
compared with downhole method 71–2, *73*
example of data obtained *71*, *72*
interpretation of data *73*, *74*, *75*
limitations 72
shear modulus determination 114
sonde used 68, *70*
S-wave
downhole survey data, example *67*
propagation velocity compared with that of P-wave 65

tailings
cyclic [liquefaction] resistance 227, *228*
residual strength 274–5
Teganuma clay
strain-dependent dynamic properties *134*
strain-dependent stiffness degradation *176*
threshold shear strain 148
cyclic stiffness degradation 172
plasticity index affecting *150*
Tia Juana silty sand 262–3
characteristic parameters listed *322*
consolidation characteristics *264*
grain size distribution curve *251*
initial state ratio *270, 271*
normalized residual strength *271*, *272*
quasi-steady state 263, *264–6*
SPT *N* curve *324*
stress paths *265–6*
parameters characterizing *271*
stress–strain relations *265–6*
time of loading 2
in classification of dynamic problems *2*
shear moduli of cohesive soils affected by *158*, *160*
strength of soils affected by *187*, *188*
Tokachioki earthquake, acceleration time histories 194, *196*